SOURCE
The Prentice Hall
ENGINEERING SOURCE

Engineering with Excel

Second Edition

Ronald W. Larsen

Department of Chemical Engineering
Montana State University–Bozeman

PEARSON

Prentice
Hall

Upper Saddle River, NJ 07458

Library of Congress Cataloging-in-Publication Data on file

Vice President and Editorial Director, ECS: *Marcia J. Horton*
Executive Editor: *Eric Svendsen*
Associate Editor: *Dee Bernhard*
Vice President and Director of Production and Manufacturing, ESM: *David W. Riccardi*
Executive Managing Editor: *Vince O'Brien*
Managing Editor: *David A. George*
Production Editor: *Craig Little*
Art Director: *Jayne Conte*
Cover Designer: *Bruce Kenselaar*
Art Editor: *Greg Dulles*
Manufacturing Manager: *Trudy Pisciotti*
Manufacturing Buyer: *Lisa McDowell*
Marketing Manager: *Holly Stark*

© 2005 Pearson Education, Inc.
Pearson Prentice Hall
Pearson Education, Inc.
Upper Saddle River, NJ 07458

Printed in the United States of America

10 9 8 7 6 5 4 3 2 1

0-13-147511-8

Pearson Education Ltd., *London*
Pearson Education Australia Pty. Ltd., *Sydney*
Pearson Education Singapore, Pte. Ltd.
Pearson Education North Asia Ltd., *Hong Kong*
Pearson Education Canada, Inc., *Toronto*
Pearson Educación de Mexico, S.A. de C.V.
Pearson Education—Japan, *Tokyo*
Pearson Education Malaysia, Pte. Ltd.
Pearson Education, Inc., *Upper Saddle River, New Jersey*

About ESource

ESource—The Prentice Hall Engineering Source—
www.prenhall.com/esource

ESource—The Prentice Hall Engineering Source gives professors the power to harness the full potential of their text and their first-year engineering course. More than just a collection of books, ESource is a unique publishing system revolving around the ESource website—www.prenhall.com/esource. ESource enables you to put your stamp on your book just as you do your course. It lets you:

Control You choose exactly which chapters are in your book and in what order they appear. Of course, you can choose the entire book if you'd like and stay with the authors' original order.

Optimize Get the most from your book and your course. ESource lets you produce the optimal text for your students needs.

Customize You can add your own material anywhere in your text's presentation, and your final product will arrive at your bookstore as a professionally formatted text. Of course, all titles in this series are available as stand-alone texts, or as bundles of two or more books sold at a discount. Contact your PH sales rep for discount information.

ESource ACCESS

Professors who choose to bundle two or more texts from the ESource series for their class, or use an ESource custom book will be providing their students with an on-line library of intro engineering content—ESource Access. We've designed ESource ACCESS to provide students a flexible, searchable, on-line resource. Free access codes come in bundles and custom books are valid for one year after initial log-on. Contact your PH sales rep for more information.

ESource Content

All the content in ESource was written by educators specifically for freshman/first-year students. Authors tried to strike a balanced level of presentation, an approach that was neither formulaic nor trivial, and one that did not focus too heavily on advanced topics that most introductory students do not encounter until later classes. Because many professors do not have extensive time to cover these topics in the classroom, authors prepared each text with the idea that many students would use it for self-instruction and independent study. Students should be able to use this content to learn the software tool or subject on their own.

While authors had the freedom to write texts in a style appropriate to their particular subject, all followed certain guidelines created to promote a consistency that makes students comfortable. Namely, every chapter opens with a clear set of **Objectives**, includes **Practice Boxes** throughout the chapter, and ends with a number of **Problems**, and a list of **Key Terms**. **Applications Boxes** are spread throughout the book with the intent of giving students a real-world perspective of engineering. **Success Boxes** provide the student with advice about college study skills, and help students avoid the common pitfalls of first-year students. In addition, this series contains an entire book titled ***Engineering Success*** by Peter Schiavone of the University of Alberta intended to expose students quickly to what it takes to be an engineering student.

Creating Your Book

Using ESource is simple. You preview the content either on-line or through examination copies of the books you can request on-line, from your PH sales rep, or by calling 1-800-526-0485. Create an on-line outline of the content you want, in the order you want, using ESource's simple interface. Insert your own material into the text flow. If you are not ready to order, ESource will save your work. You can come back at any time and change, re-arrange, or add more material to your creation. Once you're finished you'll automatically receive an ISBN. Give it to your bookstore and your book will arrive on their shelves four to six weeks after they order. Your custom desk copies with their instructor supplements will arrive at your address at the same time.

To learn more about this new system for creating the perfect textbook, go to www.prenhall.com/esource. You can either go through the on-line walkthrough of how to create a book, or experiment yourself.

Supplements

Adopters of ESource receive an instructor's CD that contains professor and student resources and **350 PowerPoint transparencies** created by Jack Leifer of University of Kentucky–Paducah for various books in the series. Professors can either follow these transparencies as pre-prepared lectures or use them as the basis for their own custom presentations.

Titles in the ESource Series

Design Concepts for Engineers, 2/e
0-13-093430-5
Mark Horenstein

Engineering Success, 2/e
0-13-041827-7
Peter Schiavone

Engineering Design and Problem Solving, 2E
0-13-093399-6
Steven K. Howell

Exploring Engineering
0-13-093442-9
Joe King

Engineering Ethics
0-13-784224-4
Charles B. Fleddermann

Introduction to Engineering Analysis, 2/e
0-13-145332-7
Kirk D. Hagen

Introduction to Engineering Communication
0-13-146102-8
Hillary Hart

Introduction to Engineering Experimentation
0-13-032835-9
Ronald W. Larsen, John T. Sears, and Royce Wilkinson

Introduction to Mechanical Engineering
0-13-019640-1
Robert Rizza

Introduction to Electrical and Computer Engineering
0-13-033363-8
Charles B. Fleddermann and Martin Bradshaw

Introduction to MATLAB 7
0-13-147492-8
Delores Etter and David C. Kuncicky with Holly Moore

MATLAB Programming
0-13-035127-X
David C. Kuncicky

Introduction to Mathcad 2000
0-13-020007-7
Ronald W. Larsen

Introduction to Mathcad 11
0-13-008177-9
Ronald W. Larsen

Introduction to Maple 8
0-13-032844-8
David I. Schwartz

Mathematics Review
0-13-011501-0
Peter Schiavone

Power Programming with VBA/Excel
0-13-047377-4
Steven C. Chapra

Introduction to Excel 2002
0-13-008175-2
David C. Kuncicky

Introduction to Excel, 2/e
0-13-016881-5
David C. Kuncicky

Engineering with Excel, 2/e
0-13-147511-8
Ronald W. Larsen

Introduction to Word 2002
0-13-008170-1
David C. Kuncicky

Introduction to PowerPoint 2002
0-13-008179-5
Jack Leifer

Graphics Concepts
0-13-030687-8
Richard M. Lueptow

Graphics Concepts with SolidWorks, 2/e
0-13-140915-8
Richard M. Lueptow and Michael Minbiole

Graphics Concepts with Pro/ENGINEER
0-13-014154-2
Richard M. Lueptow, Jim Steger, and Michael T. Snyder

Introduction to AutoCAD, 2/e
0-13-147509-6
Mark Dix and Paul Riley

Introduction to Engineering Communication
0-13-146102-8
Hillary Hart

Introduction to UNIX
0-13-095135-8
David I. Schwartz

Introduction to the Internet, 3/e
0-13-031355-6
Scott D. James

Introduction to Visual Basic 6.0
0-13-026813-5
David I. Schneider

Introduction to C
0-13-011854-0
Delores Etter

Introduction to C++
0-13-011855-9
Delores Etter

Introduction to FORTRAN 90
0-13-013146-6
Larry Nyhoff and Sanford Leestma

Introduction to Java
0-13-919416-9
Stephen J. Chapman

Introduction to Engineering Analysis
0-13-145332-7
Kirk D. Hagen

About the Authors

N o project could ever come to pass without a group of authors who have the vision and the courage to turn a stack of blank paper into a book. The authors in this series, who worked diligently to produce their books, provide the building blocks of the series.

Martin D. Bradshaw was born in Pittsburg, KS in 1936, grew up in Kansas and the surrounding states of Arkansas and Missouri, graduating from Newton High School, Newton, KS in 1954. He received the B.S.E.E. and M.S.E.E. degrees from the University of Wichita in 1958 and 1961, respectively. A Ford Foundation fellowship at Carnegie Institute of Technology followed from 1961 to 1963 and he received the Ph.D. degree in electrical engineering in 1964. He spent his entire academic career with the Department of Electrical and Computer Engineering at the University of New Mexico (1961-1963 and 1991-1996). He served as the Assistant Dean for Special Programs with the UNM College of Engineering from 1974 to 1976 and as the Associate Chairman for the EECE Department from 1993 to 1996. During the period 1987-1991 he was a consultant with his own company, EE Problem Solvers. During 1978 he spent a sabbatical year with the State Electricity Commission of Victoria, Melbourne, Australia. From 1979 to 1981 he served an IPA assignment as a Project Officer at the U.S. Air Force Weapons Laboratory, Kirkland AFB, Albuquerque, NM. He has won numerous local, regional, and national teaching awards, including the George Westinghouse Award from the ASEE in 1973. He was awarded the IEEE Centennial Medal in 2000.

Acknowledgments: Dr. Bradshaw would like to acknowledge his late mother, who gave him a great love of reading and learning, and his father, who taught him to persist until the job is finished. The encouragement of his wife, Jo, and his six children is a never-ending inspiration.

Stephen J. Chapman received a B.S. degree in Electrical Engineering from Louisiana State University (1975), the M.S.E. degree in Electrical Engineering from the University of Central Florida (1979), and pursued further graduate studies at Rice University. Mr. Chapman is currently Manager of Technical Systems for British Aerospace Australia, in Melbourne, Australia. In this position, he provides technical direction and design authority for the work of younger engineers within the company. He also continues to teach at local universities on a part-time basis.

Mr. Chapman is a Senior Member of the Institute of Electrical and Electronics Engineers (and several of its component societies). He is also a member of the Association for Computing Machinery and the Institution of Engineers (Australia).

Steven C. Chapra presently holds the Louis Berger Chair for Computing and Engineering in the Civil and Environmental Engineering Department at Tufts University. Dr. Chapra received engineering degrees from Manhattan College and the University of Michigan. Before joining the faculty at Tufts, he taught at Texas A&M University, the University of Colorado, and Imperial College, London. His research interests focus on surface water-quality modeling and advanced computer applications in environmental engineering. He has published over 50 refereed journal articles, 20 software packages and 6 books. He has received a number of awards including the 1987 ASEE Merriam/Wiley Distinguished Author Award, the 1993 Rudolph Hering Medal, and teaching awards from Texas A&M, the University of Colorado, and the Association of Environmental Engineering and Science Professors.

Acknowledgments: To the Berger Family for their many contributions to engineering education. I would also like to thank David Clough for his friendship and insights, John Walkenbach for his wonderful books, and my colleague Lee Minardi and my students Kenny William, Robert Viesca and Jennifer Edelmann for their suggestions.

Mark Dix began working with AutoCAD in 1985 as a programmer for CAD Support Associates, Inc. He helped design a system for creating estimates and bills of material directly from AutoCAD drawing databases for use in the automated conveyor industry. This system became the basis for systems still widely in use today. In 1986 he began collaborating with Paul Riley to create AutoCAD training materials, combining Riley's background in industrial design and training with Dix's background in writing, curriculum development, and programming. Mr. Dix received the M.S. degree in education from the University of Massachusetts. He is currently the Director of Dearborn Academy High School in Arlington, Massachusetts.

Delores M. Etter is a Professor of Electrical and Computer Engineering at the University of Colorado. Dr. Etter was a faculty member at the University of New Mexico and also a Visiting Professor at Stanford University. Dr. Etter was responsible for the Freshman Engineering Program at the University of New Mexico and is active in the Integrated Teaching Laboratory at the University of Colorado. She was elected a Fellow of the Institute of Electrical and Electronics Engineers for her contributions to education and for her technical leadership in digital signal processing.

Charles B. Fleddermann is a professor in the Department of Electrical and Computer Engineering at the University of New Mexico in Albuquerque, New Mexico. All of his degrees are in electrical engineering: his Bachelor's degree from the University of Notre Dame, and the Master's and Ph.D. from the University of Illinois at Urbana-Champaign. Prof. Fleddermann developed an engineering ethics course for his department in response to the ABET requirement to incorporate ethics topics into the undergraduate engineering curriculum. *Engineering Ethics* was written as a vehicle for presenting ethical theory, analysis, and problem solving to engineering undergraduates in a concise and readily accessible way.

Acknowledgments: I would like to thank Profs. Charles Harris and Michael Rabins of Texas A & M University whose NSF sponsored workshops on engineering ethics got me started thinking in this field. Special thanks to my wife Liz, who proofread the manuscript for this book, provided many useful suggestions, and who helped me learn how to teach "soft" topics to engineers.

Kirk D. Hagen is a professor at Weber State University in Ogden, Utah. He has taught introductory-level engineering courses and upper-division thermal science courses at WSU since 1993. He received his B.S. degree in physics from Weber State College and his M.S. degree in mechanical engineering from Utah State University, after which he worked as a thermal designer/analyst in the aerospace and electronics industries. After several years of engineering practice, he resumed his formal education, earning his Ph.D. in mechanical engineering at the University of Utah. Hagen is the author of an undergraduate heat transfer text.

Mark N. Horenstein is a Professor in the Department of Electrical and Computer Engineering at Boston University. He has degrees in Electrical Engineering from M.I.T. and U.C. Berkeley and has been involved in teaching engineering design for the greater part of his academic career. He devised and developed the senior design project class taken by all electrical and computer engineering students at Boston University. In this class, the students work for a virtual engineering company developing products and systems for real-world engineering and social-service clients.

Acknowledgments: I would like to thank Prof. James Bethune, the architect of the Peak Performance event at Boston University, for his permission to highlight the competition in my text. Several of the ideas relating to brainstorming and teamwork were derived from a workshop on engineering design offered by Prof. Charles Lovas of Southern Methodist University. The principles of estimation were derived in part from a freshman engineering problem posed by Prof. Thomas Kincaid of Boston University.

 **Steven Howell** is the Chairman and a Professor of Mechanical Engineering at Lawrence Technological University. Prior to joining LTU in 2001, Dr. Howell led a knowledge-based engineering project for Visteon Automotive Systems and taught computer-aided design classes for Ford Motor Company engineers. Dr. Howell also has a total of 15 years experience as an engineering faculty member at Northern Arizona University, the University of the Pacific, and the University of Zimbabwe. While at Northern Arizona University, he helped develop and implement an award-winning interdisciplinary series of design courses simulating a corporate engineering-design environment.

 **Douglas W. Hull** is a graduate student in the Department of Mechanical Engineering at Carnegie Mellon University in Pittsburgh, Pennsylvania. He is the author of *Mastering Mechanics I Using Matlab 5*, and contributed to *Mechanics of Materials* by Bedford and Liechti. His research in the Sensor Based Planning lab involves motion planning for hyper-redundant manipulators, also known as serpentine robots.

 **Scott D. James** is a staff lecturer at Kettering University (formerly GMI Engineering & Management Institute) in Flint, Michigan. He is currently pursuing a Ph.D. in Systems Engineering with an emphasis on software engineering and computer-integrated manufacturing. He chose teaching as a profession after several years in the computer industry. "I thought that it was really important to know what it was like outside of academia. I wanted to provide students with classes that were

up to date and provide the information that is really used and needed."

Acknowledgments: Scott would like to acknowledge his family for the time to work on the text and his students and peers at Kettering who offered helpful critiques of the materials that eventually became the book.

 **Joe King** received the B.S. and M.S. degrees from the University of California at Davis. He is a Professor of Computer Engineering at the University of the Pacific, Stockton, CA, where he teaches courses in digital design, computer design, artificial intelligence, and computer networking. Since joining the UOP faculty, Professor King has spent yearlong sabbaticals teaching in Zimbabwe, Singapore, and Finland. A licensed engineer in the state of California, King's industrial experience includes major design projects with Lawrence Livermore National Laboratory, as well as independent consulting projects. Prof. King has had a number of books published with titles including M*atlab*, MathCAD, Exploring Engineering, and Engineering and Society.

 **David C. Kuncicky** is a native Floridian. He earned his Baccalaureate in psychology, Master's in computer science, and Ph.D. in computer science from Florida State University. He has served as a faculty member in the Department of Electrical Engineering at the FAMU–FSU College of Engineering and the Department of Computer Science at Florida State University. He has taught computer science and computer engineering courses for over 15 years. He has published research in the areas of intelligent hybrid systems and neural networks. He is currently the Director of Engineering at Bioreason, Inc. in Sante Fe, New Mexico.

Acknowledgments: Thanks to Steffie and Helen for putting up with my late nights and long weekends at the computer. Finally, thanks to Susan Bassett for having faith in my abilities, and for providing continued tutelage and support.

**Ron Larsen** is a Professor of Chemical Engineering at Montana State University, and received his Ph.D. from the Pennsylvania State University. He was initially attracted to engineering by the challenges the profession offers, but also appreciates that engineering is a serving profession. Some of the greatest challenges he has faced while teaching have involved non-traditional teaching methods, including evening courses for practicing engineers and teaching through an interpreter at the Mongolian National University. These experiences have provided tremendous opportunities to learn new ways to communicate technical material. Dr. Larsen views modern software as one of the new tools that will radically alter the way engineers work, and his book *Introduction to MathCAD* was written to help young engineers prepare to meet the challenges of an ever-changing workplace.

Acknowledgments: To my students at Montana State University who have endured the rough drafts and typos, and who still allow me to experiment with their classes—my sincere thanks.

**Sanford Leestma** is a Professor of Mathematics and Computer Science at Calvin College, and received his Ph.D. from New Mexico State University. He has been the long-time co-author of successful textbooks on Fortran, Pascal, and data structures in Pascal. His current research interest are in the areas of algorithms and numerical computation.

**Jack Leifer** is an Assistant Professor in the Department of Mechanical Engineering at the University of Kentucky Extended Campus Program in Paducah, and was previously with the Department of Mathematical Sciences and Engineering at the University of South Carolina–Aiken. He received his Ph.D. in Mechanical Engineering from the University of Texas at Austin in December 1995. His current research interests include the analysis of ultra-light and inflatable (Gossamer) space structures.

Acknowledgments: I'd like to thank my colleagues at USC–Aiken, especially Professors Mike May and Laurene Fausett, for their encouragement and feedback; and my parents, Felice and Morton Leifer, for being there and providing support (as always) as I completed this book.

**Richard M. Lueptow** is the Charles Deering McCormick Professor of Teaching Excellence and Associate Professor of Mechanical Engineering at Northwestern University. He is a native of Wisconsin and received his doctorate from the Massachusetts Institute of Technology in 1986. He teaches design, fluid mechanics, an spectral analysis techniques. Rich has an active research program on rotating filtration, Taylor Couette flow, granular flow, fire suppression, and acoustics. He has five patents and over 40 refereed journal and proceedings papers along with many other articles, abstracts, and presentations.

Acknowledgments: Thanks to my talented and hard-working co-authors as well as the many colleagues and students who took the tutorial for a "test drive." Special thanks to Mike Minbiole for his major contributions to Graphics Concepts with SolidWorks. Thanks also to Northwestern University for the time to work on a book. Most of all, thanks to my loving wife, Maiya, and my children, Hannah and Kyle, for supporting me in this endeavor. (Photo courtesy of Evanston Photographic Studios, Inc.)

Holly Moore is a professor of engineering at Salt Lake Community College, where she teaches courses in thermal science, materials science engineering, and engineering computing. Dr. Moore received the B.S. degree in chemistry, the M.S. degree in chemical engineering from South Dakota School of Mines and Technology, and the Ph.D. degree in chemical engineering from the University of Utah. She spent 10 years working in the aerospace industry, designing and analyzing solid rocket boosters for both defense and space programs. She has also been active in the development of hands-on elementary science materials for the state of Utah.

Acknowledgments: Holly would like to recognize the tremendous influence of her father, Professor George

Moore, who taught in the Department of Electrical Engineering at the South Dakota School of Mines and Technology for almost 20 years. Professor Moore earned his college education after a successful career in the United States Air Force, and was a living reminder that you are never too old to learn.

Larry Nyhoff is a Professor of Mathematics and Computer Science at Calvin College. After doing bachelor's work at Calvin, and Master's work at Michigan, he received a Ph.D. from Michigan State and also did graduate work in computer science at Western Michigan. Dr. Nyhoff has taught at Calvin for the past 34 years—mathematics at first and computer science for the past several years.

Paul Riley is an author, instructor, and designer specializing in graphics and design for multimedia. He is a founding partner of CAD Support Associates, a contract service and professional training organization for computer-aided design. His 15 years of business experience and 20 years of teaching experience are supported by degrees in education and computer science. Paul has taught AutoCAD at the University of Massachusetts at Lowell and is presently teaching AutoCAD at Mt. Ida College in Newton, Massachusetts. He has developed a program, Computer-aided Design for Professionals that is highly regarded by corporate clients and has been an ongoing success since 1982.

Robert Rizza is an Assistant Professor of Mechanical Engineering at North Dakota State University, where he teaches courses in mechanics and computer-aided design. A native of Chicago, he received the Ph.D. degree from the Illinois Institute of Technology. He is also the author of *Getting Started with Pro/ENGINEER*. Dr. Rizza has worked on a diverse range of engineering projects including projects from the railroad, bioengineering, and aerospace industries. His current research interests include the fracture of composite materials, repair of cracked aircraft components, and loosening of prostheses.

Peter Schiavone is a professor and student advisor in the Department of Mechanical Engineering at the University of Alberta, Canada. He received his Ph.D. from the University of Strathclyde, U.K. in 1988. He has authored several books in the area of student academic success as well as numerous papers in international scientific research journals. Dr. Schiavone has worked in private industry in several different areas of engineering including aerospace and systems engineering. He founded the first Mathematics Resource Center at the University of Alberta, a unit designed specifically to teach new students the necessary *survival skills* in mathematics and the physical sciences required for success in first-year engineering. This led to the Students' Union Gold Key Award for outstanding contributions to the university. Dr. Schiavone lectures regularly to freshman engineering students and to new engineering professors on engineering success, in particular about maximizing students' academic performance.

Acknowledgements: Thanks to Richard Felder for being such an inspiration; to my wife Linda for sharing my dreams and believing in me; and to Francesca and Antonio for putting up with Dad when working on the text.

David I. Schneider holds an A.B. degree from Oberlin College and a Ph.D. degree in Mathematics from MIT. He has taught for 34 years, primarily at the University of Maryland. Dr. Schneider has authored 28 books, with one-half of them computer programming books. He has developed three customized software packages that are supplied as supplements to over 55 mathematics textbooks. His involvement with computers dates back to 1962, when he programmed a special purpose computer at MIT's Lincoln Laboratory to correct errors in a communications system.

**David I. Schwartz** is an Assistant Professor in the Computer Science Department at Cornell University and earned his B.S., M.S., and Ph.D. degrees in Civil Engineering from State University of New York at Buffalo. Throughout his graduate studies, Schwartz combined principles of computer science to applications of civil engineering. He became interested in helping students learn how to apply software tools for solving a variety of engineering problems. He teaches his students to learn incrementally and practice frequently to gain the maturity to tackle other subjects. In his spare time, Schwartz plays drums in a variety of bands.

Acknowledgments: I dedicate my books to my family, friends, and students who all helped in so many ways.

Many thanks go to the schools of Civil Engineering and Engineering & Applied Science at State University of New York at Buffalo where I originally developed and tested my UNIX and Maple books. I greatly appreciate the opportunity to explore my goals and all the help from everyone at the Computer Science Department at Cornell.

**John T. Sears** received the Ph.D. degree from Princeton University. Currently, he is a Professor and the head of the Department of Chemical Engineering at Montana State University. After leaving Princeton he worked in research at Brookhaven National Laboratory and Esso Research and Engineering, until he took a position at West Virginia University. He came to MSU in 1982, where he has served as the Director of the College of Engineering Minority Program and Interim Director for BioFilm Engineering. Prof. Sears has written a book on air pollution and economic development, and over 45 articles in engineering and engineering education.

**Michael T. Snyder** is President of Internet startup company Appointments 123.com. He is a native of Chicago, and he received his Bachelor of Science degree in Mechanical Engineering from the University of Notre Dame. Mike also graduated with honors from Northwestern

University's Kellogg Graduate School of Management in 1999 with his Masters of Management degree. Before Appointments123.com, Mike was a mechanical engineer in new product development for Motorola Cellular and Acco Office Products. He has received four patents for his mechanical design work. "Pro/ ENGINEER was an invaluable design tool for me, and I am glad to help students learn the basics of Pro/ ENGINEER."

Acknowledgments: Thanks to Rich Lueptow and Jim Steger for inviting me to be a part of this great project. Of course, thanks to my wife Gretchen for her support in my various projects.

**Jim Steger** is currently Chief Technical Officer and cofounder of an Internet applications company. He graduated with a Bachelor of Science degree in Mechanical Engineering from Northwestern University. His prior work included mechanical engineering assignments at Motorola and Acco Brands. At Motorola, Jim worked on part design for two-way radios and was one of the lead mechanical engineers on a cellular phone product line. At Acco Brands, Jim was the sole engineer on numerous office product designs. His Worx stapler has won design awards in the United States and in Europe. Jim has been a Pro/ENGINEER user for over six years.

Acknowledgments: Many thanks to my co-authors, especially Rich Lueptow for his leadership on this project. I would also like to thank my family for their continuous support.

**Royce Wilkinson** received his undergraduate degree in chemistry from Rose-Hulman Institute of Technology in 1991 and the Ph.D. degree in chemistry from Montana State University in 1998 with research in natural product isolation from fungi. He currently resides in Bozeman, MT and is involved in HIV drug research. His research interests center on biological molecules and their interactions in the search for pharmaceutical advances.

ESource Reviewers

We would like to thank everyone who helped us with or has reviewed texts in this series.

Christopher Rowe, *Vanderbilt University*
Steve Yurgartis, *Clarkson University*
Heidi A. Diefes-Dux, *Purdue University*
Howard Silver, *Fairleigh Dickenson University*
Jean C. Malzahn Kampe, *Virginia Polytechnic Institute and State University*
Malcolm Heimer, *Florida International University*
Stanley Reeves, *Auburn University*
John Demel, *Ohio State University*
Shahnam Navee, *Georgia Southern University*
Heshem Shaalem, *Georgia Southern University*
Terry L. Kohutek, *Texas A & M University*
Liz Rozell, *Bakersfield College*
Mary C. Lynch, *University of Florida*
Ted Pawlicki, *University of Rochester*
James N. Jensen, *SUNY at Buffalo*
Tom Horton, *University of Virginia*
Eileen Young, *Bristol Community College*
James D. Nelson, *Louisiana Tech University*
Jerry Dunn, *Texas Tech University*
Howard M. Fulmer, *Villanova University*
Naeem Abdurrahman, *University of Texas, Austin*
Stephen Allan, *Utah State University*
Anil Bajaj, *Purdue University*
Grant Baker, *University of Alaska–Anchorage*
William Beckwith, *Clemson University*
Haym Benaroya, *Rutgers University*
John Biddle, *California State Polytechnic University*
Tom Bledsaw, *ITT Technical Institute*
Fred Boadu, *Duk University*
Tom Bryson, *University of Missouri, Rolla*
Ramzi Bualuan, *University of Notre Dame*
Dan Budny, *Purdue University*
Betty Burr, *University of Houston*
Dale Calkins, *University of Washington*
Harish Cherukuri, *University of North Carolina –Charlotte*
Arthur Clausing, *University of Illinois*
Barry Crittendon, *Virginia Polytechnic and State University*
James Devine, *University of South Florida*

Ron Eaglin, *University of Central Florida*
Dale Elifrits, *University of Missouri, Rolla*
Patrick Fitzhorn, *Colorado State University*
Susan Freeman, *Northeastern University*
Frank Gerlitz, *Washtenaw College*
Frank Gerlitz, *Washtenaw Community College*
John Glover, *University of Houston*
John Graham, *University of North Carolina–Charlotte*
Ashish Gupta, *SUNY at Buffalo*
Otto Gygax, *Oregon State University*
Malcom Heimer, *Florida International University*
Donald Herling, *Oregon State University*
Thomas Hill, *SUNY at Buffalo*
A.S. Hodel, *Auburn University*
James N. Jensen, *SUNY at Buffalo*
Vern Johnson, *University of Arizona*
Autar Kaw, *University of South Florida*
Kathleen Kitto, *Western Washington University*
Kenneth Klika, *University of Akron*
Terry L. Kohutek, *Texas A&M University*
Melvin J. Maron, *University of Louisville*
Robert Montgomery, *Purdue University*
Mark Nagurka, *Marquette University*
Romarathnam Narasimhan, *University of Miami*
Soronadi Nnaji, *Florida A&M University*
Sheila O'Connor, *Wichita State University*
Michael Peshkin, *Northwestern University*
Dr. John Ray, *University of Memphis*
Larry Richards, *University of Virginia*
Marc H. Richman, *Brown University*
Randy Shih, *Oregon Institute of Technology*
Avi Singhal, *Arizona State University*
Tim Sykes, *Houston Community College*
Neil R. Thompson, *University of Waterloo*
Raman Menon Unnikrishnan, *Rochester Institute of Technology*
Michael S. Wells, *Tennessee Tech University*
Joseph Wujek, *University of California, Berkeley*
Edward Young, *University of South Carolina*
Garry Young, *Oklahoma State University*
Mandochehr Zoghi, *University of Dayton*

Contents

7

USING MACROS IN EXCEL   257

8

PROGRAMMING IN EXCEL WITH VBA   299

1

Introduction to Excel

1.1 INTRODUCTION

When electronic spreadsheets first became available, engineers immediately found uses for them. They discovered that many engineering tasks can be solved quickly and easily within the framework of a spreadsheet. Powerful modern spreadsheets such as Microsoft Excel®[1] allow ever more complex problems to be solved right on an engineer's desktop. This text focuses on using Microsoft Excel, currently the most popular spreadsheet in the world, to perform common engineering calculations.

1.1.1 Nomenclature

Key terms are shown in italics:	Press [F2] to edit the *active cell*.
Formulas and functions are shown in Courier font:	=B3*C4
Menu selections are indicated with slash marks:	To open a file, select File/Open.
Individual keystrokes are enclosed in brackets:	Press the [enter] key.
Key combinations are enclosed in brackets:	Press [Ctrl-C] to copy the contents of the cell.

1.1.2 Examples and Application Problems

We will use sample problems throughout the text to illustrate how Excel is applied to solve engineering problems. There are three levels of problems included in this text:

* Demonstration Examples: These are usually very simple, well-known subjects used to demonstrate specific processes in Excel. They are usually single-step examples.

OBJECTIVES

After reading this chapter, you will know

* What a spreadsheet is
* How the Excel screen is laid out
* The fundamentals of using Excel

---

[1]Excel is a trademark of Microsoft Corporation, Inc.

- Sample Problems: These problems are slightly more involved and usually involve many steps. They are designed to illustrate how to apply specific Excel functions or capabilities to engineering problems.
- Application Problems: These are intended to be larger problems, ones closer to the type of problem engineering students will see as homework problems.

The scope of this text is intended to include all engineering disciplines at the undergraduate level. There has been an attempt to cover a fairly broad range of subjects in the example and application problems and to stay away from problems that require discipline-specific knowledge.

1.1.3   What Is a Spreadsheet?

A *spreadsheet* is a piece of paper containing a grid designed to hold values. The values are written into the *cells* formed by the grid and arranged into vertical columns and horizontal rows. Spreadsheets have been used for many years by people in the business community to present financial statements in an orderly way. With the advent of the personal computer in the 1970s, the paper spreadsheet was migrated to the computer and became an *electronic spreadsheet*. The rows and columns of values are still there, and the values are still housed in cells. The layout of an electronic spreadsheet is simple, making it very easy to learn to use. People can start up a program such as Excel for the first time and start solving problems within minutes.

The primary virtue of the early spreadsheets was *automatic recalculation*: Any change in a value or formula in the spreadsheet caused the rest of the spreadsheet to be recalculated. This meant that errors found in a spreadsheet could be fixed easily without having to recalculate the rest of the spreadsheet by hand. Even more importantly, the electronic spreadsheet could be used to investigate the effect of a change in one value on the rest of the spreadsheet. Engineers who wanted to know what would happen if, for example, the load on a bridge was increased by 2, 3, or 4% quickly found electronic spreadsheets very useful.

Since the 1970s, the computing power offered by electronic spreadsheets on personal computers has increased dramatically. Graphing capabilities were added early on and have improved dramatically over time. Built-in functions were added to speed up common calculations. Microsoft added a programming language to Excel that can be accessed from the spreadsheet when needed. Also, the computing speed and storage capacity of personal computers has increased to such an extent that a single personal computer with some good software (including, but not limited to, a spreadsheet such as Excel) can handle most of the day-to-day tasks encountered by most engineers.

1.1.4   Why Use a Spreadsheet?

Spreadsheets are great for some tasks, but some problems fit the grid structure of a spreadsheet better than others. When your problem centers on columns of numbers, such as data sets recorded from instruments, it fits in a spreadsheet very well. The analysis of tabular data fits the spreadsheet layout quite well. But if your problem requires the symbolic manipulation of complex mathematical equations, a spreadsheet is not the best place to solve that problem.

The spreadsheet's heritage as a business tool becomes apparent as you use it. For example, there are any number of ways to display data in pie and bar charts, but the available X–Y chart options, generally more applicable to science and engineering, are more limited. There are many built-in functions to accomplish tasks such as calculating rates of return on investments (and engineers can find those useful), but there is no built-in function that calculates torque.

Spreadsheets are easy to use and can handle a wide range of problems. Many, if not most, of the problems for which engineers used to write computer programs are now solved by using electronic spreadsheets or other programs on their personal computers. A supercomputer might be able to "crunch the numbers" faster, but when the "crunch" time is tiny compared with the time required to write the program and create a report based on its results, the spreadsheet's ease of use and ability to print results in finished form (or easily move results to a word processor) can make the total time required to solve a problem by using a spreadsheet much shorter than that with conventional programming methods.

Spreadsheets are great for

- performing the same calculations repeatedly (e.g., analyzing data from multiple experimental runs),
- working with tabular information (e.g., finding enthalpies in a steam table— once you've entered the steam table into the spreadsheet),
- producing graphs—spreadsheets provide an easy way to get a plot of your data,
- performing parametric analyses, or "what if" studies—for example, "What would happen if the flow rate were doubled?", and
- presenting results in readable form.

There was a time when spreadsheets were not the best way to handle computationally intense calculations such as iterative solutions to complex problems, but dramatic improvements in the computational speed of personal computers has eliminated a large part of this shortcoming, and improvements in the solution methods used by Excel have also helped. Excel can now handle many very large problems that just a few years ago would not have been considered suitable for implementation on a spreadsheet.

But there are still a couple of things that spreadsheets do not do well. Programs such as Mathematica[2] and Maple[3] are designed to handle symbolic manipulation of mathematical equations; Excel is not. Electronic spreadsheets also display only the results of calculations (just as their paper ancestors did), rather than the equations used to calculate the results. You must take special care, when developing spreadsheets, to indicate how the solution was found. Other computational software programs, such as Mathcad,[4] more directly show the solution process as well as the result.

1.2   STARTING EXCEL

Excel can be purchased by itself or as part of the Microsoft Office® family of products. During installation, an Excel option is added to the Start menu, and, if Office is installed, a "New Office Document" option is added as well. Excel can be started by using either option.

---

[2]Wolfram Research, Inc., Champaign, IL, USA.
[3]Waterloo Maple Inc., Ontario, Canada.
[4]MathSoft Inc., Cambridge, MA, USA.

1.2.1 Starting Excel by Using the Excel Icon on the Start Menu

Use Start/Programs/Microsoft Excel:

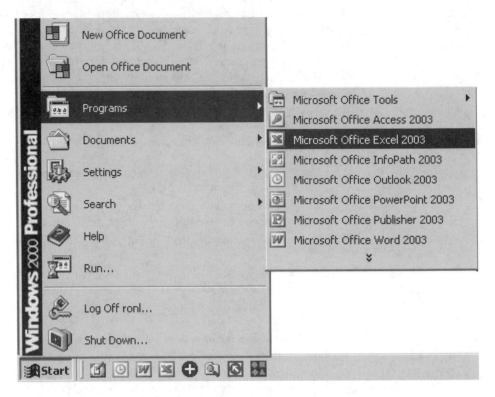

Click the Microsoft Excel icon. (The location of the icon within the Start Menu varies.) Excel will start up, and you will see a title screen; then the grid pattern of the spreadsheet will appear.

1.2.2 Starting Excel by Using New Office Document

Use Start/New Office Document. (This option will appear on the Start menu only if Microsoft Office has been installed on your computer.)

Then select Blank Workbook from the New Office Document dialog.
No matter how Excel is started, the blank workbook should look similar to the following:

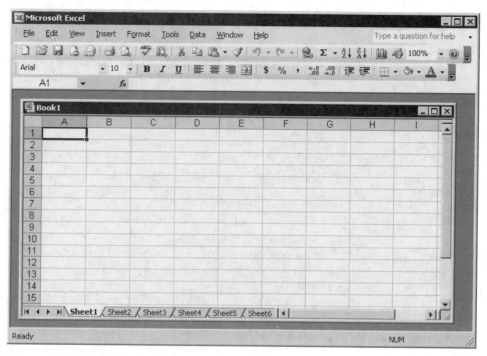

There are a lot of features available in the Excel program, but only a few will be addressed here.

1.3 A LITTLE WINDOWS®

The look of the windows depends on the version of Windows® you are running. Images shown here are from Excel 2003 running under Windows 2000, but users of other versions should see few differences.

1.3.1 Title Bar

The top line of a window is called the *title bar* and contains the name of the program running in the window—"Microsoft Excel", in the preceding figure. If the window is smaller than the full screen, you can drag the title bar with the mouse to move the window across the screen.

1.3.2 Control Buttons

Close   The top right corner of the window contains the control buttons. At the very right is the Close button. Click this button to close the window and terminate the program.

Minimize and Maximize Buttons   The buttons in the top right corner of the window (to the left of the close button) are used to minimize and maximize the window. A window can be minimized, maximized, or restored.

- A minimized window is simply an icon representing the window in the taskbar at the bottom of the screen. Clicking on the icon will restore the window to the size it was before it was minimized.
- A maximized window covers the entire screen. The size of a maximized window cannot be changed by dragging the edge of the window with the mouse.

When a window covers only a portion of the screen, the size and shape of the window can usually be changed by grabbing the window's border with the mouse and dragging. Some types of windows, such as error-message windows, might not allow you to move or resize them.

1.3.3   Menu Bar

Just below the title bar is the *main menu*. Clicking on any of the items in the main menu displays a list of additional menu selections.

1.3.4   Standard Toolbar

Below the menu bar is a row of buttons called the *standard toolbar*. The actions of these buttons generally duplicate menu items. They are provided to allow you to do common tasks quickly and easily. The number and choice of toolbar buttons varies, depending on how your computer has been configured, so menu selections will generally be used in this text.

Excel is very flexible and lets you modify the appearance of the window to suit your needs. The *toolbar* buttons and other features described here might not appear on your screen. Use View/Toolbars to select the toolbars you want to use. To customize the toolbars to display buttons you want to use, select View/Toolbars/Customize.

Customizing a Toolbar

The right end of each Toolbar is a link to a pop-up menu that allows you to customize the Toolbar.

When you click the end marker, a pop-up menu appears. Clicking on the Add or Remove Buttons option opens a second menu.

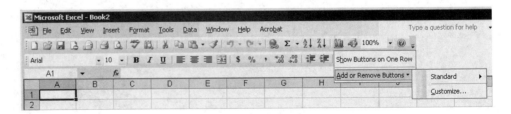

The Standard option allows you to activate or deactivate any of the standard Toolbar buttons. The Customize . . . option opens a dialog box that allows you to add additional buttons to the Toolbar.

The Standard menu is shown next. Simply click the available buttons to toggle them on and off.

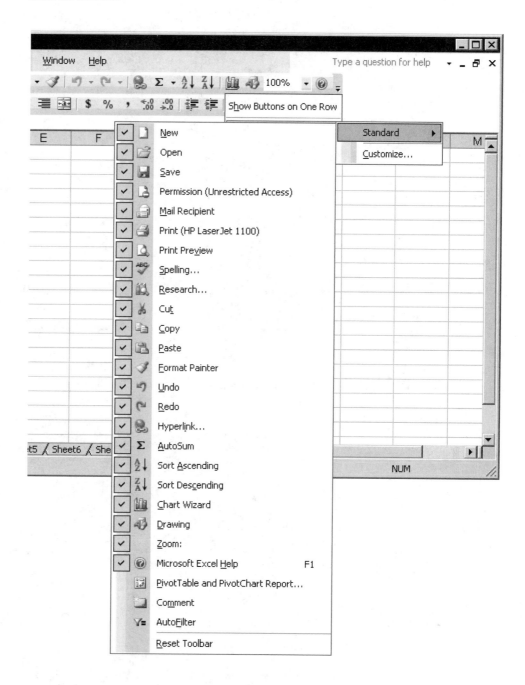

Selecting Customize . . . opens the Customize dialog. There are three panels available. The Toolbars panel allows you to activate or deactivate entire toolbars.

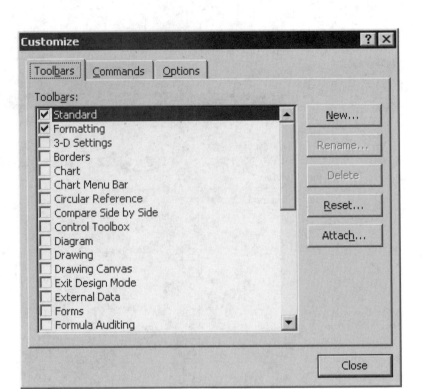

The Commands panel allows you to add specialized buttons to any toolbar. Simply find the button you want and drag it to a toolbar.

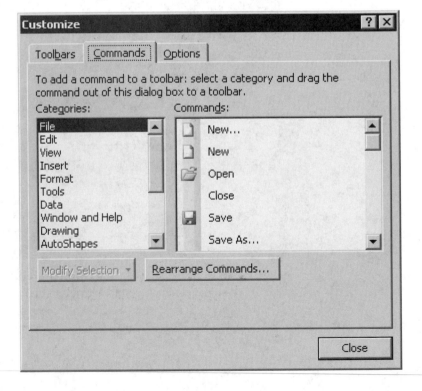

The Options panel allows you to specify how you want the toolbars arranged and whether large icons should be used on the toolbars.

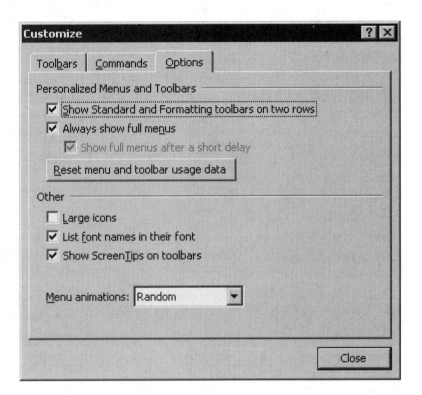

1.3.5   Formatting Toolbar

The *formatting toolbar* typically is displayed just beneath the standard toolbar. The formatting toolbar contains fields that indicate the property settings for the selected or *active cell*. In the figure shown next, the heavy border around cell A1 shows that it is the active cell. The fields on the left side of the formatting toolbar indicate that text entered into cell A1 would be displayed in the Arial font sized to 10 point and not bolded, italicized, or underlined:

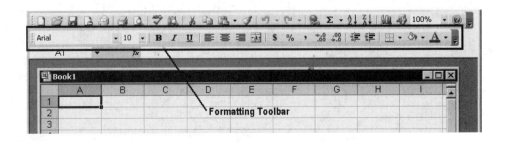

Each of the property indicators is shown in either a drop-down list box or a button that toggles a feature. For example, you can click on the Arial indicator to change the font used to display the cell's contents or click on the Bold button to toggle bold characters on or off.

1.3.6 Formula Bar

Just below the Formatting Toolbar (usually) is the *formula bar*. The box at the left shows the cell designation or address of the active cell, in this case, cell A1. Excel also uses this box as a drop-down list to display (and allow you to reuse) recently used functions.

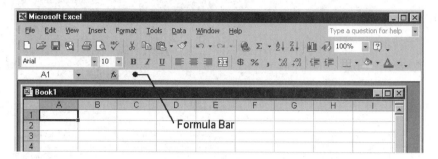

To the right of the cell indicator is a box (next to the f_x symbol) that shows the contents of the active cell. In the example just shown, cell A1 is empty.

> *Note: If the formula bar is not displayed, you can activate it by selecting View/Formula Bar.*

1.3.7 Spreadsheets and Workbooks

Near the bottom of the Excel window, just below the spreadsheet grid, you'll notice "Sheet1" on something that looks like the tab on a file folder. This indicates that spreadsheet 1 is currently in use. By default, there are multiple spreadsheets available in the window, named Sheet1, Sheet2, and so on. You can add additional sheets if needed by right-clicking a tab. Right-clicking on any tab pops up a small menu of options, including adding, renaming, deleting, or moving sheets. You can also change the default number of sheets included in a new workbook by using the Options dialog on the Tools menu. Specifically, menu commands Tools/Options . . . open the Options dialog.

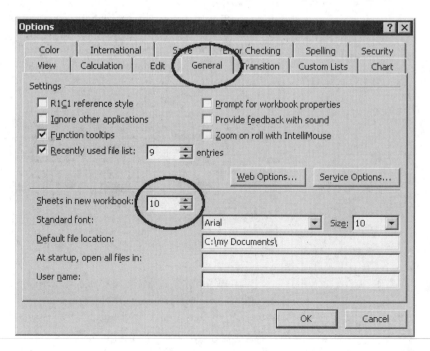

On the dialog's General tab, you can set the number of sheets in a new workbook.

A collection of spreadsheets is called a *workbook*. You can keep each spreadsheet in a separate file if you want to, but workbooks are a convenient way to collect related spreadsheets:

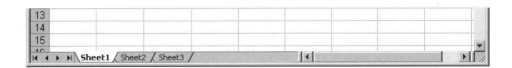

You can also have multiple workbooks open at once. Each workbook is displayed in its own window within the main Excel window. Each workbook has its own control buttons and can be minimized and maximized, just like the main window. In the following figure, only one workbook is being displayed (called Book 1 by default), and the window is neither minimized nor maximized:

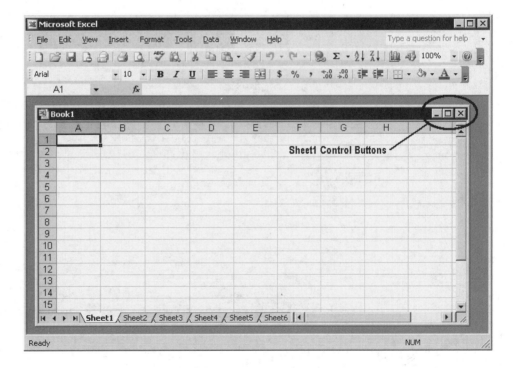

Clicking the spreadsheet's maximize button (or double-clicking the title bar) expands the workbook window to fill the available space in the main window, as illustrated in the next figure. A maximized workbook's control buttons (as shown in the figure) appear directly below the main window's control buttons, and the workbook name is added to the title bar:

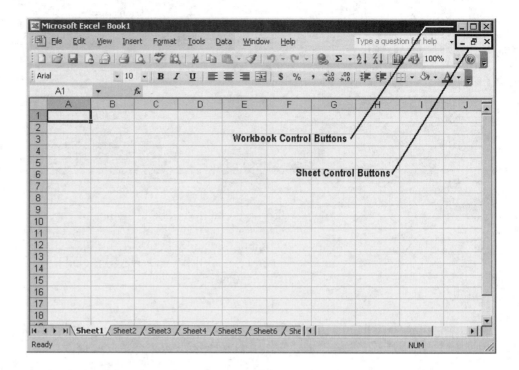

1.3.8 Status Bar

At the very bottom of the main window is the *status bar*. In the figure, the status bar indicates that the [Num Lock] key is set (i.e., that the numeric keypad is set for numeric entry) and that the spreadsheet is ready for data entry. When the spreadsheet is performing a large number of calculations, the ready indicator can change to BKGD, indicating that Excel is doing the calculations in the background while allowing you to continue working.

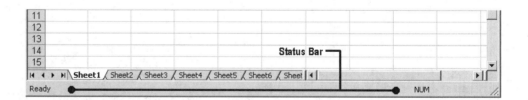

1.4 SPREADSHEET BASICS

1.4.1 Labels, Numbers, Formulas

When you start Excel, you see an empty grid on the screen. Each element, or cell, is denoted by its *cell address*, made up of a column letter and a row number. For example, the cell in the second column from the left and the third row from the top would be called cell B3.

The active cell is indicated by a heavy border (cell B3 in the next figure) and is identified in the Name Box at the left side of the Formula Bar. You can enter data only into the active cell. The location of the active cell can be changed with the arrow keys or by clicking a cell with the mouse pointer.

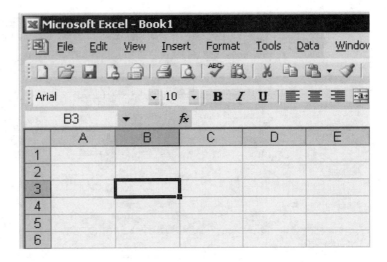

Each cell can contain a *label* (a word or words), a *value*, or a *formula*. (The folks at Microsoft use the term "formula" rather than "equation".)

- Typing anything beginning with a letter or a single quote shows up as a label in the spreadsheet.
- If you enter a number, that value appears in the active cell.
- Formulas may be entered by using an initial equal sign to let Excel know that a formula follows, but the *result* of the formula is actually displayed. For example, if you enter =2+4, the formula =2+4 would be stored as the contents of the active cell, but the result, 6, would be displayed.

The ability to use formulas is one of the keys to successful spreadsheets. What makes formulas useful is the ability to use cell addresses (e.g., B3) as variables in the formulas. The following example uses a spreadsheet to compute a velocity from a distance and a time interval.

EXAMPLE 1.1

A car drives 90 miles in 1.5 hours. What is the average velocity in miles per hour?

This is a pretty trivial example, and illustrates another consideration in the use of a spreadsheet: Single, simple calculations are what calculators are for. A spreadsheet is overkill for this example, but we'll use it as the starting point for a more detailed problem.

1. Enter labels as follows:
    - "Distance" in cell B3 and the units "miles" in cell B4.
    - "Time" in cell C3 and the units "hours" in cell C4.
    - "Velocity" in cell D3, with units "mph" in cell D4.
2. Enter values as follows:
    - 90 in cell B5 (i.e., in the Distance column (column B)).
    - 1.5 in cell C5 (i.e., in the Time column (column C)).
3. Enter the formula =B5/C5 in cell D5.

Note that, in Step 6, an equal sign was used to keep the spreadsheet from interpreting B5/C5 as a label; without the equal sign, the first character would be a letter:

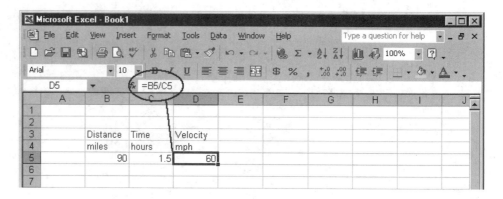

Cell D5 contains the formula =B5/C5, as indicated on the formula bar in the preceding figure, but displays the result of the formula (60 miles per hour).

Note: The formula bar will be included in most screen images in this text to show the contents of the active cell.

1.4.2 Entering Formulas

Normally, when entering formulas, you don't actually type =B5/C5. Instead, for the previous example, after moving the *cell cursor* (active cell indicator) to the desired location (D5), you would enter the equal sign and then click the mouse on cell B5 (or use the cursor keys to select cell B5). At this point, =B5 would appear on the entry line in the formula bar and in the cell being edited:

Pressing the [/] key tells the spreadsheet that you are done pointing at B5, so you can then use the mouse or arrow keys to point at cell C5.

Pressing [enter] concludes the entry of the formula in cell D5.

Note: While you are editing a formula, Excel shows the cells being used in the formula in color-coded boxes.

1.4.3   Copying Cell Ranges to Perform Repetitive Calculations

If you have a column of distances and another of times, you do not need to enter the formula for calculating the velocity over and over again; instead simply copy the formula in cell D5 to the *range of cells* D6 through D9 (written D6:D9):

	D5	▼	*fx* =B5/C5	
	A	B	C	D	E
1					
2					
3		Distance	Time	Velocity	
4		miles	hours	mph	
5		90	1.5	60	
6		120	6		
7		75	3		
8		80	2		
9		135	9		
10					

The procedure for copying the contents of cell D5 to D6:D9 is as follows:

1. Select the cell (or cells, if more than one) containing the formula(s) to be copied. For one cell, simply select the cell using the mouse or arrow keys. When multiple cells are to be copied, select the source range of cells by clicking on the first cell in the range. Then use the mouse to drag the cursor to the other end of the range (or hold down the shift key while using the mouse or arrow keys to select the cell at the other end of the range). Do <u>not</u> click on the little box at the lower-right corner of a cell. This is the *fill handle*, which is useful for several things, but not for selecting a range of cells. Its use will be described later.

    Note: There is a fast way to select a column of values by using the [end] key. Select the first cell in the range, hold the [shift] key down, press [end], and then press either the up or down arrow key. Excel will select all contiguous filled cells in the column, stopping at the first empty cell. To select a contiguous row of cells, select an end cell, hold the [shift] key down, press [end], and then press either the left or right arrow key.

    In this example, we simply need to select cell D5.

2. Copy the contents of cell D5 to the Windows clipboard by selecting Edit/Copy or by using the copy button on the standard toolbar. Excel uses a dashed border to show the cell that has been copied.

	D5	▼	*fx* =B5/C5	
	A	B	C	D	E
1					
2					
3		Distance	Time	Velocity	
4		miles	hours	ft/sec	
5		90	1.5	60	
6		120	6		
7		75	3		
8		80	2		
9		135	9		
10					

3. Select the beginning cell of the destination range, cell D6.

4. Select the entire destination range by dragging the mouse to cell D9 (or hold the shift key and click on cell D9). Excel uses a heavy border to show the selected range of cells.

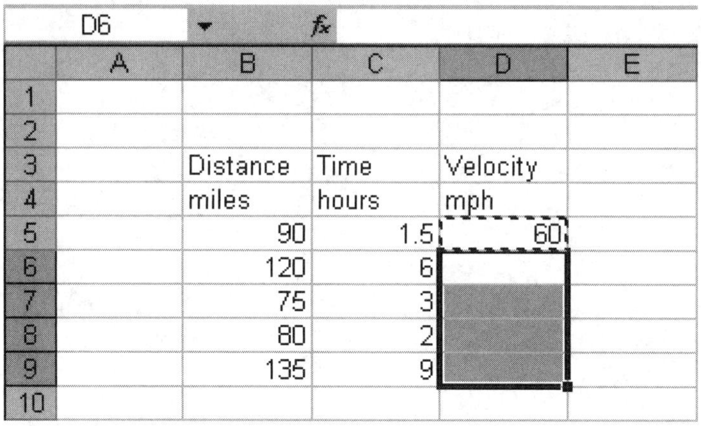

	D6	▼		*fx*	
	A	B	C	D	E
1					
2					
3		Distance	Time	Velocity	
4		miles	hours	mph	
5		90	1.5	60	
6		120	6		
7		75	3		
8		80	2		
9		135	9		
10					

5. Paste the information stored on the Windows clipboard into the destination cells by selecting Edit/Paste or by using the Paste button on the standard toolbar.

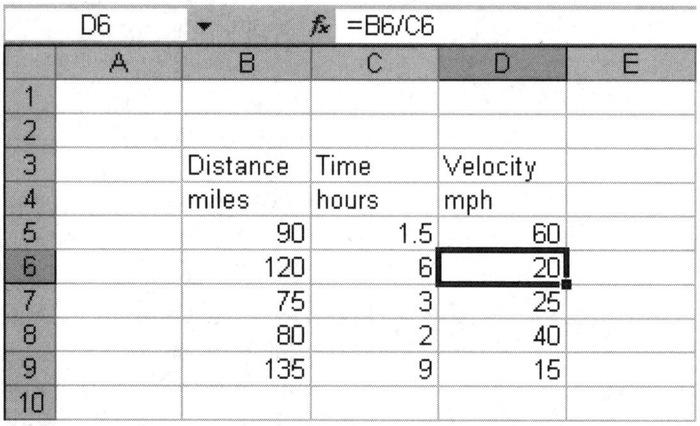

	D6	▼		*fx* =B6/C6	
	A	B	C	D	E
1					
2					
3		Distance	Time	Velocity	
4		miles	hours	mph	
5		90	1.5	60	
6		120	6	20	
7		75	3	25	
8		80	2	40	
9		135	9	15	
10					

However, there's an even easier way to copy the cells ...

1.4.4   Using the Fill Handle

The small square at the bottom right corner of the active cell border is called the *fill handle*. If the cell contains a formula and you grab and drag the fill handle, the formula will be copied.

After selecting the source cell, grab the fill handle of the selected cell with the mouse and drag, as shown in the following figure:

	A	B	C	D	E
1					
2					
3		Distance	Time	Velocity	
4		miles	hours	mph	
5		90	1.5	60	
6		120	6		
7		75	3		
8		80	2		
9		135	9		
10					

The destination range will be outlined as you drag. Continue dragging until the entire range (cell range D5:D9) is outlined.

	A	B	C	D	E
1					
2					
3		Distance	Time	Velocity	
4		miles	hours	mph	
5		90	1.5	60	
6		120	6		
7		75	3		
8		80	2		
9		135	9		
10					

Then release the mouse. Excel will copy the contents of the original cell to the entire range.

	A	B	C	D	E
1					
2					
3		Distance	Time	Velocity	
4		miles	hours	mph	
5		90	1.5	60	
6		120	6	20	
7		75	3	25	
8		80	2	40	
9		135	9	15	
10					
11					

The little square icon next to the fill handle after the copy is a link to a pop-up menu that allows you to modify how the copy process is carried out. If you click on the icon, a pop-up menu will appear with some options for how the copy using the fill handle should be completed. The Copy Cells option (default) copies both the contents and the formatting of the source cell. Other options allow you to copy only the contents or only the formatting.

	A	B	C	D	E	F	G
1							
2							
3		Distance	Time	Velocity			
4		miles	hours	mph			
5		90	1.5	60			
6		120	6	20			
7		75	3	25			
8		80	2	40			
9		135	9	15			
10							
11							
12					○ Copy Cells		
13					○ Fill Formatting Only		
14					○ Fill Without Formatting		
15							

The fill handle can be used for other purposes as well:

- To fill a range with a series of values incremented by one, place the first value (not formula) in a cell and then drag the fill handle to create the range of values.

- If you want a range of values with an increment other than one, enter the first two values of the series in adjacent cells, select both cells as the source, and drag the fill handle.

- If you want a nonlinear fill, drag the fill handle by using the right mouse button. When you release the mouse, a menu will be displayed giving a variety of fill options.

1.4.5 Relative and Absolute Addressing

At this point, the contents of cell D5 have been copied, with significant modifications, to cells D6:D9. If you move the cell pointer to D6, you will discover that D6 contains the formula =B6/C6.

D6		f_x =B6/C6		
	A	B	C	D	E
1					
2					
3		Distance	Time	Velocity	
4		miles	hours	ft/sec	
5		90	1.5	60	
6		120	6	20	
7		75	3	25	
8		80	2	40	
9		135	9	15	
10					

As the formula was copied from row 5 to row 6, the row numbers in the formula were incremented by one. Similarly, as the formula was copied from row 5 to row 7, the row numbers in the formula were incremented by two. So, the velocity calculated in cell D7 uses the distance and time values from row 7, as desired. This automatic incrementing of cell addresses during the copy process is called *relative cell addressing* and is an important feature of electronic spreadsheets.

Note: If you had copied the formula in D5 to cell E5, the column letters would have been incremented. Copying down increments row numbers, copying across increments column letters.

Sometimes you don't want relative addressing. You can make any address *absolute* in a formula by including dollar signs in the address, as $B$5. The nomenclature $B$5 tells Excel not to automatically increment either the B or the 5 while copying. Similarly, $B5 tells the program to increment the 5, but not the B during a copy, and B$5 tells it to increment the B, but not the 5. During a copy, any row or column designation preceded by a $ is left alone. This is useful when you have a constant that you wish to apply in a number of places.

EXAMPLE 1.2

If you wanted to display the velocity in feet per second, you could build the conversion factor, 1.467 (ft/sec)/(mile/hr) into the spreadsheet and use its absolute address in the formulas.

To keep the spreadsheet organized, I generally place parameter values, such as the unit conversion factor, near the top, as shown next. The factor can then be used in the formulas in column D. It is only necessary to reenter or edit one of the formulas in column D. Once one formula has been changed, it can be copied to the rest of column D.

	D6	▼		f_x =B6/C6*$B$1	
	A	**B**	**C**	**D**	**E**
1	**Factor:**	1.467	(ft/sec) / (mile/hr)		
2					
3		Distance	Time	Velocity	
4		miles	hours	ft/sec	
5		90	1.5	88.0	
6		120	6	29.3	
7		75	3	36.7	
8		80	2	58.7	
9		135	9	22.0	
10					

Notice that the constant appears in the equation as $B$1, telling the program to always use the value in cell B1 and increment neither the row nor the column during the copy process.

APPLICATIONS: FLUID STATICS

Pressures at the Bottom of Columns of Liquid. The pressure at the bottom of a column of fluid is caused by the mass of the fluid, m, being acted on by the acceleration due to gravity, g. The resulting force is called the weight of the fluid, F_W, and can be calculated as

$$F_W = mg, \tag{1.1}$$

which is a specific version of Newton's law when the acceleration is due to the earth's gravity. The pressure is the force divided by the area of the bottom of the column, A:

$$P = \frac{F_W}{A} = \frac{mg}{A} \tag{1.2}$$

The mass of the fluid can be replaced by its density times the fluid volume,

$$P = \frac{F_W}{A} = \frac{mg}{A} = \frac{\rho V g}{A}, \tag{1.3}$$

and the volume is the area of the column times its height, h:

$$P = \frac{F_W}{A} = \frac{mg}{A} = \frac{\rho V g}{A} = \frac{\rho A h g}{A} = \rho h g. \tag{1.4}$$

Calculate the pressure at the bottom of a column of mercury 760 mm high. ($SG_{mercury} = 13.6$). [Do not consider any imposed pressure (e.g., air pressure) on the top of the column in this problem.

C8			f_x =C7*C3*C5		
	A	B	C	D	E
1	**Pressure at the Bottom of a Column of Fluid**				
2					
3	**Grav. Acceleration:**		9.8	m/s^2	
4	**Height:**		760	mm	
5	**Height (SI):**		0.76	m	
6	**Specific Gravity:**		13.6		
7	**Density:**		13600	kg/m3	
8	**Pressure:**		1.013E+05	Pa	
9	**Pressure:**		1.000	atm	
10					

PRACTICE!

Calculate the pressure at the bottom of the columns of fluid specified. [Do not consider any imposed pressure (e.g., air pressure) on the top of the column in this problem.]

a. A column of water 10 meters high ($SG_{water} = 1$). [Answer: 0.97 atm]

b. A column of water 11,000 meters high. This is the depth of the Marianas Trench, the deepest spot known in the earth's oceans ($SG_{water} = 1.03$ —ignore the variation in water density with pressure). [Answer: 1090 atm]

Letting Excel Add the Dollar Signs You don't actually have to enter the dollar signs by hand. If you enter B1 and then press the [F4] key, Excel will automatically enter the dollar signs for you. Pressing [F4] once converts B1 to $B$1. Pressing the [F4] key multiple times changes the number and arrangement of the dollar signs. In practice, you simply press [F4] until you get the dollar signs where you want them. If you use the mouse to point to a cell in a formula, you must press [F4] right after you click on the cell. You can also use [F4] while editing a formula; just move the edit cursor to the cell address that needs the dollar signs and press [F4].

1.4.6 Using Named Cells

Excel allows *named cells*, allowing you to assign descriptive names to individual cells or ranges of cells. Then you can use the names, rather than the cell addresses, in formulas. For example, the cell containing the conversion factor in the previous example might be given the name, ConvFactor. This name could then be used in the velocity formulas instead of $B$1.

To give a single cell a name, click on the cell (cell B1 in this example) and enter the name in the *Name Box* at the left side of the formula bar.

ConvFactor			f_x 1.467			
	A	B	C	D	E	F
1	**Factor:**	1.467	(ft/sec) / (mile/hr)			
2						

The name can then be used in place of the cell address in the velocity formulas, as in the formula in cell D5 in the following figure:

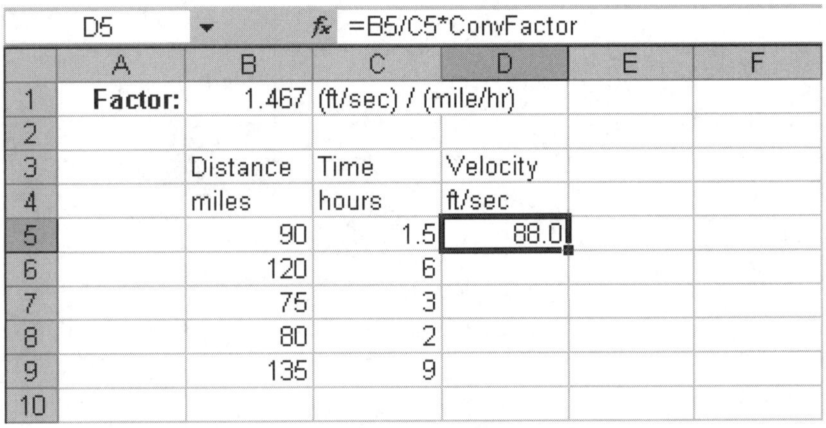

	D5	▼		*fx*	=B5/C5*ConvFactor		
	A	B	C	D	E	F
1	**Factor:**	1.467	(ft/sec) / (mile/hr)			
2						
3		Distance	Time	Velocity		
4		miles	hours	ft/sec		
5		90	1.5	88.0		
6		120	6			
7		75	3			
8		80	2			
9		135	9			
10						

When the formula in cell D5 is copied to cells D6 through D9, the cell name is included in the formulas, as shown in cell D6 in the following figure:

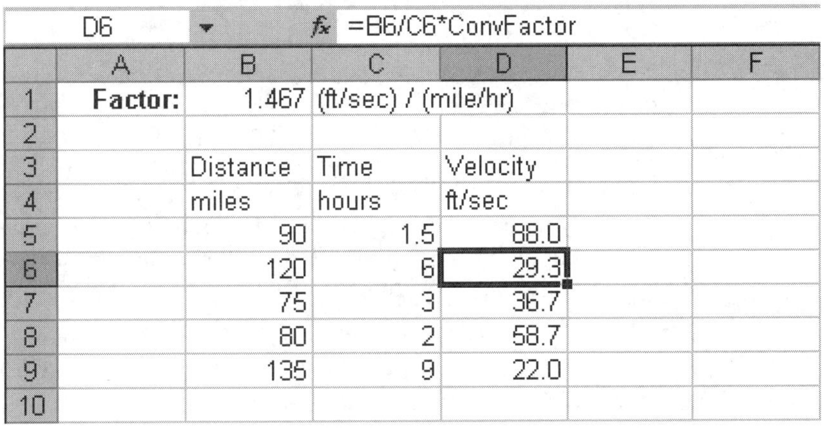

	D6	▼		*fx*	=B6/C6*ConvFactor		
	A	B	C	D	E	F
1	**Factor:**	1.467	(ft/sec) / (mile/hr)			
2						
3		Distance	Time	Velocity		
4		miles	hours	ft/sec		
5		90	1.5	88.0		
6		120	6	29.3		
7		75	3	36.7		
8		80	2	58.7		
9		135	9	22.0		
10						

Notice that `ConvFactor` copied in the same way that `$B$1` had copied previously. A cell name acts as an *absolute cell address* and is not modified when the formula is copied.

It is also possible to assign names to cell ranges, such as the columns of distance values. To assign a name to a cell range, first select the range and then enter the name in the name box:

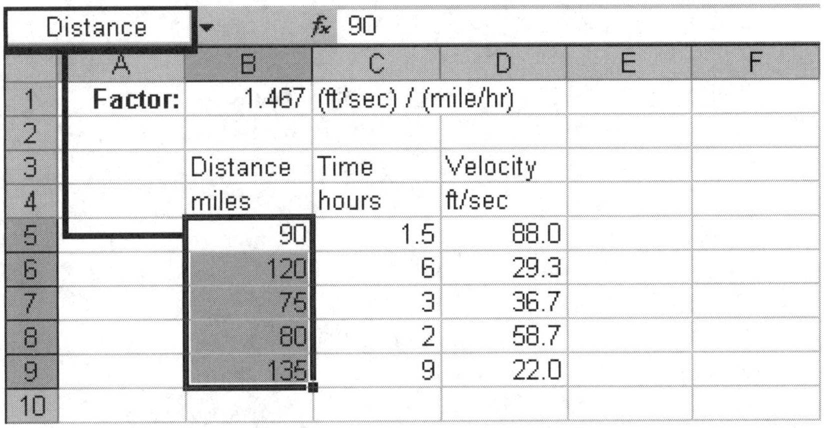

	Distance	▼		*fx*	90		
	A	B	C	D	E	F
1	**Factor:**	1.467	(ft/sec) / (mile/hr)			
2						
3		Distance	Time	Velocity		
4		miles	hours	ft/sec		
5		90	1.5	88.0		
6		120	6	29.3		
7		75	3	36.7		
8		80	2	58.7		
9		135	9	22.0		
10						

The name can then be used in place of the cell range, as in the =AVERAGE(Distance) function in cell B11. (AVERAGE is Excel's built in function for calculating the average of a set of values.)

B11	▾		f_x =AVERAGE(Distance)			
	A	B	C	D	E	F
1	**Factor:**	1.467	(ft/sec) / (mile/hr)			
2						
3		Distance	Time	Velocity		
4		miles	hours	ft/sec		
5		90	1.5	88.0		
6		120	6	29.3		
7		75	3	36.7		
8		80	2	58.7		
9		135	9	22.0		
10						
11	**Avg:**	100				
12						

If you have assigned a name and decide to remove it, use menu commands Insert/Name/Define ... to see a list of the defined names in the worksheet, select the name to remove, and click the Delete button.

1.4.7 Editing the Active Cell

You can enter *edit mode* to change the contents of the active cell either by double-clicking the cell, or by selecting a cell and then pressing [F2]. Be sure to press [enter] when you are done editing the cell contents, or your next mouse click will change your formula.

Note: Changing a formula that was previously copied to another range of cells does not cause the contents of the other cells to be similarly modified. Recopy the range if the change should be made to each cell.

You can edit either on the formula bar or right in the active cell if this option is activated. To activate or deactivate editing in the active cell, select Tools/Options ... to

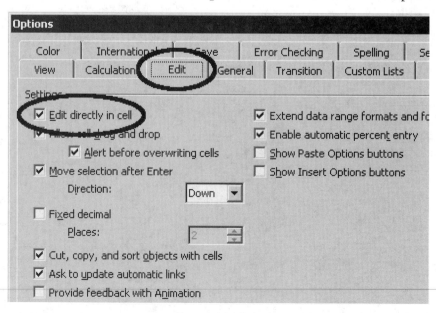

open the Options dialog, and select the Edit tab. Under the Settings heading, check or uncheck the box corresponding to "Edit directly in cell."

1.4.8   Using Built-In Functions

Spreadsheets have *built-in functions* to do lots of handy things. Originally, they handled business functions, but newer versions of spreadsheet programs also have many functions useful to engineers. For example, you can compute the arithmetic average and standard deviation of the values in the distance, time, and velocity columns, using Excel's AVERAGE and STDEV functions. These functions work on a range of values, so we'll have to tell the spreadsheet which values to include in the computation of the average and standard deviation. Here we enter =AVERAGE(B5:B9) in cell B11, and =STDEV(B5:B9) in cell B12:

B11	▼	*fx* =AVERAGE(B5:B9)		
	A	B	C	D	E
1	**Factor:**	1.467	(ft/sec) / (mile/hr)		
2					
3		Distance	Time	Velocity	
4		miles	hours	ft/sec	
5		90	1.5	88.0	
6		120	6	29.3	
7		75	3	36.7	
8		80	2	58.7	
9		135	9	22.0	
10					
11	**Avg:**	100			
12	**St Dev:**	26.2			
13					

Again, you don't normally enter the entire formula from the keyboard. Typically, you would type the equal sign, the function name, and the opening parenthesis [=AVERAGE(] and then use the mouse to indicate the range of values to be included in the formula. Then you would enter the closing parenthesis and press [enter] to complete the formula. The formula for standard deviation, =STDEV(B5:B9) was entered into cell B12 in this manner.

C12	▼	*fx* =STDEV(C5:C9)		
	A	B	C	D	E
1	**Factor:**	1.467	(ft/sec) / (mile/hr)		
2					
3		Distance	Time	Velocity	
4		miles	hours	ft/sec	
5		90	1.5	88.0	
6		120	6	29.3	
7		75	3	36.7	
8		80	2	58.7	
9		135	9	22.0	
10					
11	**Avg:**	100	4.3	46.9	
12	**St Dev:**	26.2	3.2	26.8	
13					

To compute the averages and standard deviations of all three columns, simply copy the source range, B11:B12, to the destination range C11:D11. (Yes, D11, not D12—you tell the spreadsheet where to copy the top left corner of the source range, not the entire source range.) The final spreadsheet is shown in the following figure (cell C12 has been selected to show the contents):

1.4.9 Error Messages in Excel

Sometimes Excel cannot recognize the function you are trying to use (e.g., when the function name is misspelled) or cannot perform the requested math operation (e.g., because of a divide by zero). When this happens, Excel puts a brief error message in the cell to let you know that something went wrong. Some common error messages are listed in the following table:

MESSAGE	MEANING
#DIV/0	Attempted to divide by zero.
#N/A	Not available. There is an NA() function in Excel that returns #N/A, meaning that the result is "not available." Some Excel functions return #N/A for certain errors, such as failures in using a lookup table. Attempts to do math with #N/A values also return #N/A.
#NAME?	Excel could not recognize the name of the function you tried to use.
#NUM!	Not a valid number. A function or math operation returned an invalid numeric value (e.g., a value too large to be displayed). This error message is also displayed if a function fails to find a solution. (For example, the IRR() function uses an iterative solution to find the internal rate of return and may fail.)
#REF!	An invalid cell reference was encountered. For example, this error message is returned from the VLOOKUP() function if the column index number (offset) points at a column outside of the table range.
#VALUE!	This error can occur when the wrong type of argument is passed to a function, or when you try to use math operators on something other than a value (such as a text string).

When these error messages are displayed, they indicate that Excel detected an error in the formula contained in the cell. The solution involves fixing the formula or the values in the other cells that the formula references.

Sometimes Excel will detect a typographical error when you press [enter] to save a formula in a cell. When Excel detects this type of error, it pops up a message box indicating that there is an error in the formula. For example, the following formula for area of a circle has unbalanced parentheses:

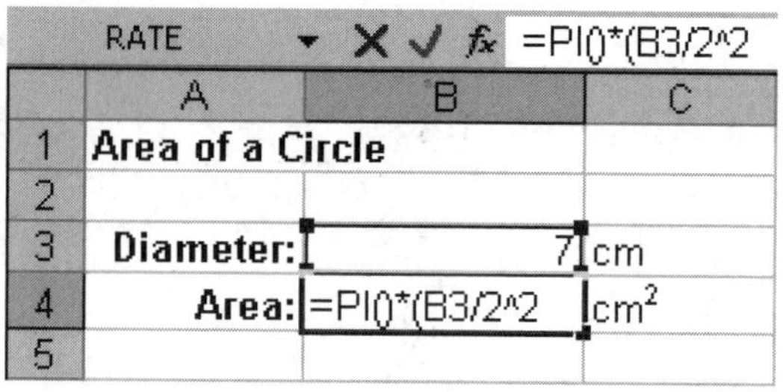

```
=PI()*(B3/2^2  should be  =PI()*(B3/2)^2
```

If you attempt to enter this formula in a cell, Excel will detect the missing parenthesis and display a message box indicating the problem and proposing a correction:

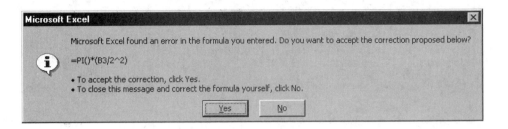

If the proposed correction gives the desired result, simply click the Yes button and let Excel correct the formula. However, if the proposed solution is not correct, click the No button and fix the formula by hand. In this example, the proposed correction would place the closing parenthesis in the wrong place, so the formula must be corrected by hand.

B4	▼	f_x =PI()*(B3/2)^2		
	A	B	C	D
1	Area of a Circle			
2				
3	Diameter:	7	cm	
4	Area:	38.5	cm^2	
5				

1.4.10 Formatting Cells

Excel provides a wide range of formatting options for the displayed contents of cells. To change the way the cell contents are displayed, either use the formatting toolbar (for operations such as font changes, justification, changing the number of displayed digits, or using the percentage format) or use the Format Cells dialog. To open the Format Cells dialog, select the cell or range of cells to format. Then either use menu commands Format/Cells … or right click and select Format Cells … from the pop-up menu:

Clicking on Format Cells … opens the Format Cells dialog:

The Format Cells dialog contains six panels:

- Number: Perhaps mislabeled, the number panel allows you to control how the contents of a cell are displayed, regardless of whether the cell contains a number. The named formats available for displaying numbers include the following:

  - General: Excel tries to display numbers in a very readable form, switching to scientific notation for very large or small values.

  - Number: The number format allows you to set the number of displayed decimal places. It also allows you to select how negative values should be displayed, using either the minus signs common to scientists and engineers or the parentheses of accounting style.

  - Currency: The currency format is much like the number format, but allows you to select a currency symbol, such as $ or €.

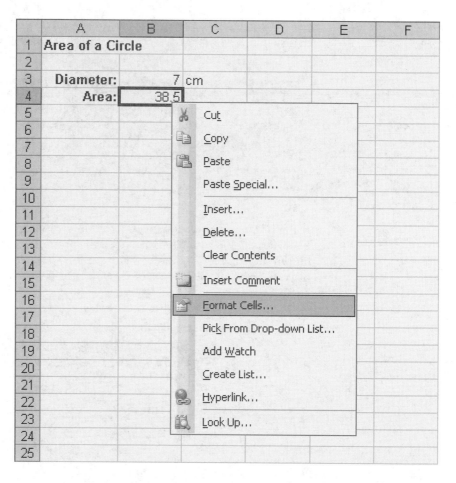

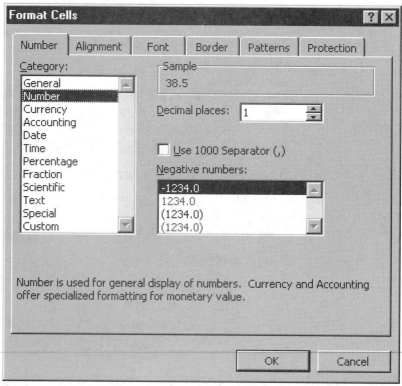

- Accounting: This allows you to indicate the number of decimal places to display and a currency symbol. This format aligns both the currency symbols and the decimal places.

- Percentage: The percentage format multiplies the value displayed in the cell by 100 and adds a percent symbol. For example, if a cell contained the value 0.75, the percentage format would cause the cell contents to be displayed as 75.00% (assuming that you requested two decimal places).

- Fraction: Selecting this option displays the value in the cell as a fraction. You select the number of digits to be used when computing the nearest fraction. For example, if a cell contained the value 0.75, it would be displayed as $^3/_4$ with fraction format, and so would 0.76 if you asked for only one digit to be used. If you requested two digits in the fractions, 0.75 would still be displayed as $^3/_4$, but 0.76 would be displayed as 19/25.

- Scientific: This displays the cell contents in scientific notation. You choose the number of decimal places to include. For example, if a cell contained the value 0.75, it would be displayed as 7.50E-01, using scientific format with two decimal places.

- Custom: This option allows you to define how you want the value to be displayed. For example, the custom format "#,##0.00" tells Excel to always display at least three digits: two after the decimal point (.00) and one before the decimal point (0.). The pound symbol tells Excel to display digits in these positions if the value is large enough. The comma in the format string tells Excel to include a comma as a thousands separator if the number is 1,000 or greater. (If the number exceeds 1,000,000, then two commas will be included.)

The Number panel also includes formats for dates, times, text, and special formats specific to geographical regions. The various styles used for telephone numbers and postal codes are typical special formats.

- Alignment: The alignment panel allows you to control the placement of the displayed contents in the cell. From this panel, you can set the horizontal and vertical alignments and control the text direction and wrapping. You can also merge the selected cells from this panel.

- Font: The font panel provides access to the typical text attributes, such as typeface, font style, size, and color. It also allows you to make the selected text either a subscript or a superscript.

Note: When a cell contains text, you can edit the cell contents (double click on the cell or press [F2]), select a portion of the text (such as a letter), and use the font panel (select Format/Cells ... and use the Font tab) to subscript or superscript just the selected portion of the cell contents.

- Border: The border panel allows you to control the way lines are drawn around selected groups of cells on the spreadsheet. You can choose from a variety of line styles to put a border just around the outside of the selected cells, just inside lines, or both.

- Patterns: The patterns panel allows you to set the background color and pattern of a selected group of cells. The color options displayed on the patterns panel are for the background color. If you select a background color (other than "none"), then the patterns become available. The Pattern drop-down list contains a second set of color options for the pattern color.

- Protection: When you set or clear the locked or hidden check boxes on the protection panel, you do not actually lock or hide anything. Instead, you are telling Excel how to handle the selected cells after the entire spreadsheet is password protected (Tools/Protection/Protect sheet...) If you clear the "locked" check box for a selected group of cells, users will be able to edit those cells after the spreadsheet is protected. If you check the "hidden" check box for a selected group of cells, any formulas in those cells will work, but the formulas will not be displayed in the formula bar after the spreadsheet is protected.

1.4.11 Hyperlinks In Excel

Cells can contain hyperlinks to

- other locations in the same workbook (bookmarks),
- other files on your computer,
- other files on your network,
- locations on the World Wide Web.

To put a hyperlink in a cell, right click in the cell and select Hyperlink from the pop-up menu:

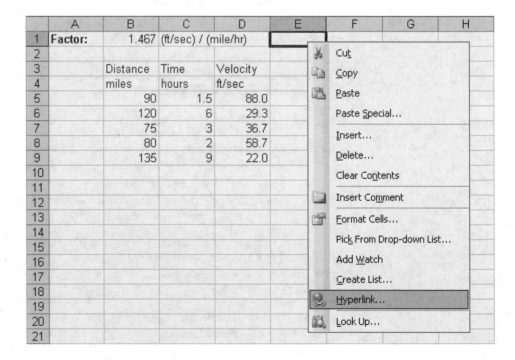

Inserting a hyperlink to another location in the workbook (a bookmark) In the Insert Hyperlink dialog box, enter the text to display in the cell ("See Note"), then click the Bookmark button.

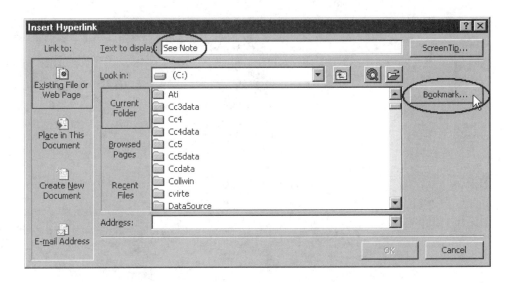

Indicate the location to jump to when the hyperlink is clicked. Here, the hyperlink has been connected to cell A2 on Sheet2:

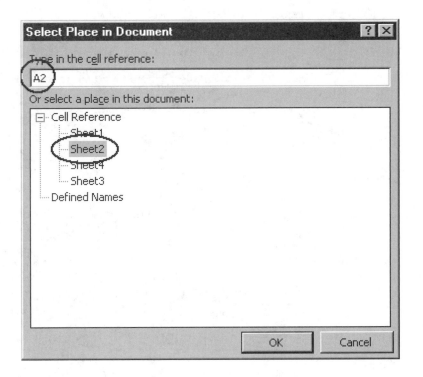

Click OK to return to the Insert Hyperlink dialog. Click the ScreenTip button if you want information about the hyperlink to be displayed when the mouse hovers over the link:

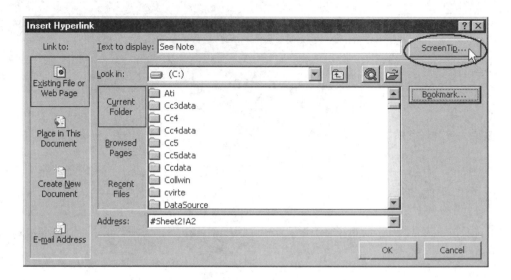

Enter the text to be displayed when the mouse hovers over the link. Click OK to return to the Insert Hyperlink dialog box:

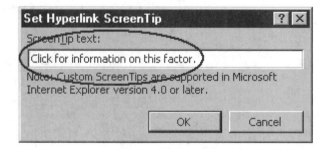

When you click the OK button on the Insert Hyperlink dialog box, a link labeled "See Note" is created from cell E1 on Sheet1 to cell A2 on Sheet2, as shown in the following figure:

	A	B	C	D	E	F	G	H
1	**Factor:**	1.467	(ft/sec) / (mile/hr)		See Note			
2								
3		Distance	Time	Velocity				
4		miles	hours	ft/sec				
5		90	1.5	88.0				
6		120	6	29.3				
7		75	3	36.7				
8		80	2	58.7				
9		135	9	22.0				
10								
11	**Avg:**	100	4.3	46.9				
12	**St Dev:**	26.2	3.2	26.8				
13								

Clicking on the link causes cell A2 on Sheet2 to be displayed. Information about the conversion factor has been entered on Sheet2 so that it is displayed when the hyperlink is clicked:

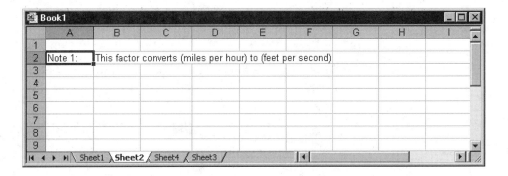

Inserting a hyperlink to a location on the World Wide Web   To insert a hyperlink to the World Wide Web, right click in a cell and select Hyperlink from the pop-up menu. Then enter the Web address or click the Browse the Web... button to start your browser to find the Web page you want to.

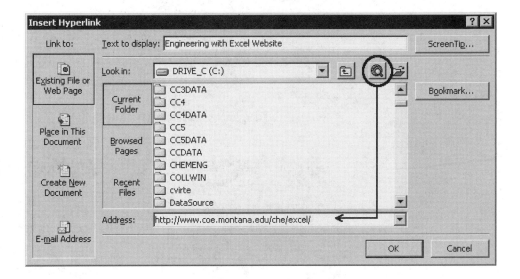

In this example, we are creating a link to this text's website, at *http://www.coe. montana.edu/che/excel*. This website provides access to all of the data sets used in this text, plus solutions to the Practice! problems.

When you have browsed to the desired Web page, return to the Insert Hyperlink dialog box without closing your browser. The address of the Web page displayed in the browser will appear in the Insert Hyperlink dialog's Address field.

Note: You can simply type a Web address into a cell. Excel will recognize that it is a Web address and create a hyperlink automatically.

1.4.12 Splitting and Freezing Panes

Splitting Panes means dividing the displayed portion of the spreadsheet in two or four pieces, each of which can still scroll. **Freezing panes** is very similar, except that the top or left panes will not scroll—they're frozen. The most common uses for freezing and splitting panes is allowing column headings to continue to be displayed as you scroll through a long list of data values and allowing values near the top of the spreadsheet to be visible and readily available as formulas are entered further down the spreadsheet.

Splitting the Display   To split the displayed portion of the spreadsheet into four panes, click the cell that will become the top-left corner of the lower right pane:

	A	B	C	D	E	F
1	**Factor:**	1.467	(ft/sec) / (mile/hr)		See Note	
2						
3		Distance	Time	Velocity		
4		miles	hours	ft/sec		
5		90	1.5	88.0		
6		120	6	29.3		
7		75	3	36.7		
8		80	2	58.7		
9		135	9	22.0		
10						
11	**Avg:**	100	4.3	46.9		
12	**St Dev:**	26.2	3.2	26.8		
13						

Then split the display by selecting Window/Split:

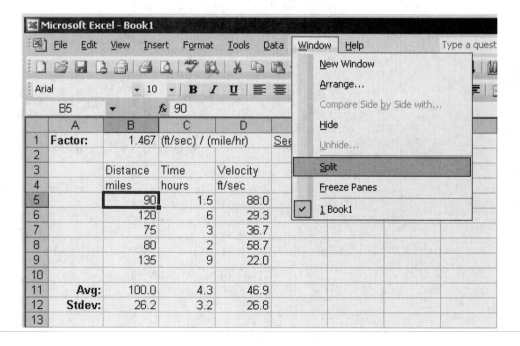

The display will include heavy bars showing the location of the splits:

	A	B	C	D	E	F
1	**Factor:**	1.467	(ft/sec) / (mile/hr)		See Note	
2						
3		Distance	Time	Velocity		
4		miles	hours	ft/sec		
5		90	1.5	88.0		
6		120	6	29.3		
7		75	3	36.7		
8		80	2	58.7		
9		135	9	22.0		
10						
11	**Avg:**	100	4.3	46.9		
12	**St Dev:**	26.2	3.2	26.8		
13						

The split bars can be moved to change the size of the panes, and double clicking on either of the split bars removes that split. You can also select Window/Remove Split to restore the display to a single pane.

To divide the display into two panes, select an entire row or column by clicking on the row or column label. Then select Window/Split. The split will place the selected row in the lower pane or the selected column in the right pane.

Freezing the Display   The process for freezing portions of the display is nearly the same. To divide the display into four panes, click the cell that will become the top left corner of the lower right pane and select Window/Freeze Panes:

	A	B	C	D	E	F
1	**Factor:**	1.467	(ft/sec) / (mile/hr)		See Note	
2						
3		Distance	Time	Velocity		
4		miles	hours	ft/sec		
5		90	1.5	88.0		
6		120	6	29.3		
7		75	3	36.7		
8		80	2	58.7		
9		135	9	22.0		
10						
11	**Avg:**	100	4.3	46.9		
12	**St Dev:**	26.2	3.2	26.8		
13						

The edges of the panes are indicated by thin lines.

Only the lower right pane will scroll in both directions. The lower panes scroll up and down, and the right panes scroll left and right. The top left pane does not scroll at all. To unfreeze the panes, you must select Window/Unfreeze Panes.

To divide the display into two panes, select an entire row or column by clicking on the row or column label. Then select Window/Freeze Panes. Freezing the panes will place the selected row in the lower pane or the selected column in the right pane. Only the lower or right panes will scroll.

APPLICATIONS: ORGANIZE THE INFORMATION ON YOUR SPREADSHEETS

Computer programs, including spreadsheets, typically contain the following standard elements:

- Titles
- Input Values: values entered by the user when the spreadsheet is used
- Parameter Values: values needed by the formulas, but not varied as part of the input
- Formulas: the equations that calculate results based on the input values and parameter values
- Results: the computed values

Your spreadsheets will be easier to create and use if you group these elements together (when possible) and develop a standard placement of these items on your spreadsheets. These elements can be placed anywhere, but, for small problems, a common placement puts titles at the top followed by input values and parameters further down the page. Formulas and results are placed even further down the page.

	A	B	C	D	E	F
1						
2		**Titles**				
3						
4						
5		**Input Values**				
6						
7						
8						
9		**Parameter Values**				
10						
11						
12						
13		**Formulas**				
14						
15						
16						
17		**Results**				
18						
19						
20						

One feature of this approach is that the information flows down the page, making the spreadsheet easier to read. One drawback of this approach is that the results are separated from the input values, making it hard to prepare a report showing the input values and the calculated results together.

The following layout puts the input values next to the results, but can be harder to read, because the information flow is not simply from top to bottom:

The following variation is often used when the input values include a data set:

When the problem requires a lot of space on the spreadsheet (several screens), it can be convenient to put the input values, parameter values, formulas, and results on different spreadsheets within the same workbook:

	A	B	C	D	E	F	G
1							
2		**Title Screen**					
3							
4		**Author:**					
5		**Date:**					
6							
7							

Titles / Input Values / Parameter Values / Formulas / Results /

The specific layout you use in your spreadsheets is up to you, but consistency will make your spreadsheets easier for you to create and others to use.

1.5 PRINTING THE SPREADSHEET

Printing an Excel spreadsheet is typically a two-step process. First, set the area to be printed; and then print. You might also want to set printing options from the Print Preview screen.

1.5.1 Setting the Print Area

Select the region of the spreadsheet that should be printed:

	A	B	C	D	E	F
1	**Factor:**	1.467	(ft/sec) / (mile/hr)		See Note	
2						
3		Distance	Time	Velocity		
4		miles	hours	ft/sec		
5		90	1.5	88.0		
6		120	6	29.3		
7		75	3	36.7		
8		80	2	58.7		
9		135	9	22.0		
10						
11	**Avg:**	100	4.3	46.9		
12	**St Dev:**	26.2	3.2	26.8		
13						

Then select File/Print Area/Set Print Area:

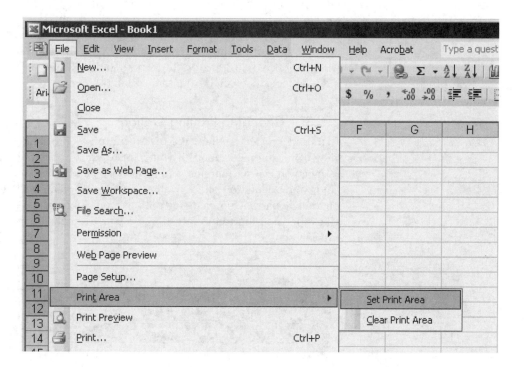

Once a print area has been set, Excel shows the region that will be printed with dashed lines. (If multiple pages are to be printed, Excel also shows the page breaks with dashed lines.)

	A	B	C	D	E	F
1	**Factor:**	1.467	(ft/sec) / (mile/hr)		See Note	
2						
3		Distance	Time	Velocity		
4		miles	hours	ft/sec		
5		90	1.5	88.0		
6		120	6	29.3		
7		75	3	36.7		
8		80	2	58.7		
9		135	9	22.0		
10						
11	**Avg:**	100	4.3	46.9		
12	**St Dev:**	26.2	3.2	26.8		
13						

1.5.2   Printing Using Current Options

Once the desired print area has been set, you can click the print button on the standard toolbar or select File/Print . . . from the main menu. If you use File/Print . . . , the print dialog will be displayed, allowing you to select a printer or change printer properties. If you use the print button on the toolbar, your output will go straight to the printer that was previously selected from the workbook (or to the default printer, if no printer has been previously selected from the workbook).

1.5.3   Changing Printing Options

Use Print Preview (either the Preview button on the standard toolbar or File/Print Preview from the main menu) if you want to change printing options. From the preview screen, you can set margins, add or remove headers or footers, change from portrait to landscape orientation, show gridlines or row and column labels, or force the output to fit on a single page. After changing options, you can print directly from the preview screen. Click the Close button to return to the spreadsheet. The following is an example of the preview screen:

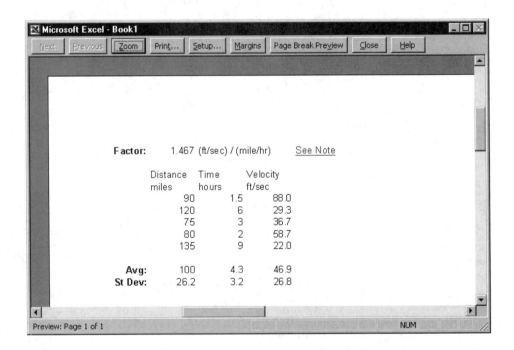

Clicking the Setup button opens the Page Setup dialog box, which provides access to most of the adjustable print features:

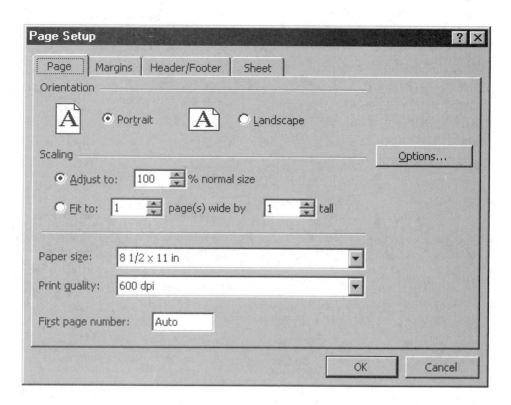

The options you set on the preview screen are saved with your workbook; you don't have to set the print options each time (unless you change workbooks).

1.6 SAVING, OPENING WORKBOOKS, AND EXITING

1.6.1 Saving the Workbook

To save the workbook as a file, press the Save button on the standard toolbar or use File/Save. If the workbook has not been saved before, you will be asked to enter a file name in the Save As dialog box.

Note: In Windows, it is generally preferable not to add an extension to the file name. Excel will allow you to add an extension, but the file name will not appear in any future selection box unless Excel's default extension is used. If you do not include a file-name extension, Excel will add its default extension, .xls.

1.6.2 Opening a Previously Saved Workbook

Select File/Open . . . , and select the workbook file you wish to use from the Open File dialog box.

1.6.3 Exiting Excel

Select File/Exit. If there are open unsaved workbooks, Excel will ask if you want to save your changes before exiting.

APPLICATIONS: FLUID STATICS

Manometer I

Manometers connected to two different locations can be used to measure pressure differences between the two locations. Three manometer configurations are as follows:

    a. Sealed-End Manometer: This is used to determine the absolute pressure at P_L.

    b. Gauge Manometer: This is used to determine the gauge pressure at P_L. (TBP stands for "today's barometric pressure.")

    c. Differential Manometer: This is used to determine the pressure difference between P_L and P_R.

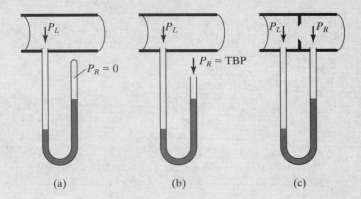

      (a)   (b)   (c)

The general manometer equation applies to all of these configurations:

$$P_L + \rho_L g h_L = P_R + \rho_R g h_R + \rho_M g R. \tag{1.5}$$

In this equation,

P_L is the pressure at the top of the left arm of the manometer,
P_R is the pressure at the top of the right arm of the manometer,
g is the acceleration due to gravity,
ρ_L is the density of the fluid in the upper portion of the left manometer arm,
ρ_R is the density of the fluid in the upper portion of the right manometer arm,
ρ_M is the density of the manometer fluid in the lower portion of the manometer,
h_L is the height of the column of fluid in the upper portion of the left manometer arm,
h_R is the height of the column of fluid in the upper portion of the right manometer arm,
R is the manometer reading, as shown in the figure below:

Simplifications

- For the sealed-end manometer, the space above the manometer fluid on the right side has been evacuated, so $P_R = 0$ and $\rho_R = 0$. The general manometer equation then simplifies to

$$P_L = \rho_M g R - \rho_L g h_L. \tag{1.6}$$

- If a manometer arm contains a gas or vacuum (sealed end on the right side), the density is considered negligible compared with a liquid density, and the term containing the gas density drops out.
- On an open-end manometer, setting today's barometric pressure to zero ($P_R = 0$) causes the calculated pressure to be a gauge pressure.
- For a differential manometer, the difference in column heights on the left and right sides is related to the manometer reading ($h_L - h_R = R$), and the general manometer equation simplifies to

$$P_L - P_R = R_g(\rho_M - \rho_F), \tag{1.7}$$

where ρ_F is the density of the fluid in the left and right arms of the manometer ($\rho_F = \rho_L = \rho_R$).

Calculate the following pressures and express your result in atmospheres:

a. Water ($\rho_L = 1,000$ kg/m^3) flows in the pipe connected to a sealed-end manometer. The height of the water column on the left side is $h_L = 35$ cm. The manometer reading is 30 cm of mercury ($\rho_M = 13,600$ kg/m^3). What is the absolute pressure, P_L?

	A	B	C	D	E	F
1	**a) Sealed-End Manometer**					
2						
3		ρ_L	1000	kg/m^3		
4		ρ_M	13600	kg/m^3		
5		g	9.8	m/s^2		
6		h_L	0.35	m		
7		R	0.30	m		
8						
9		P_L	36554	kg/m s^2 = N/m^2 = Pa		
10		P_L	0.3608	atm		
11						
12	**Formulas:**					
13		C9:	=C4*C5*C7-C3*C5*C6			
14		C10:	=C9/101325			
15						

b. Air ($\rho_L \approx 0$) flows in the pipe connected to an open-ended manometer. The manometer reading is 12 cm of mercury ($\rho_M = 13{,}600$ kg/m^3). The barometric pressure when the manometer was read was 740 mm Hg. What is the *gauge* pressure, P_L?

	A	B	C	D	E	F
1	**b) Open-End Manometer**					
2						
3		ρ_M	13600	kg/m^3		
4		g	9.8	m/s^2		
5		R	0.12	m		
6						
7		P_L	15993.6	kg/m s^2 = N/m^2 = Pa		
8		P_L	0.1578	atm (gauge)		
9						
10	**Formulas:**					
11		C7:	=C3*C4*C5			
12		C8:	=C7/101325			
13						

c. Air ($\rho_L \approx 0$) flows in the pipe connected to an open-ended manometer. The manometer reading is 12 cm of mercury ($\rho_M = 13{,}600$ kg/m^3). The barometric pressure when the manometer was read was 740 mm Hg. What is the *absolute* pressure, P_L?

	A	B	C	D	E	F
1	**c) Open-End Manometer**					
2						
3		ρ_M	13600	kg/m^3		
4		g	9.8	m/s^2		
5		R	0.12	m		
6						
7		P_L	15993.6	kg/m s^2 = N/m^2 = Pa		
8		P_L	0.1578	atm (gauge)		
9		P_R	740	mm Hg		
10		P_R	0.9737	atm (TBP)		
11						
12		P_L	1.1315	atm (abs.)		
13						
14	**Formulas:**					
15		C7:	=C3*C4*C5			
16		C8:	=C7/101325			
17		C10:	=C9/760			
18		C12:	=C8+C10			
19						

d. Water ($\rho_L = \rho_R = 1,000$ kg/m^3) flows in a pipe connected to a differential manometer. The manometer reading is 30 cm of mercury ($\rho_M = 13,600$ kg/m^3). What is the pressure difference, $P_L - P_R$?

	A	B	C	D	E	F
1	d) Differential Manometer					
2						
3		ρ_L	1000	kg/m^3		
4		ρ_M	13600	kg/m^3		
5		g	9.8	m/s^2		
6		R	0.30	m		
7						
8		P$_L$-P$_R$	37044	kg/m s^2 = N/m^2 = Pa		
9		P$_L$-P$_R$	0.3656	atm		
10						
11	Formulas:					
12		C8:	=C6*C5*(C4-C3)			
13		C9:	=C8/101325			
14						

KEY TERMS

Absolute cell addressing
Active cell
Automatic recalculation
Built-in functions
Cell address
Cell cursor
Cells
Edit mode
Electronic spreadsheet
Fill handle

Formatting toolbar
Formula
Formula bar
Freeze panes
Label
Main menu
Name box
Named cells
Range of cells
Relative cell addressing

Split panes
Spreadsheet
Standard toolbar
Status Bar
Title bar
Toolbar
Value
Workbook

SUMMARY

Cell Contents   Cells can contain labels (text), numbers, and equations (formulas). Excel interprets what you type into a cell as follows:

- If the entry begins with an equal sign (or a plus sign followed by something other than a number), it is considered a formula.
- If the entry begins with a number, a currency symbol followed by a number, or a minus sign, Excel assumes that you are entering a value.
- If your entry cannot be interpreted as a formula or a value, Excel treats it as a label.

Cell References: Relative and Absolute Addressing   In formulas, cells are identified by using the cell's address, such as G18. Dollar signs are used to prevent the row index (18) and the column index (G) from incrementing if the formula is copied to other cells.

For example—

=$G18+2   Here, the column index G will not change if the formula is copied. If the formula is copied down the column, the 18 will automatically be incremented as the formula is entered into each new cell.

=G$18+2   In this case, the row index 18 will not change if the formula is copied. If the formula is copied across the row, the G will automatically be incremented as the formula is entered into each new cell.

=$G$18+2   In this case, neither the column index G nor the row index 18 will change if the formula is copied.

Selecting Cell Ranges

- To select a single cell, simply click on the cell, or use the arrow keys to locate the cell. The selected cell becomes the active cell.
- To select multiple cells in a rectangular region, click on a corner cell and move the mouse cursor to the opposite corner of the rectangular region. Or use the arrow keys to select the first corner cell and hold the [Shift] key down while moving the cell cursor to the opposite corner of the rectangular region.

Copying Cell Ranges by Using the Clipboard

1. Select the source cell or range of cells.
2. Copy the contents of the source cell(s) to the clipboard by selecting Edit/Copy.
3. Select the destination cell or range of cells.
4. Paste the clipboard contents into the destination cell(s) by selecting Edit/Paste.

Copying Cell Ranges by Using the Fill Handle

1. Select the source cell or range of cells.
2. Click the fill handle with the mouse and drag it to copy the selected cell contents to the desired range.

Editing a Cell's Contents   Double click the cell or select the cell and press [F2].

Printing a Spreadsheet

1. Select the area to be printed and then select File/Print Area/Set Print Area.
2. Press the Print button or the Print Preview button on the toolbar select or File/Print . . . or File/ Print Preview.

Setting Print Options   Select File/Print Preview and click the Setup button.

Saving a Workbook   Press the Save button on the toolbar, or select File/Save.

Opening a Previously Saved Workbook   Press the Open button on the toolbar, or select File/Open. . . .

Exiting Excel   Select File/Exit, or click the close button at the top right corner of the Excel window.

Problems

Basic Fluid Flow I

1. Water flows through a nominal 100-mm (4-in), schedule 40 steel pipe (actual inside diameter $D_i = 102.3$ mm) at average velocity $v = 2.1$ m/s.

    a. Calculate the area available for flow through the pipe,

    $$A_{\text{flow}} = \frac{\pi}{4} D_i^2,$$ (1.8)

    and the volumetric flow rate

    $$\dot{V} = v A_{\text{flow}}.$$ (1.9)

    b. What pipe diameter should be used to reduce the fluid velocity to 1.3 m/s?

Fluid Statics: Pressure at the Bottom of the Pool

2. Swimmers can feel the pressure increase on their ears as they swim to the bottom of the pool.

    a. Calculate the pressure in atmospheres (absolute) at the bottom of a swimming pool (depth = 3 m) on a day when the barometric pressure is 0.95 atm.

    b. If the swimming pool were filled with salt water instead of fresh water, would the pressure at 3 m be the same, higher, or lower? Why?

Resistance Temperature Detector I

3. A *resistance temperature detector* (RTD) is a device that uses a metal wire or film to measure temperature. The electrical resistance of the metal depends on temperature, so, by measuring the resistance of the metal, you can calculate the temperature. The equation relating temperature to resistance is

    $$R_T = R_0[1 + \alpha T],$$ (1.10)

    where

R_T	is the resistance at the unknown temperature T,
R_0	is the resistance at 0°C (known, one of the RTD specifications), and
α	is the linear temperature coefficient (known, one of the RTD specifications).

    The common grade of platinum used for RTDs has the α value 0.00385 ohm/ohm/°C (sometimes written simply as $0.00385°C^{-1}$).

    a. For a 100-ohm platinum RTD (i.e., $R_0 = 100$-ohm) with the α value $0.00385°C^{-1}$, calculate the resistance that would be measured when the RTD is at the following temperatures:

    (i)   100°C.
    (ii) 250°C.
    (iii) 400°C.

b.   Calculate the temperature of the RTD when the resistance of the RTD is

(i)   68 ohms.

(ii)   109.7 ohms.

(iii)   152 ohms.

Resistance Temperature Detector II

4.   The *linear temperature coefficient* α, of a *resistance temperature detector* (RTD) is a physical property of the metal used to make the RTD, one that indicates how the electrical resistance of the metal changes as the temperature increases. The equation relating temperature to resistance is

$$R_T = R_0[1 + \alpha T], \tag{1.11}$$

where

R_T is the resistance at the unknown temperature T.

R_0 is the resistance at $0°C$ (known, one of the RTD specifications), and

α is the linear temperature coefficient (known, one of the RTD specifications).

The common grade of platinum used for RTDs has the α value 0.00385 ohm/ohm/°C (sometimes written simply as $0.00385°C^{-1}$). However, older RTDs used a different grade of platinum and operated with $\alpha = 0.003902°C^{-1}$, and laboratory grade RTDs use very high-purity platinum with $\alpha = 0.003923°C^{-1}$. If the wrong α value is used to compute temperatures from RTD readings, the computed temperatures will be incorrect.

A laboratory grade RTD has accidentally been installed in an industrial furnace. The computer system that controls the temperature in the furnace assumes that the α value will be $0.00385°C^{-1}$. If the control system (using $\alpha = 0.00385°C^{-1}$) indicates that the temperature is 500°C, what is the true temperature (using $\alpha = 0.003923°C^{-1}$)?

Ideal Gas Equation I

5.   A glass cylinder fitted with a movable piston contains 5 g of ammonia gas (NH_3). When the gas is at room temperature (25°C), the piston is 2 cm from the bottom of the container. The pressure on the gas is 1 atm.

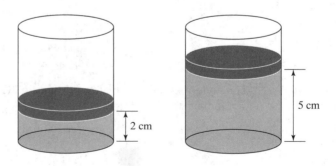

The ideal gas law is

$$PV = nRT, \tag{1.12}$$

where the symbols are defined as follows:

TABLE 1.1

	Description	Typical Units
P	absolute pressure of the gas	atm
V	volume of gas in the container	liters (L)
N	number of moles of gas in the container	gram moles (gmol)
T	absolute temperature	Kelvins (K)

273 + 25
298

If R is the ideal gas constant, which has a value of 0.08206 L · atm/gmol · K,

    a.   What is the volume of the glass cylinder (in liters)?

If the gas is then heated at constant pressure until the piston is 5 cm from the bottom of the container,

    b.   What is the final temperature of the gas (in Kelvins)?

(*Hint*: Assume that ammonia behaves as an ideal gas under these conditions and that the molecular weight of ammonia is 17 g/mol.)

Ideal Gas Equation II

6.   Nitrogen gas is added to a glass cylinder 2.5 cm in radius, fitted with a movable piston. Gas is added until the piston is lifted 5 cm off the bottom of the glass. At that time, the pressure on the gas is 1 atm, and the gas and cylinder are at room temperature (25°C).

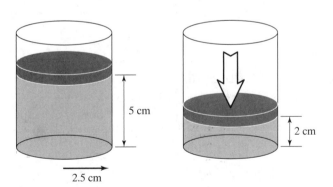

    a.   How many moles of nitrogen were added to the cylinder?

If pressure is then applied to the piston, compressing the gas until the piston is 2 cm from the bottom of the cylinder,

    b.   When the temperature returns to 25°C, what is the pressure in the cylinder (in atm)?

(*Hint*: Assume that nitrogen behaves as an ideal gas under these conditions, and that the molecular weight of nitrogen (N_2) is 28 g/mol.)

Pulley I

7.   A 200-kg mass is suspended from a pulley. Both ends of the cord around the pulley are connected to the support. The cord ends hang straight down from the support.

What is the tension in each cord?

Pulley II

8.   A 200-kg mass is suspended from a pulley as shown in the accompanying figure. One end of the cord around the pulley is connected to the overhead support, and the other wraps around a second pulley. All cords hang vertically.

   a.   What force must be applied to the free end of the cord to keep the mass from moving?

   b.   If the applied force is 1.3 times that calculated in part (a), at what rate is the mass accelerating (neglect the mass of the pulley system and friction)?

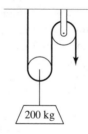

Pulley III

9.   The block and tackle shown in the accompanying figure has two pulleys on the lower portion and three on the upper portion. What forces must be applied on the right and left side cords to keep the 200-kg mass from moving? (Assume a frictionless system.)

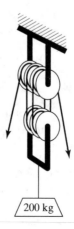

Basic Fluid Flow II

10. The Reynolds number is a dimensionless quantity that is commonly used in fluid flow problems. For flow in pipes, a Reynolds number less than 2,100 lets you know that the flow will be laminar. If the Reynolds number is greater than 10,000, the flow will certainly be turbulent. For Reynolds numbers between 2,100 and 10,000, the flow type can't be known with certainty, although turbulent flow is more likely, especially at the higher end of the range. The Reynolds number is defined as

$$N_{\mathrm{Re}} = \frac{D_i \bar{V} \rho}{\mu}, \tag{1.13}$$

where the symbols are defined as follows:

	DESCRIPTION	TYPICAL UNITS
D_i	inside diameter of pipe	mm
V	average velocity of fluid in pipe	m/s
ρ	fluid density	kg/m^3
μ	absolute viscosity of fluid	Pa·s

Calculate the Reynolds numbers for each of the following cases and state whether the type of flow will be laminar, turbulent, or unknown (*caution: watch your units to be sure that they all cancel when you calculate the Reynolds number.*):

a. Water at room temperature ($\rho = 998$ kg/m^3, $\mu = 9.82 \times 10^{-4}$ Pa·s) flows through a nominal 100-mm, schedule 40 steel pipe ($D_i = 102.3$ mm) at an average velocity of 2.1 m/s.

b. Water at 90°C ($\rho = 958$ kg/m^3, $\mu = 3.1 \times 10^{-4}$ Pa·s) flows through a nominal 100-mm, schedule 40 steel pipe ($D_i = 102.3$ mm) at an average velocity of 2.1 m/s.

c. Honey at room temperature ($\rho = 1400$ kg/m^3, $\mu = 10$ Pa·s) flows through a nominal 100-mm, schedule 40 steel pipe ($D_i = 102.3$ mm) at an average velocity of 2.1 m/s.

d. Warm air ($T = 100$°C, $\rho = 0.95$ kg/m^3, $\mu = 2.1 \times 10^{-5}$ Pa·s) flows through a round duct with an inside diameter of 0.61 m at an average velocity of 6.3 m/s.

2

Graphing with Excel

2.1 INTRODUCTION

An Excel spreadsheet is a convenient place to generate graphs. The process is quick and easy, and Excel gives you a lot of control over the appearance of the graph. Once a graph is created, you can then begin analyzing your data by adding a trendline to a graphed data set with just a few mouse clicks. The majority of Excel's trendlines are regression lines, and the equation of the best-fit curve through your data set is immediately available. Excel graphs are great tools for visualizing and analyzing data.

2.2 GETTING READY TO GRAPH

You must have some data in the spreadsheet before creating a graph, and those data can come from a variety of sources. Creating a graph from values that were calculated within the spreadsheet is probably the most common, but data can also be imported from data files, copied from other programs (such as getting data from the Internet by using a browser), or perhaps even read directly from an experiment that uses automated data acquisition. Once the data, from whatever source, are in your spreadsheet, there are some things you can do to make graphing quicker and easier. That is the subject of this section on getting ready to create a graph.

Excel provides a *Chart Wizard* to speed the process of creating a graph. The wizard will try to analyze your data and automatically fill in a number of required fields to help create the graph. You can assist the process by laying out your data in a standard form. A typical set of data for plotting might look like this:

OBJECTIVES

After reading this chapter, you will know

• How to import data for graphing
• How to create an XY Scatter plot
• How to modify plot formatting
• How to add trend lines to graphs
• How to add error bars to graphs

	A	B	C
1	**Temperature vs. Time Data**		
2			
3	Time (sec.)	Temp. (°C)	
4	0	54.23	
5	1	45.75	
6	2	28.41	
7	3	28.30	
8	4	26.45	
9	5	17.36	
10	6	17.64	
11	7	9.51	
12	8	5.76	
13	9	8.55	
14	10	6.58	
15	11	4.62	
16	12	2.73	
17	13	2.91	
18	14	0.32	
19	15	1.68	
20			

Excel is fairly flexible, but the typical data layout for an *X–Y* graph, such as the preceding, includes the following elements:

- The data to be plotted on the *x*-axis (Time) are stored in a single column.
    - No blank rows exist between the first *x* value and the last *x* value.
- The data to be plotted on the *y*-axis (Temp.) are stored in a single column to the right of the column of *x* values.
    - No blank rows exist between the first *y* value and the last *y* value.
    - The series name may be included in the cell directly above the first *y* value. If the top cell in the *y* values column contains text rather than a number, Excel will use that text as the name of the series and will include that name in the graph's legend.

Keeping data in columns is the most common practice, but rows will also work. The same data set shown above would look as follows if it were stored in rows. (Only the first seven values of each row have been shown.)

	A	B	C	D	E	F	G	H
1	**Temperature vs. Time Data**							
2								
3	Time (sec.)	0	1	2	3	4	5	6
4	Temp. (°C)	54.23	45.75	28.41	28.30	26.45	17.36	17.64
5								

By putting your data into a standard form, you make it possible for the Chart Wizard to assist in the creation of graphs, so creating graphs in Excel becomes even easier.

2.3   CREATING AN XY SCATTER PLOT

Once you have a block of data in the spreadsheet, it can be plotted by following these steps:

1.   Select the Data Range
2.   Start the Chart Wizard
3.   Select the Chart Type
4.   Check the Data Series
5.   Format Basic Graph Features
6.   Select the Location for the Graph

Select the Data Range

The majority of graphs used by engineers are XY Scatter plots, so preparing this type of graph is covered here. Other types of graphs available in Excel are described in Section 2.7.

For an XY Scatter plot, select the two columns (or rows) of data to be graphed and call up Excel's Chart Wizard to assist you with this task. Excel assumes that the first column (or row) of data contains the x values and that the second column (or row) contains the y values.

To select the data to be plotted, simply click on the cell containing the first x value and hold the left mouse button down as you move the mouse pointer to the cell containing the last y value. In the temperature-versus-time data shown here, cells A4 through B19 have been selected for graphing.

	A	B	C
1	**Temperature vs. Time Data**		
2			
3	Time (sec.)	Temp. (°C)	
4	0	54.23	
5	1	45.75	
6	2	28.41	
7	3	28.30	
8	4	26.45	
9	5	17.36	
10	6	17.64	
11	7	9.51	
12	8	5.76	
13	9	8.55	
14	10	6.58	
15	11	4.62	
16	12	2.73	
17	13	2.91	
18	14	0.32	
19	15	1.68	
20			

Including Series Names

Excel will interpret any text in the top row of the column of *y* values as the name of the series and will automatically use this name in the legend that is created for the graph. If you want to use this feature, then include the column headings in the selected range (A3:B19 in this example).

	A	B	C
1	**Temperature vs. Time Data**		
2			
3	Time (sec.)	Temp. (°C)	
4	0	54.23	
5	1	45.75	
6	2	28.41	
7	3	28.30	
8	4	26.45	
9	5	17.36	
10	6	17.64	
11	7	9.51	
12	8	5.76	
13	9	8.55	
14	10	6.58	
15	11	4.62	
16	12	2.73	
17	13	2.91	
18	14	0.32	
19	15	1.68	
20			

Note: The degree symbol may be entered in several ways:

1. *From the numeric keypad (not the numbers at the top of the keyboard), press [Alt-0176]; that is, hold down the [Alt] key while pressing 0176 on the numeric keypad.*
2. *Use Insert/Symbol, and select the degree symbol from the list of available symbols.*

Start the Chart Wizard

Click on the Chart Wizard button on the standard toolbar or select Insert/Chart . . . .

Select a Chart Type

The Chart Wizard will first ask you to select a graph type. The most common graph type for scientific work is the XY (Scatter) graph, which has been highlighted in the menu illustrated next. You can also choose how you want the points connected. In the following figure, *smooth curves with data-point markers* has been selected:

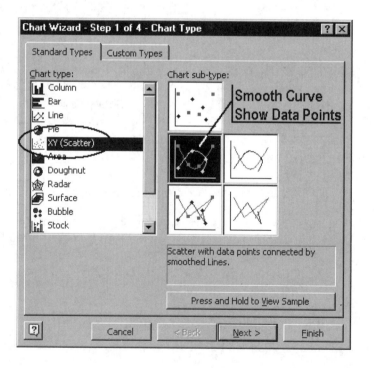

Click Next > to continue.

Check the Data Series

The Chart Wizard indicates the range of cells containing the data and shows a preview of what your graph will look like, using all the default parameters. On this screen, click on the Series tab: .

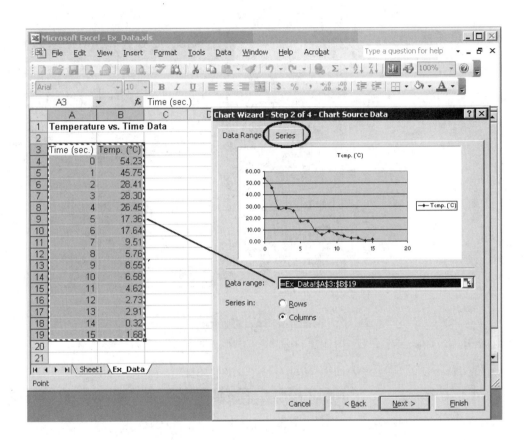

On the Series panel, notice that the contents of cell B3 are being used as the name of the series. This is because we included the y-column heading in the range selected for graphing. The series name is also used, by default, in the legend and as the graph title. If you want to use a different name for the series (and in the legend), enter the new name in the Name field. The graph title may be changed in the next step of the Chart Wizard.

Press Next $>$ to continue.

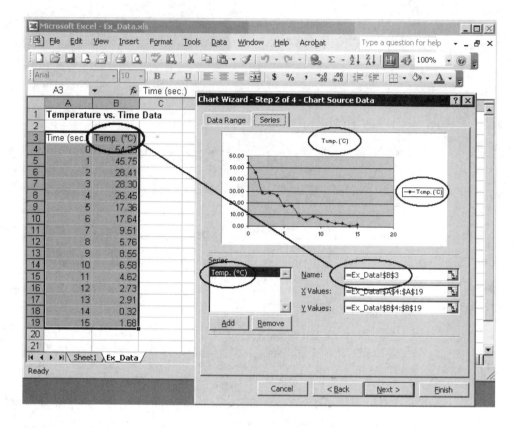

Formatting the Graph

We can now enter titles for the graph:

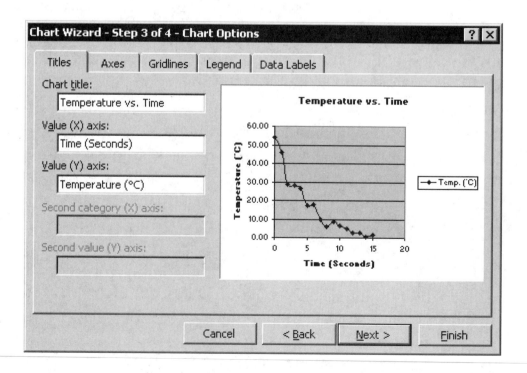

Note: The degree symbol must be entered from the numeric keypad as [Alt-0176], since Insert/Symbol is not available.

You can also use this screen to change a variety of other chart features. For example, the *legend* is taking up a lot of space, so let's move it to the bottom of the chart. To do this, click on the Legend tab, and select "Bottom."

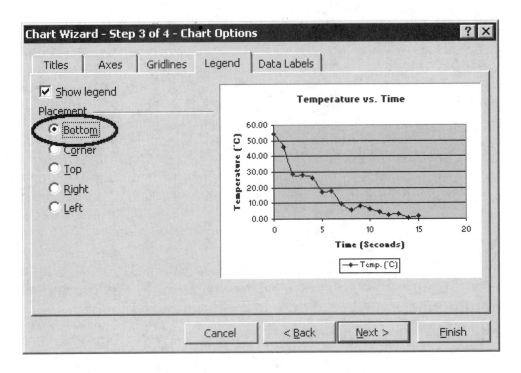

If you did not want a legend, you would clear the check box in front of the "Show legend" label.

When you are done changing chart options, press Next > to continue.

Placing the Graph on the Worksheet

The final Chart Wizard screen asks where you want to put your graph. The default is to place the graph on the current spreadsheet, called Ex_Data in this example. Click "Finish" to accept the default:

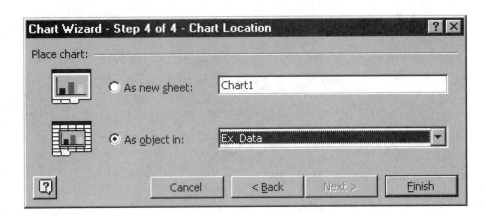

At this point, the Chart Wizard places your graph on the spreadsheet:

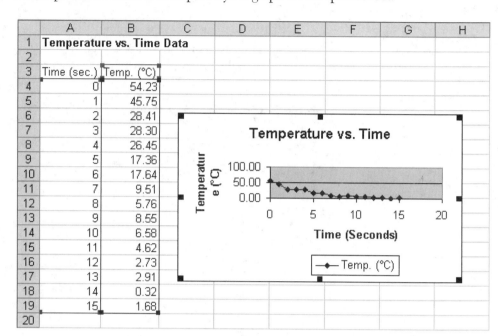

The size and location are set by default and can look unappealing, but they can be changed. Also, you can continue to edit the graph after it has been placed on the spreadsheet.

2.4 EDITING AN EXISTING GRAPH

2.4.1 Modifying Graph Features

You can still change the appearance of the graph. Click anywhere on the graph to enter edit mode. A graph in edit mode is indicated by squares (called *handles*) located around the border. Click and drag on any white space inside the graph to move the graph

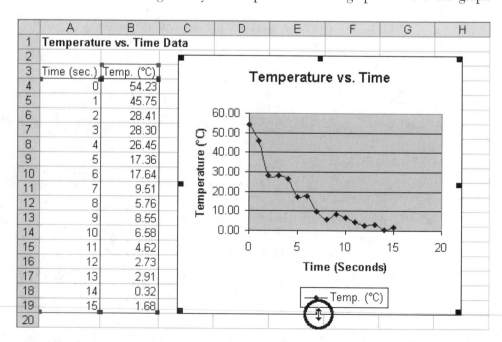

around on the spreadsheet. Or grab a handle with the mouse and drag it to change the size of the graph as follows:

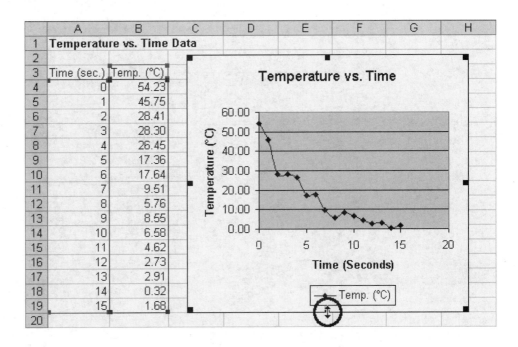

There are three ways to change an existing graph (after entering edit mode):

1. Right click on the graph to bring up a menu of options.
2. Double click on a particular feature of the graph (e.g., an axis, a curve, or the legend) to bring up a dialog box for that feature. For example, double clicking the graph's background brings up the following dialog box:

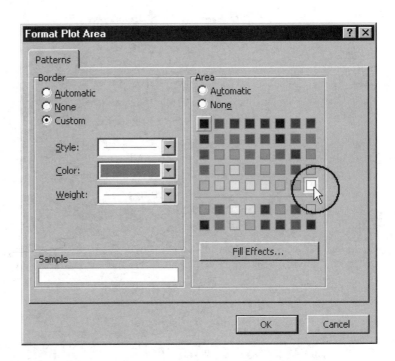

To get rid of the shaded background, change the properties of the *plot area*. Click on the white color sample to change the background color to white.

3. Select a feature from the Chart toolbar. To activate the Chart toolbar, select View/Toolbars/Chart. The Chart toolbar is displayed only when a graph is in edit mode:

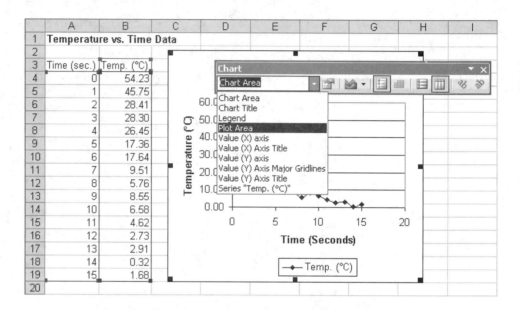

There are many properties of the chart, axes, and curves that may be set. As a typical illustration, consider changing the scaling on the *x*-axis from automatic (the default) to manual with a maximum value of 15. To change the scaling on the *x*-axis, double click it. The Format Axis dialog box will appear:

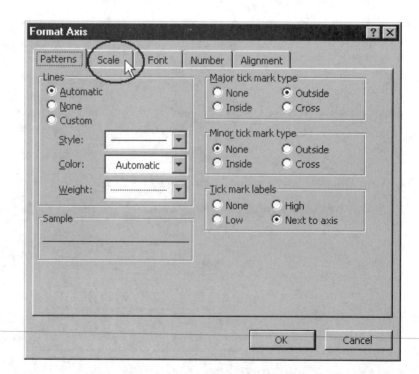

The Format Axis dialog box has a number of panels. In the Scale section, the maximum value displayed on the *x*-axis was changed from 20 to 15. When you enter a value for an axis limit, the Auto (*automatic scaling*) check box for that limit is automatically cleared:

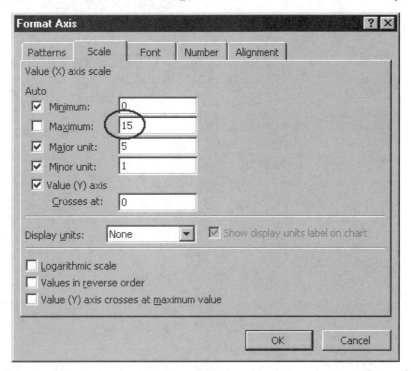

Note that this is also the dialog box that is used to change the axis to log-scale if that is desired.

At this point, the graph should resemble the following:

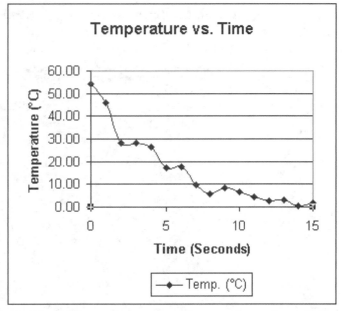

Notice that the values on the *y*-axis are shown with two decimal places. The *y* values in column B were formatted to show two decimal places, and Excel carried that formatting into the label values displayed on the *y*-axis. To change the number of displayed decimal places on axis labels, click on the axis to select it, and then use the Increase Decimal or Decrease Decimal button on the Format Toolbar.

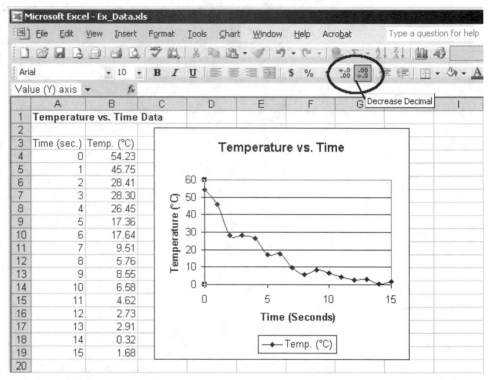

2.4.2   Adding a TrendLine

A *trendline* is a curve that has been fit to the data. Excel will automatically put several types of trendline on charts. To see the trendline clearly, we need to get rid of the original line connecting the dots.

Removing the Smooth Curve Connecting the Data Points   To get rid of the current line, double click it. The Format Data Series dialog box is displayed. Select the Patterns panel. (It is the default panel and normally is displayed.) Set the Line pattern to None:

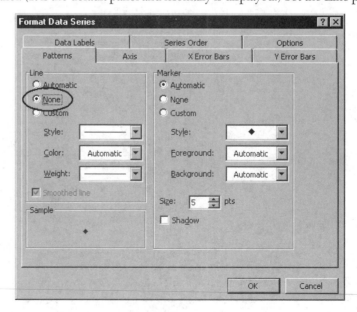

The graph should now resemble the following:

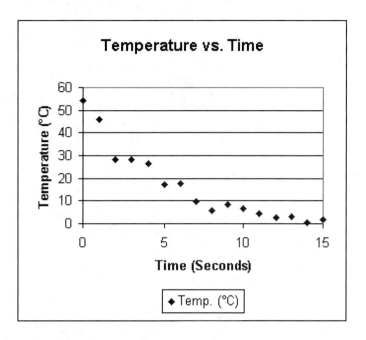

Adding a Trendline   To add a trendline, right click any of the data markers and select Add Trendline . . . from the pop-up menu, as shown:

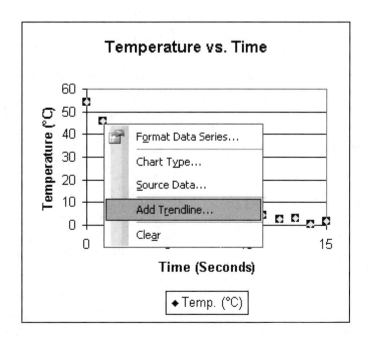

The Add Trendline dialog box is then displayed:

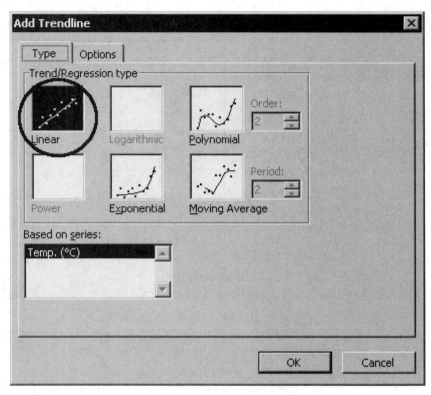

We'll start with a linear trendline. It's a very poor choice for this data, but will show how Excel fits trendlines to data.

Note: The options tab can be used if you want the equation of the trendline to be displayed on the graph.

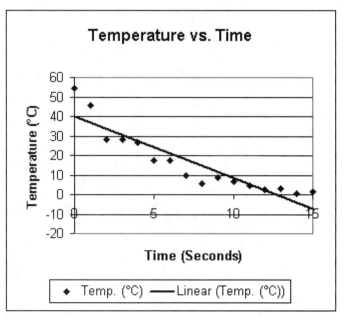

The linear trendline is the straight line that best fits the data. It doesn't fit very well, because the relationship between time and temperature shown in the data is not actually linear. Double click the trendline to edit its properties to change the type of trendline used, as follows:

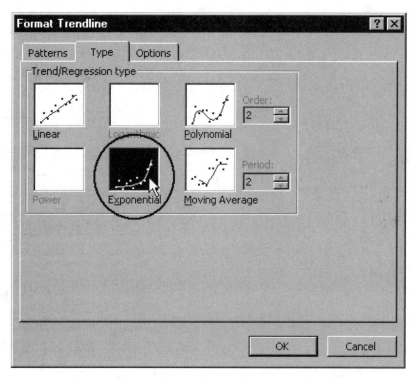

Use the Options panel to display the equation of the trendline and the R^2 value (coefficient of determination).

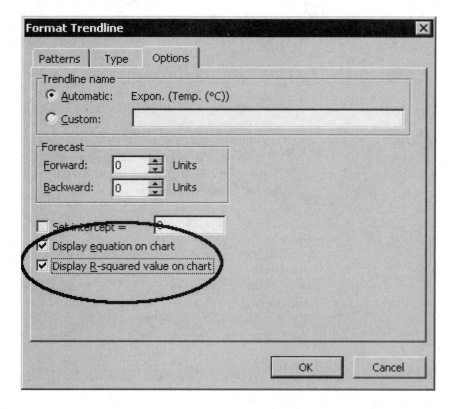

We'll select an exponential trendline, because the decline in temperature over time looks something like an exponential decay curve. The exponential trendline looks like a better fit to the data:

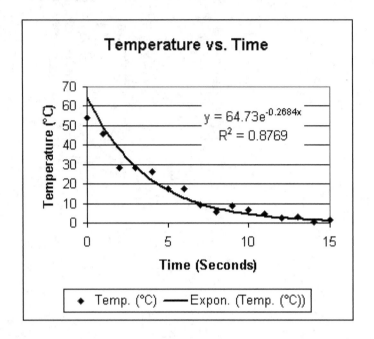

2.4.3   Adding Error Bars

Excel can automatically add *x*- or *y-error bars*, calculated from the data in several ways:

- Fixed Value: A fixed value is added to and subtracted from each data value and plotted as an error bar.
- Fixed Percentage: A fixed percentage is multiplied by each data value, and the result is added to and subtracted from the data value and plotted as an error bar.
- Standard Deviation: The selected standard deviation of all of the values in the data set is computed and then added to and subtracted from each data value and plotted as an error bar.
- Standard Error: The standard error of all of the values in the data set is computed and then added to and subtracted from each data value and plotted as an error bar.

Additionally, you can calculate the values you want plotted as error bars by using the Custom error bar option.

To demonstrate how to add error bars to a graph, fixed-percentage (30%) error bars will be added to the temperature values (*y* values).

To bring up the Format Data Series dialog, double click any of the data markers on the graph or select Series "Temp.(°C)" from the chart toolbar's drop-down list and click the Format Data Series button on the toolbar. Select the first display option to display error bars both above and below the data marker, and set the fixed percentage to 30%.

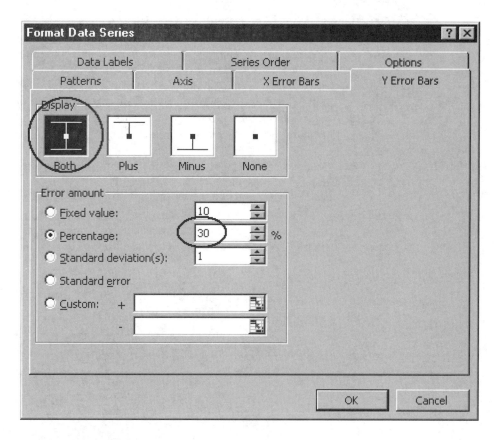

This process yields the following result:

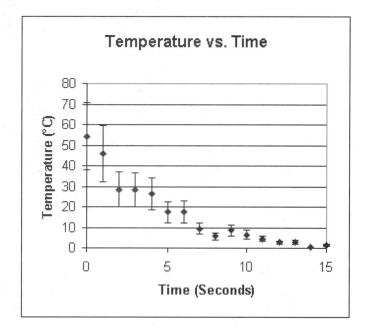

The trendline was removed from this graph to improve the visibility of the markers; the legend was removed to allow the plot area to be expanded.

2.5 CREATING GRAPHS WITH MULTIPLE CURVES

There is a particularly simple way to create an XY scatter graph with multiple curves *if only one column (or row) of x values is needed*: You simply select all of the columns (or rows) of data [one column (or row) of *x* values, multiple columns (or rows) of *y* values], and follow the Chart Wizard steps for creating the graph. Excel will assume that the leftmost column (or top row) contains the *x* values and that all other columns (rows) contain *y* values. Each column (or row) of *y* values becomes a curve on the graph.

EXAMPLE 2.1

To demonstrate plotting multiple series on a single plot, we will graph sine and cosine curves over the range from 0 to 2π radians. The angle values in degrees are included in the data set but will not be plotted.

If the data are available in the spreadsheet, begin creating the graph by selecting the three columns of data. Include the column headings in the range of data values to be plotted if you want to make Excel use the headings to name the series.

	A	B	C	D	E
1	Sine and Cosine Data				
2					
3	θ (degrees)	θ (radians)	Sin(θ)	Cos(θ)	
4	0	0.000	0.000	1.000	
5	20	0.349	0.342	0.940	
6	40	0.698	0.643	0.766	
7	60	1.047	0.866	0.500	
8	80	1.396	0.985	0.174	
9	100	1.745	0.985	-0.174	
10	120	2.094	0.866	-0.500	
11	140	2.443	0.643	-0.766	
12	160	2.793	0.342	-0.940	
13	180	3.142	0.000	-1.000	
14	200	3.491	-0.342	-0.940	
15	220	3.840	-0.643	-0.766	
16	240	4.189	-0.866	-0.500	
17	260	4.538	-0.985	-0.174	
18	280	4.887	-0.985	0.174	
19	300	5.236	-0.866	0.500	
20	320	5.585	-0.643	0.766	
21	340	5.934	-0.342	0.940	
22	360	6.283	0.000	1.000	
23					

Note: The Greek letter θ (theta) was inserted into the spreadsheet by using Insert/Symbol. Including the column headings within the range of data to be plotted will make the Greek letters also appear in the graph's legend.

Start the Chart Wizard, and select an XY (Scatter) plot. The wizard will assume that the left column (column B, containing θ values in radians) is to be plotted on the *x*-axis and that the other columns [Sin(θ) and Cos(θ)] should be plotted as the *y* values for series 1 and series 2.

In the Wizard's Step 2 of 4—Chart Source Data—you can see that two series appear in the Series list. Because the $\text{Sin}(\theta)$ and $\text{Cos}(\theta)$ column headings were included in the selected data, the series are named $\text{Sin}(\theta)$ and $\text{Cos}(\theta)$.

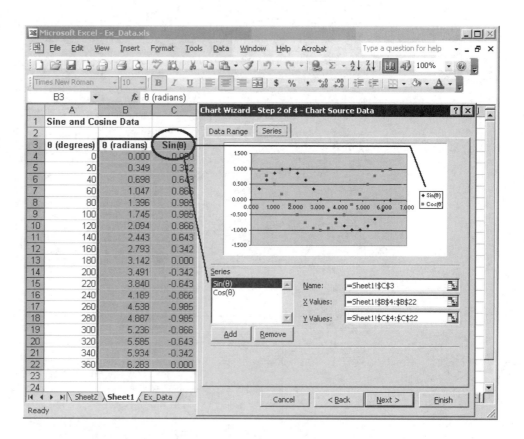

Note: If column headings were not included in the data range to be graphed, the series would have been given the default names: "Series 1" and "Series 2".

In the Chart Wizard's next step, the *x*-axis title was added. Insert/Symbol is not available while the Chart Wizard is running, so the word "Theta" was used in the axis label. This will be changed after the graph is created.

In the final step of the Chart Wizard, the chart was placed on the spreadsheet. Then the following cosmetic changes were made:

- The legend was moved over a blank spot in the graph.
- The plot area background color was changed to white.
- The line style for the Cosine series was changed to a dashed line.
- The number of displayed decimal places on the *y*-axis labels was reduced to one.
- The minimum and maximum values on the *x*-axis were set at 0 and 6.283 (2π), respectively.
- Labels on the *x*-axis were turned off (i.e., tick mark labels set to "None" on the Patterns panel of the Format Axis dialog).

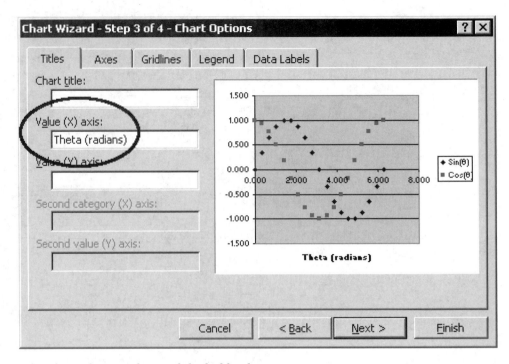

After these changes, the graph looks like this:

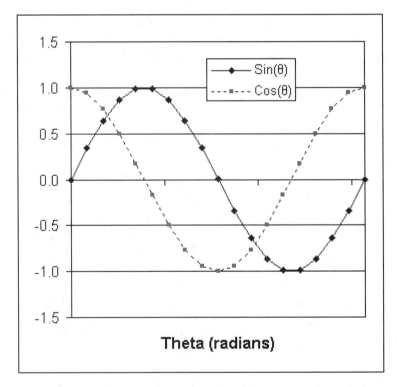

Now the x-axis title can be edited to replace "Theta" by the Greek symbol θ. To do this, you have to cheat a little, because Insert/Symbol is not available while editing the title. To work around this, first copy the symbol θ from any of the column headings (cells A3 through D3) to the Windows clipboard. Then the symbol can be pasted into the title while it is being edited.

To edit the title, click on the title to select it, then click it again to enter edit mode.

Note: This is two single clicks, not a double click. Leave a slight pause between the clicks. Double clicking on the title opens the Format Axis Title dialog, which is useful, but not for editing the contents of the axis title.

Once you are editing the x-axis title, select the word "Theta", then Edit/Paste (or [Ctrl-V]) the Greek letter θ from the Windows clipboard into the axis title. Click anywhere outside the axis title to leave edit mode.

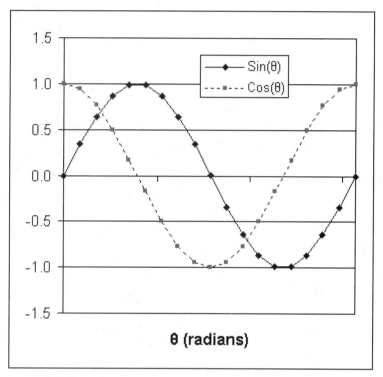

Adding an Additional Curve

If you have already created a graph and decide to add another curve, you can. For example, to add a $\sin^2(\theta)$ curve to the graph created in the previous example, we would need to add a new series of y values, but use the same x values as for the sine and cosine curves.

Once the $\sin^2(\theta)$ data are available in the spreadsheet (having been added in column E), you would begin by right clicking on the existing graph and selecting Source Data from the pop-up menu.

	A	B	C	D	E	F	G	H	I	J	K	L
1	Sine and Cosine Data											
2												
3	θ (degrees)	θ (radians)	Sin(θ)	Cos(θ)	Sin²(θ)		1.5	Format Chart Area...				
4	0	0.000	0.000	1.000	0.000			Chart Type...		Sin(θ)		
5	20	0.349	0.342	0.940	0.117							
6	40	0.698	0.643	0.766	0.413		1.0	Source Data...		Cos(θ)		
7	60	1.047	0.866	0.500	0.750			Chart Options...				
8	80	1.396	0.985	0.174	0.970		0.5	Location...				
9	100	1.745	0.985	-0.174	0.970							
10	120	2.094	0.866	-0.500	0.750			3-D View...				
11	140	2.443	0.643	-0.766	0.413		0.0	Chart Window				
12	160	2.793	0.342	-0.940	0.117			Cut				
13	180	3.142	0.000	-1.000	0.000							
14	200	3.491	-0.342	-0.940	0.117		-0.5	Copy				
15	220	3.840	-0.643	-0.766	0.413			Paste				
16	240	4.189	-0.866	-0.500	0.750			Clear				
17	260	4.538	-0.985	-0.174	0.970		-1.0					
18	280	4.887	-0.985	0.174	0.970			Bring to Front				
19	300	5.236	-0.866	0.500	0.750			Send to Back				
20	320	5.585	-0.643	0.766	0.413		-1.5					
21	340	5.934	-0.342	0.940	0.117			Assign Macro...		ns)		
22	360	6.283	0.000	1.000	0.000							
23												

On the Series panel of the Source Data dialog box, we need to add a new series to the list of Series. Set the Name of the new series to "Sin2(θ)" by pointing at cell E3 (specifically, `Sheet1!$E$3`). Use the small button at the right of the Name: field to return to the spreadsheet to point out the cell containing the series name.

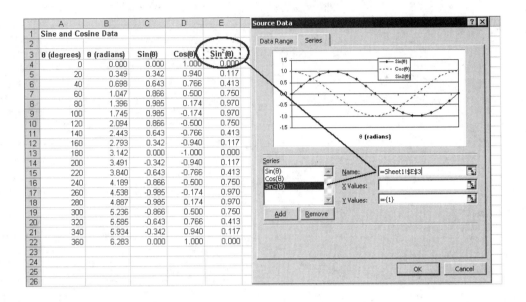

Next, tell Excel where to find the *x* values for the new curve. Again, the small button at the right of the X Values field will take you back to the spreadsheet so that you can select the *x* values.

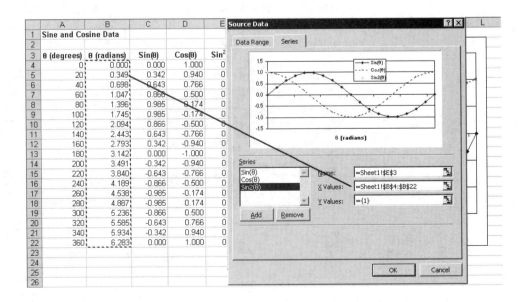

Next, tell Excel where to find the y values for the new curve.

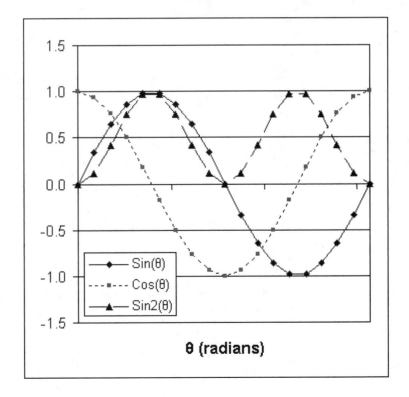

At this point, the new series has the name "Sin2(θ)" (which came from cell E3), a specified range of cells containing x values (=Sheet1!$B$4:$B$22), and a range of y values (=Sheet1!$E$4:$E$22). Click the OK button to add the new curve to the existing graph.

The graph now includes three curves. The following modifications have been made:

- The legend has been moved to the lower left corner of the plot area.
- The color and line style of the new curve were changed from the default values.

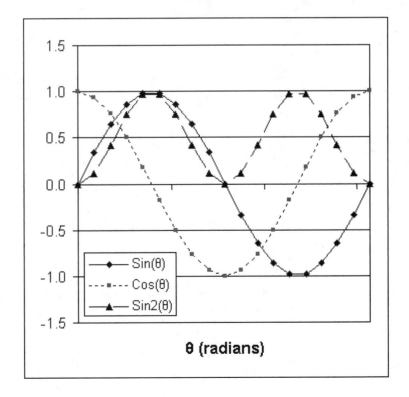

2.6   PRINTING THE GRAPH

There are two ways to get a graph onto paper:

1.   Print the graph only.
2.   Print the spreadsheet containing the graph.

To print only the graph, first click the graph to select it. (Handles will be displayed around the edge of the graph.) If you use the Print or Print Preview button on the standard toolbar or select File/Print... or File/Print Preview from the main menu, the graph will be printed or displayed on the preview page appropriately. The graph will be resized to fit the page. Use the Setup button on the preview page and select the Chart panel on the Page Setup dialog box to change the size of the printed graph.

To print the spreadsheet containing the graph, simply include the graph when you set the print area by selecting File/Print Area/Set Print Area. Then, when you use the Print or Print Preview button on the standard toolbar or select File/Print... or File/Print Preview, the spreadsheet with the graph will be printed or displayed on the preview page appropriately.

2.7   TYPES OF GRAPHS

Excel provides a number of chart types other than XY (Scatter) plots. XY (Scatter) plots are the most common type of graph for engineering work; however, several of the other chart types are also used. Other standard chart types include

*   *Line Graphs*
*   *Column* and *Bar Graphs*
*   *Pie Charts*
*   *Surface Plots*

WARNING: *A common error is to use a Line Graph when an XY Scatter plot is more appropriate. You will get away with this if the* x *values in your data set are uniformly spaced. But if the* x *values are not uniformly spaced, your curve will appear distorted on the Line Graph.*

EXAMPLE 2.2

The *x* values in the spreadsheet shown next are calculated as

$$x_{i+1} = 1.2 \cdot x_i. \tag{2.1}$$

This creates nonuniformly spaced *x* values. The *y* values are calculated from the *x* values as

$$y_i = 3 \cdot x_i, \tag{2.2}$$

so the relationship between *x* and *y* is linear. On an XY Scatter plot, the linear relationship is evident:

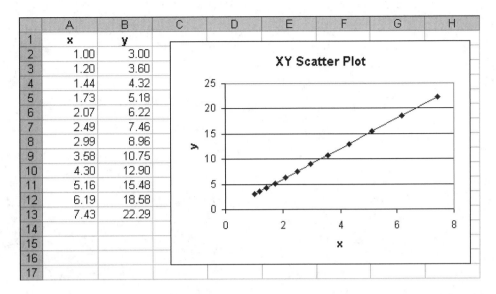

However, on a Line Graph, the relationship between x and y appears to be nonlinear:

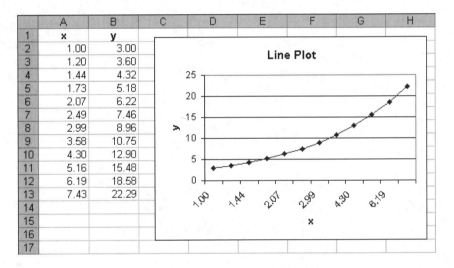

This is an artifact of the Line Graph, because the x values were not used to plot the points on the graph. Excel's Line Graph simply causes the y values to be spread out evenly across the chart (effectively assuming uniform x spacing). If you include two columns of values when you create the chart, Excel assumes that the left column is a column of labels to use on the x-axis. This can be misleading on a Line Graph, because the x values are displayed on the x-axis, but they were not used to plot the data points.

Excel's Line, Column, Bar, and Pie Charts all require only y values to create the graph. If you provide two columns (or rows) of data, the left column (or top row) will be used as labels for the chart. If you need to plot x and y values on a graph, you must use an XY Scatter plot.

Surface Plots in Excel   A surface plot takes the values in a two-dimensional range of cells and displays the values graphically.

EXAMPLE 2.3

The following surface plot shows the value of $F(x,y) = \sin(x)\cos(y)$ for $-1 \le x \le 2$ and $-1 \le y \le 2$:

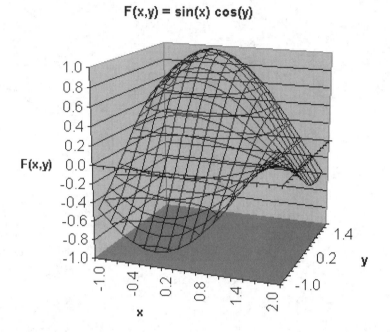

To create the plot, x values ranging from -1 to 2 were calculated in column A, and y values ranging from -1 to 2 were calculated in row 2:

C2	▼	f_x =B2+0.2															
	A	B	C	D	E	F	G	H	I	J	K	L	M	N	O	P	Q	R
1		y >>>																
2	x	-1.0	-0.8	-0.6	-0.4	-0.2	0.0	0.2	0.4	0.6	0.8	1.0	1.2	1.4	1.6	1.8	2.0	
3	-1.0																	
4	-0.8																	
5	-0.6																	
6	-0.4																	
7	-0.2																	
8	0.0																	
9	0.2																	
10	0.4																	
11	0.6																	
12	0.8																	
13	1.0																	
14	1.2																	
15	1.4																	
16	1.6																	
17	1.8																	
18	2.0																	
19																		

The first value for $F(x,y)$ is calculated in cell B3 by means of the formula B3:=SIN($A3)*COS(B$2), as the following shows:

| B3 | ▼ | | f_x | =SIN($A3)*COS(B$2) | | | | | | | | | | | | | |

	A	B	C	D	E	F	G	H	I	J	K	L	M	N	O	P	Q	R
1		y >>>																
2	x	-1.0	-0.8	-0.6	-0.4	-0.2	0.0	0.2	0.4	0.6	0.8	1.0	1.2	1.4	1.6	1.8	2.0	
3	-1.0	-0.45																
4	-0.8																	
5	-0.6																	
6	-0.4																	
7	-0.2																	
8	0.0																	
9	0.2																	
10	0.4																	
11	0.6																	
12	0.8																	
13	1.0																	
14	1.2																	
15	1.4																	
16	1.6																	
17	1.8																	
18	2.0																	
19																		

The dollar signs on the A in SIN($A3) and the 2 in COS(B$2) allow the formula to be copied to the other cells in the range. After the formula is copied, the SIN() functions will always reference x values in column A, and COS() functions will always reference y values in row 2.

Next, the formula in cell B3 is copied to all of the cells in the range B3:Q18, to fill the two-dimensional array.

	A	B	C	D	E	F	G	H	I	J	K	L	M	N	O	P	Q	R
1		y >>>																
2	x	-1.0	-0.8	-0.6	-0.4	-0.2	0.0	0.2	0.4	0.6	0.8	1.0	1.2	1.4	1.6	1.8	2.0	
3	-1.0	-0.45	-0.59	-0.69	-0.78	-0.82	-0.84	-0.82	-0.78	-0.69	-0.59	-0.45	-0.30	-0.14	0.02	0.19	0.35	
4	-0.8	-0.39	-0.50	-0.59	-0.66	-0.70	-0.72	-0.70	-0.66	-0.59	-0.50	-0.39	-0.26	-0.12	0.02	0.16	0.30	
5	-0.6	-0.31	-0.39	-0.47	-0.52	-0.55	-0.56	-0.55	-0.52	-0.47	-0.39	-0.31	-0.20	-0.10	0.02	0.13	0.23	
6	-0.4	-0.21	-0.27	-0.32	-0.36	-0.38	-0.39	-0.38	-0.36	-0.32	-0.27	-0.21	-0.14	-0.07	0.01	0.09	0.16	
7	-0.2	-0.11	-0.14	-0.16	-0.18	-0.19	-0.20	-0.19	-0.18	-0.16	-0.14	-0.11	-0.07	-0.03	0.01	0.05	0.08	
8	0.0	0.00	0.00	0.00	0.00	0.00	0.00	0.00	0.00	0.00	0.00	0.00	0.00	0.00	0.00	0.00	0.00	
9	0.2	0.11	0.14	0.16	0.18	0.19	0.20	0.19	0.18	0.16	0.14	0.11	0.07	0.03	-0.01	-0.05	-0.08	
10	0.4	0.21	0.27	0.32	0.36	0.38	0.39	0.38	0.36	0.32	0.27	0.21	0.14	0.07	-0.01	-0.09	-0.16	
11	0.6	0.31	0.39	0.47	0.52	0.55	0.56	0.55	0.52	0.47	0.39	0.31	0.20	0.10	-0.02	-0.13	-0.23	
12	0.8	0.39	0.50	0.59	0.66	0.70	0.72	0.70	0.66	0.59	0.50	0.39	0.26	0.12	-0.02	-0.16	-0.30	
13	1.0	0.45	0.59	0.69	0.78	0.82	0.84	0.82	0.78	0.69	0.59	0.45	0.30	0.14	-0.02	-0.19	-0.35	
14	1.2	0.50	0.65	0.77	0.86	0.91	0.93	0.91	0.86	0.77	0.65	0.50	0.34	0.16	-0.03	-0.21	-0.39	
15	1.4	0.53	0.69	0.81	0.91	0.97	0.99	0.97	0.91	0.81	0.69	0.53	0.36	0.17	-0.03	-0.22	-0.41	
16	1.6	0.54	0.70	0.82	0.92	0.98	1.00	0.98	0.92	0.82	0.70	0.54	0.36	0.17	-0.03	-0.23	-0.42	
17	1.8	0.53	0.68	0.80	0.90	0.95	0.97	0.95	0.90	0.80	0.68	0.53	0.35	0.17	-0.03	-0.22	-0.41	
18	2.0	0.49	0.63	0.75	0.84	0.89	0.91	0.89	0.84	0.75	0.63	0.49	0.33	0.15	-0.03	-0.21	-0.38	
19																		

Then the array of values is selected before starting the Chart Wizard.

	Microsoft Excel - ch02.xls																	

File   Edit   View   Insert   Format   Tools   Data   Window   Help   Type a question for help

Arial   ▼ 10 ▼   B I U $ % , Chart Wizard   100%

| B3 | ▼ | | f_x | =SIN($A3)*COS(B$2) | | | | | | | | | | | | | |

	A	B	C	D	E	F	G	H	I	J	K	L	M	N	O	P	Q	R
1		y>>>																
2	x	-1.0	-0.8	-0.6	-0.4	-0.2	0	0.2	0.4	0.6	0.8	1.0	1.2	1.4	1.6	1.8	2.0	
3	-1.0	-0.45	-0.59	-0.69	-0.78	-0.82	-0.84	-0.82	-0.78	-0.69	-0.59	-0.45	-0.30	-0.14	0.02	0.19	0.35	
4	-0.8	-0.39	-0.50	-0.59	-0.66	-0.70	-0.72	-0.70	-0.66	-0.59	-0.50	-0.39	-0.26	-0.12	0.02	0.16	0.30	
5	-0.6	-0.31	-0.39	-0.47	-0.52	-0.55	-0.56	-0.55	-0.52	-0.47	-0.39	-0.31	-0.20	-0.10	0.02	0.13	0.23	
6	-0.4	-0.21	-0.27	-0.32	-0.36	-0.38	-0.39	-0.38	-0.36	-0.32	-0.27	-0.21	-0.14	-0.07	0.01	0.09	0.16	
7	-0.2	-0.11	-0.14	-0.16	-0.18	-0.19	-0.20	-0.19	-0.18	-0.16	-0.14	-0.11	-0.07	-0.03	0.01	0.05	0.08	
8	0.0	0.00	0.00	0.00	0.00	0.00	0.00	0.00	0.00	0.00	0.00	0.00	0.00	0.00	0.00	0.00	0.00	
9	0.2	0.11	0.14	0.16	0.18	0.19	0.20	0.19	0.18	0.16	0.14	0.11	0.07	0.03	-0.01	-0.05	-0.08	
10	0.4	0.21	0.27	0.32	0.36	0.38	0.39	0.38	0.36	0.32	0.27	0.21	0.14	0.07	-0.01	-0.09	-0.16	
11	0.6	0.31	0.39	0.47	0.52	0.55	0.56	0.55	0.52	0.47	0.39	0.31	0.20	0.10	-0.02	-0.13	-0.23	
12	0.8	0.39	0.50	0.59	0.66	0.70	0.72	0.70	0.66	0.59	0.50	0.39	0.26	0.12	-0.02	-0.16	-0.30	
13	1.0	0.45	0.59	0.69	0.78	0.82	0.84	0.82	0.78	0.69	0.59	0.45	0.30	0.14	-0.02	-0.19	-0.35	
14	1.2	0.50	0.65	0.77	0.86	0.91	0.93	0.91	0.86	0.77	0.65	0.50	0.34	0.16	-0.03	-0.21	-0.39	
15	1.4	0.53	0.69	0.81	0.91	0.97	0.99	0.97	0.91	0.81	0.69	0.53	0.36	0.17	-0.03	-0.22	-0.41	
16	1.6	0.54	0.70	0.82	0.92	0.98	1.00	0.98	0.92	0.82	0.70	0.54	0.36	0.17	-0.03	-0.23	-0.42	
17	1.8	0.53	0.68	0.80	0.90	0.95	0.97	0.95	0.90	0.80	0.68	0.53	0.35	0.17	-0.03	-0.22	-0.41	
18	2.0	0.49	0.63	0.75	0.84	0.89	0.91	0.89	0.84	0.75	0.63	0.49	0.33	0.15	-0.03	-0.21	-0.38	
19																		

From the Chart Wizard, a Surface Plot is displayed that uses a *wire frame*.

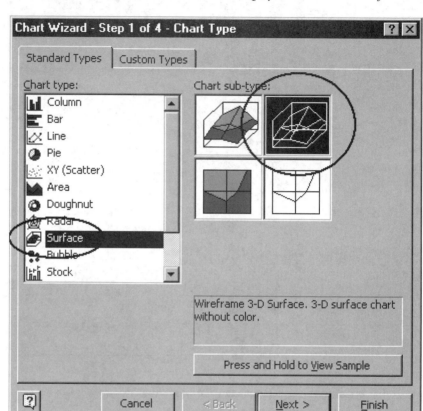

The following is the resulting plot:

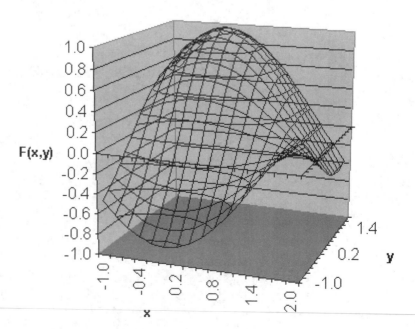

A few other things were added or changed during creation of the graph via the Chart Wizard:

- The labels (not values) displayed on the x-axis were added in Step 2 of the Chart Wizard by using the Series panel of the Source Data step. The contents of cells A3:A18 were used for these labels.
- The names of each of the data series (in Step 2 of the Chart Wizard on the Series panel of the Source Data step) were changed to "−1.0," "−0.8," "−0.6," ..., "1.8," "2.0," so that these series names would display as labels (not values) on the y-axis of the final plot.
- The x- and y-axis labels "x" and "y" were added as axis titles, along with the z-axis label "F(x,y)" and the chart title "F(x, y) = sin(x) cos(y)."

Note that the x and y values on the spreadsheet (column A and row 2) were never used to plot the points on the surface plot. The $F(x, y)$ values in cells B3 through Q18 were plotted by assuming uniform spacing in the x- and y-directions. This is a significant limitation of surface plotting in Excel.

APPLICATIONS: MATERIALS TESTING

Stress–Strain Curve I

Strength testing of materials often involves a *tensile test* in which a sample of the material is held between two **mandrels** and increasing force—actually, *stress* (i.e., force per unit area)—is applied. A stress-vs.-strain curve for a typical ductile material is shown next. The sample first stretches reversibly (*A* to *B*). Then irreversible stretching occurs (*B* to *D*). Finally, the sample breaks (point *D*).

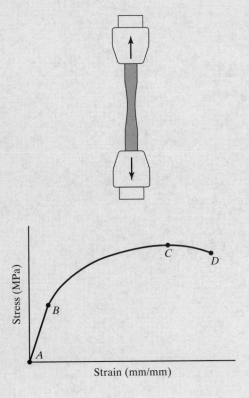

Point *C* is called the material's *ultimate stress*, or *tensile strength*, and represents the greatest stress that the material can endure (with deformation) before coming apart. The *strain* is the amount of elongation of the sample (mm) divided by the original sample length (mm).

The reversible stretching portion of the curve (*A* to *B*) is linear, and the proportionality constant relating stress and strain in this region is called *Young's modulus*, or the *modulus of elasticity*.

Tensile test data on a soft, ductile sample are tabulated as follows (and available electronically at *http:// www.coe.montana.edu/che/Excel*):

Strain (mm/mm)	Stress (MPa)
0.000	0.00
0.003	5.38
0.006	10.76
0.009	16.14
0.012	21.52
0.014	25.11
0.017	30.49
0.020	33.34
0.035	44.79
0.052	52.29
0.079	57.08
0.124	59.79
0.167	60.10
0.212	59.58
0.264	57.50
0.300	55.42

As these test data are analyzed, there are some things we will want to do:

a.   Plot the tensile test data as a stress-vs.-strain graph.
b.   Evaluate the tensile strength from the graph (or the data set).
c.   Create a second graph containing only the elastic-stretch (linear) portion of the data.
d.   Add a linear trendline to the new plot to determine the modulus of elasticity for this material.

First, the data are entered into the spreadsheet:

	A	B	C	D	E	F	G	H
1	**Stress—Strain Curve I**							
2								
3		**Strain**	**Stress**					
4		(mm/mm)	(MPa)					
5		0.000	0.00					
6		0.003	5.38					
7		0.006	10.76					
8		0.009	16.14					
9		0.012	21.52					
10		0.014	25.11					
11		0.017	30.49					
12		0.020	33.34					
13		0.035	44.79					
14		0.052	52.29					
15		0.079	57.08					
16		0.124	59.79					
17		0.167	60.10					
18		0.212	59.58					
19		0.264	57.50					
20		0.300	55.42					
21								

Then, a graph is prepared. The ultimate tensile stress can be read from the graph or the data set:

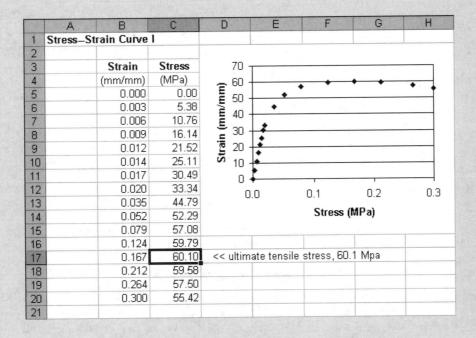

Finally, another graph is prepared, one that contains only the linear portion of the data (the first eight data points). A linear trendline with the intercept forced through the origin is added to the graph. We will use the slope from the equation for the trendline to compute the modulus of elasticity; the R^2 value, 1, provides reassurance that we have indeed plotted the linear portion of the test data:

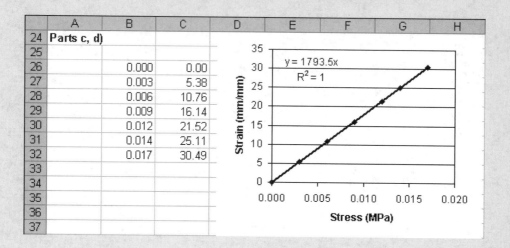

From slope of the regression line, we see that the modulus of elasticity for this material is 1793.7 MPa, or 1.79 GPa.

2.8 GRAPHING WEB DATA

The Internet's World Wide Web is becoming an increasingly common place to locate data, but those data may be in many different forms. Web pages may provide links to data files or embed the data as HTML tables. There is no single way to move Web data into Excel for graphing, but the following two methods frequently work:

1. Copy and paste; or
2. Save the data file, then import into Excel.

The first method is described in this section; the second method is introduced, then described more fully, in Section 2.9.

2.8.1 Copying and Pasting Web Data

Data sets presented on the Web are often HTML tables. It is usually possible to copy the information from such tables and paste it into Excel, but you have to copy entire rows; you can't select portions. Then, when you want to paste the data into Excel, you might need to use Paste Special... to instruct Excel to paste the values as text and ignore the HTML format information. The copy and paste process is illustrated on page 83 and 84.

The temperature vs. time data used at the beginning of this chapter is available on the text's website in two forms: as an HTML table, and as text file Ex_Data.prn. The text's website is located at

http://www.coe.montana.edu/che/excel

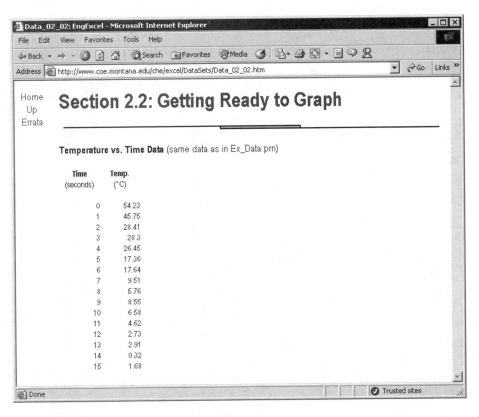

To copy the data from the HTML table, simply select all of the rows (and headings, if desired) and use Edit/Copy (or [Ctrl-C]) to copy the data to the Windows clipboard.

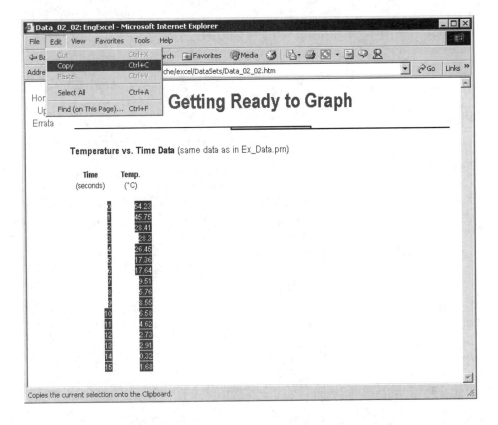

Then, you can paste the information into Excel by using Edit/Paste or [Ctrl-V].

	A	B	C	D
B3			fx 28.41	
1	0	54.23		
2	1	45.75		
3	2	28.41		
4	3	28.3		
5	4	26.45		
6	5	17.36		
7	6	17.64		
8	7	9.51		
9	8	5.76		
10	9	8.55		
11	10	6.58		
12	11	4.62		
13	12	2.73		
14	13	2.91		
15	14	0.32		
16	15	1.68		
17				

With this data set, using Edit/Paste worked fine. Sometimes, you will find it necessary to use Edit/Paste Special ... to paste the data into the spreadsheet as "Text." This tells Excel to ignore HTML formatting information and paste just the data into the cells.

2.8.2   Importing Data Files from the Web

A Web page could provide a link to a data file, such as the links to file Ex_Data.prn from the text's website. What happens when you click on a link to a .prn file depends on how your browser is configured:

1.   The browser might display the contents of the file on the screen.
2.   The browser might present an option box asking whether you want to open the file or save it.

If your browser displays the file contents on the screen, you can try copying and pasting the data. The process is exactly like that used in the previous section. If copying and pasting the data fails (or if your browser will not display the data file contents), then you might need to save the file to your computer and import the file into Excel.

If you right click on the data-file link, a pop-up menu will offer a Save Target As ... option. This option allows you to save a copy of the link's target (the data file) to your own computer.

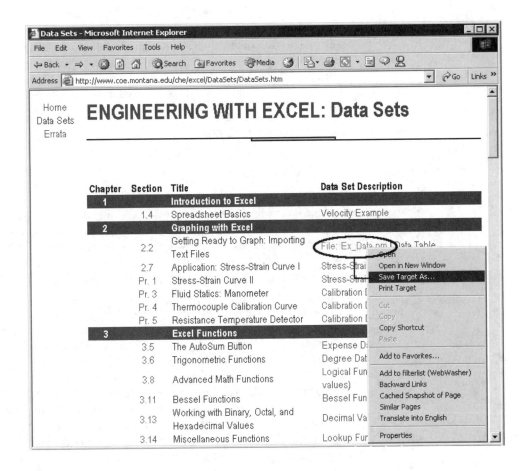

Once the data file has been saved on your own computer, it can be imported into Excel by Excel's Text Import Wizard. That process is described in the next section.

2.9   IMPORTING TEXT FILES

Text files are a common way to move data from one program to another, and Excel is good at creating graphs that use data from other programs. Importing a text file is one way to get the data to be plotted into Excel. Excel provides a Text Import Wizard to make it easy to import data from text files.

Two types of text files are used to store data:

- delimited
- fixed width

Delimited data has a special character, called a *delimiter*, between data values. Commas, spaces and tabs are the most common delimiters, but any nonnumeric character can be used. Quotes are frequently used as text-string delimiters.

The following is an example of comma-delimited data:

0, 54.23

1, 45.75

2, 28.41

Fixed width files align the data values in columns and use character position to distinguish individual data values. The comma-delimited data shown previously would look quite different in a fixed-width data file. In the following example, the data have been written to the file with 8-character fields, using 4 decimal places (a couple of header lines have been included to show the layout of the data fields):

```
Field 1 Field 2
1234567812345678

  0.0000 54.2300
  1.0000 45.7500
  2.0000 28.4100
```

Excel can read the data from either type of file, but Excel must know the format used in the data file before the data can be imported. The Text Import Wizard allows you to select the appropriate format as part of the import process. Fixed-width files were once very common, but delimited data seems to be more common at this time. Both types of data files are used regularly, and Excel's Text Import Wizard can handle either type of file.

2.9.1 Using the Text Import Wizard

The data set used in Section 2 of this chapter consists of 16 temperature values measured at 1-second intervals from 0 to 15 seconds. The values are available as a space-delimited text file called Ex_Data.prn. The file is available at the text's website, *http://www.coe.montana.edu/che/excel*. The following example assumes that the data file is available on the C: drive of the computer that is running Excel, in a folder called "ch_02".

You begin importing a text file into Excel by attempting to open the file. Select File/Open ... and enter the name of the data file—or, to select the data file from the files on the drive, change the file type (near the bottom of the Open File dialog box) to "Text Files (*.prn; *.txt; *.csv)" and browse for the file.

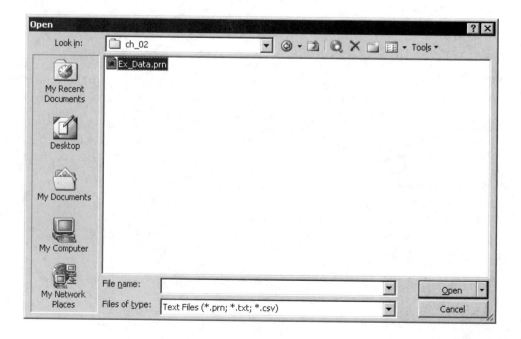

When Excel attempts to open the file and finds that it is not saved as an Excel workbook, it starts the Text Import Wizard to guide you through the import process. The steps in the process are as follows:

Select the Type of Text File First, select the data format that best fits your data, either delimited or fixed width. Excel will analyze the file contents and make a recommendation, but you should verify the data format. The data in Ex_Data.prn is space delimited, so we select Delimited.

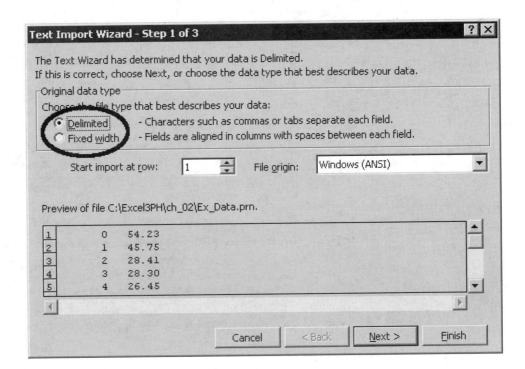

Notice that the text-import dialog allows you to begin importing data at any row (by using the Start import at row: field.) This is very useful if your data file contains heading or title information that you do not want to import into the spreadsheet.

Click Next > to go to the next step in the process.

Select the Type of Delimiter(s) If you selected "Delimited" data in the previous step, then you can choose the type(s) of delimiters. In file Ex_Data.prn, the values have leading spaces at the left of each line and spaces between two columns of values. When "Space" delimiter is checked, Excel treats the spaces in the files as delimiters and adds lines in the Data preview panel to show how the values will be separated into columns:

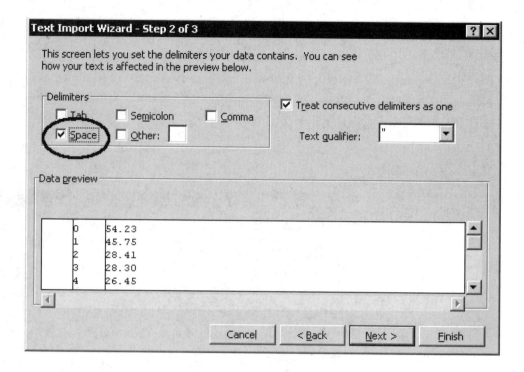

Click Next > to go to the next step in the process.

Select the Data to Be Imported and the Data Formats to be Used   The dialog box for the final import step allows you to tell Excel the number format you want used for each imported column. You can also select not to import one or more columns. To select a column, click on the column heading. Then indicate the data format to be used for the data in that column, or select "Do not import column (skip)" to elect not to import the selected column. In the following figure, I chose to skip the first column (because Excel

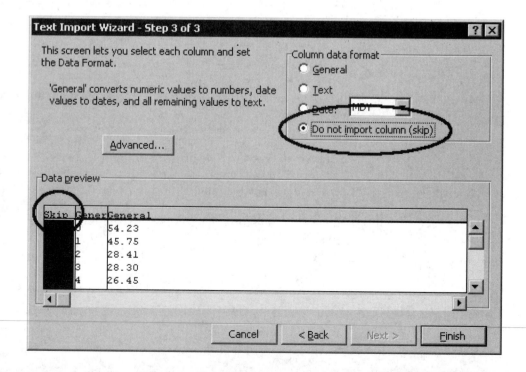

interpreted leading spaces on each line as an empty column) and to import the other two columns, saving them in the spreadsheet with a general number format.

When the correct formats have been specified for each column to be imported, click "Finish" to complete the import process. The values are placed at the top left corner of the spreadsheet:

	A	B	C
1	0	54.23	
2	1	45.75	
3	2	28.41	
4	3	28.3	
5	4	26.45	
6	5	17.36	
7	6	17.64	
8	7	9.51	
9	8	5.76	
10	9	8.55	
11	10	6.58	
12	11	4.62	
13	12	2.73	
14	13	2.91	
15	14	0.32	
16	15	1.68	
17			

You can move the cells to a different location within the same spreadsheet or copy and paste them to another spreadsheet. In the following example, the values were moved down a few rows to make room for titles:

	A	B	C
1	**Time**	**Temp.**	
2	(seconds)	(°C)	
3			
4	0	54.23	
5	1	45.75	
6	2	28.41	
7	3	28.3	
8	4	26.45	
9	5	17.36	
10	6	17.64	
11	7	9.51	
12	8	5.76	
13	9	8.55	
14	10	6.58	
15	11	4.62	
16	12	2.73	
17	13	2.91	
18	14	0.32	
19	15	1.68	
20			

Note: Since file Ex_Data.prn was opened, the workbook is called Ex_Data.prn. Excel will allow you to use the workbook just as you would a standard .xls file, but any nonalphanumeric content (e.g., graphs) will be lost if the file is saved as a .prn file. Excel will show a warning if you attempt to save the file with a .prn extension.

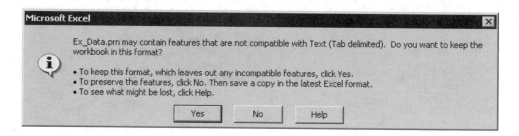

After the importing of a text file, the Excel workbook normally should be saved with the file extension .xls.

KEY TERMS

Automatic scaling	Legend	Tensile strength
Bar Graph	Line Graph	Tensile test
Column Graph	Modules of elasticity	Text file
Chart Wizard	Pie chart	Trendline
Delimited	Plot area	Ultimate stress
Error bars	Surface plot	Wire frame
Fixed width	Strain	Young's modulus
Handles	Stress	XY Scatter plot

SUMMARY

Creating an XY Scatter Plot from Existing Data

Before Starting the Chart Wizard, select the data range (*x* values in left column or top row).

The following are the Chart Wizard steps:

1. Select Chart Type XY (Scatter) and Basic Features (markers or no markers and no lines, straight lines, or smoothed lines)
2. Check the Data Series. Use Step 2 (Series panel) to add names to the series (displayed in the legend) or add additional curves to the graph.
3. Format the Graph (Add titles, turn grid lines on or off, turn the legend on or off, etc.)
4. Graph Placement. Indicate where the graph should be placed (on new spreadsheet or in the current spreadsheet).

Editing an Existing Graph

Double click the item you wish to modify. Items that can be modified include

- Axes
- Grid lines
- Legend
- Markers and Lines
- Plot Area
- Titles

Add Another Curve or Modify Existing Data Series   Right click in some white space on the graph, and select Source Data Series from the pop-up menu.

Add New Titles   Right click in some white space on the graph, and select Chart Options from the pop-up menu.

Printing a Graph

There are two ways to print a graph:

1.   Select the graph (click on it) and use the Print or Print Preview buttons on the toolbar. (Or select File/Print . . . or File/Print Preview.)
     This method causes only the graph to be printed.

2.   Select a cell range on the spreadsheet that includes the graph and then use the Print or Print Preview buttons on the toolbar. (Or select File/Print . . . or File/Print Preview.)

This causes the graph to be printed with the selected range of the spreadsheet.

Available Graph Types

- XY Scatter Plots
- Line Graphs
- Column and Bar Graphs
- Pie Charts
- Surface Plots

Problems

Stress–Strain Curve II

1.   The following tabulated data represent stress–strain data from an experiment on an unknown sample of a white metal (modulus of elasticity values for various white metals are also listed):

   a.   Graph the stress–strain data. If the data include values outside the elastic-stretch region, discard those values.

   b.   Use a linear trendline to compute the modulus of elasticity for the sample.

   c.   What type of metal was tested?

STRESS	STRAIN		MODULUS OF ELASTICITY
(MM/MM)	(MPA)	MATERIAL	(GPA)
0.0000	0	Mg Alloy	45
0.0015	168	Al Alloy	70
0.0030	336	Ag	71
0.0045	504	Ti Alloy	110
0.0060	672	Pt	170
		SS	200

Note: The modulus of elasticity depends on the type of alloy or purity of a nonalloyed material. The values listed here are typical.

Tank Temperature During a Wash-Out

2. One evening, a few friends come over for a soak, and you discover that the water in the hot tub is at 115°F (46°C)—too hot to use. As your friends turn on the cold water to cool down the tub, the engineer in you wants to know how long this is going to take, so you write an energy balance on a well-mixed tank (ignoring heat losses to the air). You end up with the following differential equation relating the temperature in the tank, T, to the temperature of the cold water flowing into the tank, T_{in}, the volume of the tank, V, and the volumetric flow rate of the cold water, $\dot{V}$:

$$\frac{dT}{dt} = \frac{\dot{V}}{V}(T_{in} - T). \tag{2.3}$$

Integrating, you get an equation for the temperature in the tank as a function of time:

$$T = T_{in} - (T_{in} - T_{init.})e^{\frac{-\dot{V}}{V}t}. \tag{2.4}$$

If the initial temperature $T_{init.}$ is 115°F, the cold water temperature is 35°F (1.7°C), and the volume and volumetric flow rate are 3,000 liters and 30 liters per minute, respectively,

a. calculate the expected water temperature at 5-minute intervals for the first 60 minutes after the flow of cold water is established;

b. plot the water temperature in the hot tub as a function of time;

c. calculate how long it should take for the water in the tub to cool to 100°F (37.8°C);

d. explain whether a hot tub is really a well-mixed tank. If it is not, will your equation predict a time that is too short or too long? Explain your reasoning.

Fluid Statics: Manometer

3. Manometers used to be common pressure-measurement devices, but, outside of laboratories, electronic pressure transducers are now more common. Manometers are sometimes still used to calibrate the pressure transducers.

In the calibration system shown in the accompanying figure, the mercury manometer on the right and the pressure transducer on the left are both connected to a piston-driven pressure source filled with hydraulic oil ($\rho = 880 \text{ kg/ m}^3$). The bulbs connected to the transducer and the right side of the manometer are both evacuated ($\rho = 0$).

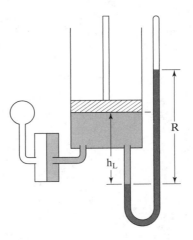

During the calibration, the piston is moved to generate a pressure on both the manometer and the transducer. The manometer reading R is recorded, along with the output of the pressure transducer A (assuming a 4- to 20-mA output current from the transducer).

Consider the following calibration data:

	CALIBRATION DATA		
PISTON SETTING	h_L (MM OIL)	MANOMETER READING (MM HG)	TRANSDUCER OUTPUT (MA)
1	300	0	4.0
2	450	150	5.6
3	600	300	7.2
4	750	450	8.8
5	900	600	10.4
6	1050	750	12.0
7	1200	900	13.6
8	1350	1050	15.2
9	1500	1200	16.8
10	1650	1350	18.4
11	1800	1500	20.0

a. Calculate pressures from the manometer readings.

b. Create a calibration table and graph showing the transducer output (mA) as a function of measured pressure.

Thermocouple Calibration Curve

4.   A type J (iron/constantan) thermocouple was calibrated by using the system illustrated in the accompanying figure. The thermocouple and a thermometer were dipped into a beaker of water on a hot plate. The power level was set at a preset level (known only as 1, 2, 3, ... on the dial) and the thermocouple readings were monitored on a computer screen. When steady state had been reached, the thermometer was read, and 10 thermocouple readings were recorded. Then the power level was increased and the process repeated.

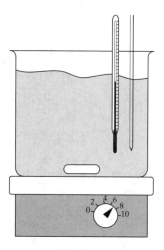

The accumulated calibration data (steady-state data only) are as follows (available electronically at *http://www.coe.montana.edu/che/Excel*):

	THERMOMETER	THERMOCOUPLE	
POWER SETTING	(°C)	AVERAGE (mV)	STD. DEV. (mV)
0	24.6	1.264	0.100
1	38.2	1.841	0.138
2	50.1	2.618	0.240
3	60.2	2.900	0.164
4	69.7	3.407	0.260
5	79.1	4.334	0.225
6	86.3	4.506	0.212
7	96.3	5.332	0.216
8	99.8	5.084	0.168

a.   Plot the thermocouple calibration curve with temperature on the *x*-axis and average thermocouple reading on the *y*-axis.

b. Add a linear trendline to the graph and have Excel display the equation for the trendline and the R^2 value.

c. Use the standard-deviation values to add error bars (± 1 std. dev.) to the graph.

d. The millivolt output of an iron/constantan thermocouple can be related to temperature by the correlation equation[1]

$$T = aV^b, \tag{2.5}$$

where

$$b\sqrt{\frac{T}{a}}$$

T is temperature in °C,
V is the thermocouple output in millivolts,
a is 19.741 for iron/constantan, and
b is 0.9742 for iron/constantan.

Use this equation to calculate predicted thermocouple outputs at each temperature, and add these to the graph as a second data series. Do the predicted values appear to agree with the average experimental values?

Resistance Temperature Detector

5. The *linear temperature coefficient α* of a *resistance temperature detector* (RTD) is a physical property of the metal used to make the RTD that indicates how the electrical resistance of the metal changes as the temperature increases. The equation relating temperature to resistance is

$$R_T = R_0[1 + \alpha T], \tag{2.6}$$

or

$$R_T = R_0 + (R_0\alpha)T \qquad (\textit{in linear-regression form}), \tag{2.7}$$

where

R_T is the resistance at the unknown temperature, T,
R_0 is the resistance at 0°C (known, one of the RTD specifications), and
α is the linear temperature coefficient (known, one of the RTD specifications).

The common grade of platinum used for RTDs has an α value equal to 0.00385 ohm/ ohm/°C (sometimes written simply as $0.00385°C^{-1}$), but older RTDs used a different grade of platinum and operated with $\alpha = 0.003902°C^{-1}$, and laboratory grade RTDs use very high-purity platinum with $\alpha = 0.003923°C^{-1}$. If the wrong α value is used to compute temperatures from RTD readings, the computed temperatures will be incorrect.

---

[1]Thermocouple correlation equation from *Transport Phenomena Data Companion*, L.P.B.M. Janssen and M.M.C.G. Warmoeskerken, Arnold DUM, London, 1987, p. 20.

The following data show the temperature vs. resistance for an RTD:

| TEMPERATURE | RESISTANCE |
°C	OHMS
0	100.0
10	103.9
20	107.8
30	111.7
40	115.6
50	119.5
60	123.4
70	127.3
80	131.2
90	135.1
100	139.0

a.   Use Excel to graph the data and add a trend line to evaluate the linear temperature coefficient.

b.   Is the RTD of laboratory grade?

3

Excel Functions

3.1 INTRODUCTION TO EXCEL FUNCTIONS

In a programming language, the term *function* is used to mean a piece of the program dedicated to a particular calculation. A function accepts input from a list of *arguments* or *parameters*, performs calculations, and then returns a value or a set of values. Excel's functions work the same way and serve the same purposes. They receive input from an argument list, perform a calculation, and return a value or a set of values.

Excel provides a wide variety of *functions* that are predefined and immediately available. This chapter presents the functions most commonly used by engineers.

Functions are used whenever you want to

- Perform the same calculations multiple times, using different input values.
- Reuse the calculation in another program without retyping it.
- Make a complex program easier to comprehend by assigning a section a particular task.

3.2 EXCEL'S BUILT-IN FUNCTIONS

Excel provides a wide assortment of *built-in functions*. Several commonly used classes of functions are

- Elementary math functions.
- Trigonometric functions.
- Advanced math functions.
- Matrix math functions (not described in this chapter).

OBJECTIVES

After reading this chapter, you will

- Know how to use Excel's built-in functions
- Know the functions commonly used by engineers
- Become familiar with the quick reference to the Excel functions included at the end of this chapter

- Functions for financial calculations (not described in this chapter).
- Functions for statistical calculations (not described in this chapter).
- Date and time functions.
- String functions.
- Lookup and reference functions.
- File-handling functions.
- Functions for working with databases (not described in this chapter).

Additionally, you can always write your own functions, directly from Excel, by using Visual Basic for Applications (VBA).

3.2.1 Function Syntax

Built-in functions are identified by a *name* and usually require an *argument list*. They use the information supplied in the argument list to compute and return a value or set of values. One of the simplest functions in Excel, but still a very useful one, is PI(). When this function is called (without any arguments inside the parentheses) it returns the value of π.

EXAMPLE 3.1

In the following figure, cell B2 is assigned the value π by the formula =PI():

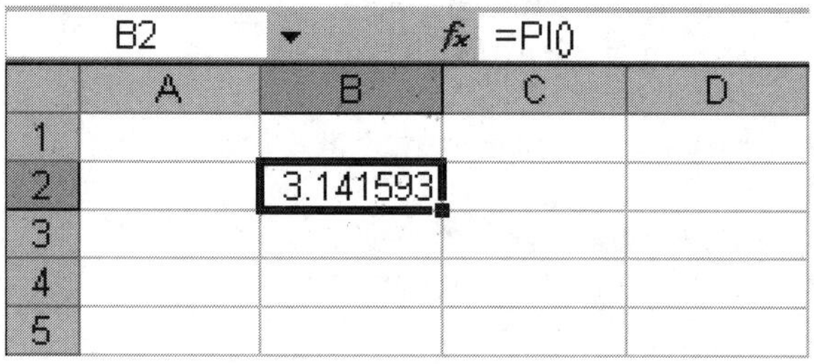

B2			f_x =PI()	
	A	B	C	D
1				
2		3.141593		
3				
4				
5				

EXAMPLE 3.2

Here the function PI() is used to calculate the area of a circle:

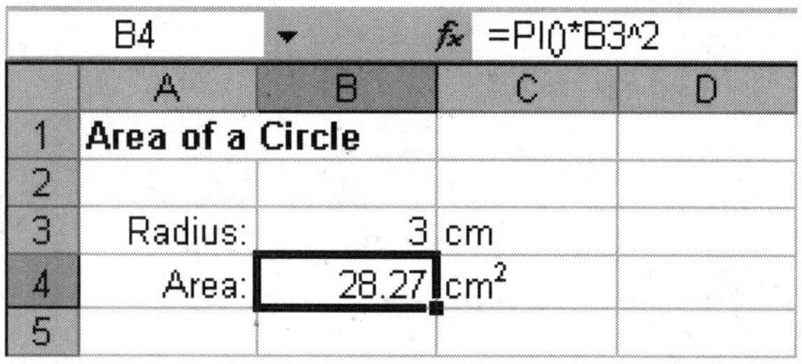

B4			f_x =PI()*B3^2	
	A	B	C	D
1	Area of a Circle			
2				
3	Radius:	3	cm	
4	Area:	28.27	cm²	
5				

Most functions take one or more values as arguments. An example of a function that takes a single value is the factorial function, FACT(x). The factorial of 4 is 24

$(4 \times 3 \times 2 \times 1 = 24)$. This can be computed by using the FACT ( ) function:

B2	▾	f_x =FACT(4)	
	A	B	C	D
1				
2		24		
3				
4				
5				

Some functions take a range of values as arguments, such as the SUM (*range*) function:

B5	▾	f_x =SUM(B1:B3)	
	A	B	C	D
1		10		
2		12		
3		14		
4				
5		36		
6				

An easy way to enter a function that takes a range of cells as an argument is to type in the *function name* and the opening parenthesis, and then use the mouse to highlight the range of cells to be used as an argument. For the preceding example, first, in cell B5, type =SUM (. Then use the mouse to select cells B1:B3. The screen should look like this:

RATE	▾ ✗ ✓	f_x =sum(B1:B3	
	A	B	C	D
1		10		
2		12		
3		⬚		
4				
5		=sum(B1:B3		
6		SUM(**number1**, [number2], ...)		
7				

The dashed line indicates the range of cells currently selected, and the selected range is included in the formula in cell B5. The *ScreenTip* below the formula lets you know what arguments the function requires. Once the complete range has been selected, either enter the closing parenthesis and press [enter] or simply press [enter], and Excel will automatically add the final parenthesis.

The cells to be included in the range do not have to be next to each other on the worksheet, as the following example shows:

RATE	▾	✕ ✓	f_x	=sum(B1:B3,D2:D3
	A	B	C	D	E
1		10			
2		12		16	
3		14		18	
4					
5		=sum(B1:B3,D2:D3			
6		SUM(number1, **[number2]**, [number3], ...)			
7					

To select noncontiguous cells, you first select the first portion of the cell range (B1:B3) with the mouse. Then either enter a comma or hold down the control key while you select the second range (D2:D3) with the mouse:

B5		▾	f_x	=SUM(B1:B3,D2:D3)
	A	B	C	D	E
1		10			
2		12		16	
3		14		18	
4					
5		70			
6					

PROFESSIONAL SUCCESS

When do you use a calculator to solve a problem, and when should you use a computer? The following three questions will help you make this decision:

1. Is the calculation long and involved?
2. Will you need to perform the same calculation numerous times?
3. Do you need to document the results for the future, either to give them to someone else or for your own reference?

A "yes" answer to any of these questions suggests that you consider using a computer. Moreover, a "yes" to the second question suggests that you might want to write a reusable function to solve the problem.

3.3   USING THE CONVERT( ) FUNCTION TO CONVERT UNITS

Excel provides an interesting function called CONVERT() to convert units. It is part of the Analysis ToolPak, which must be installed and activated before the CONVERT() function is available.

3.3.1   Checking to See Whether CONVERT( ) Is Available

If Data Analysis . . . appears on the Tools menu, the Analysis ToolPak has been installed and activated, and the CONVERT() function should be available:

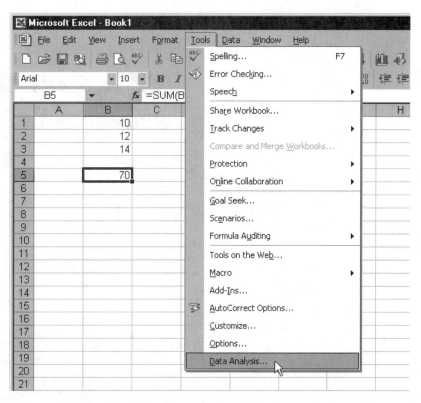

If Data Analysis ... does not appear on the Tools menu, then the Analysis ToolPak either has not been installed or has not been activated.

To see whether the Analysis ToolPak has been installed, select Tools/Add-Ins. ...

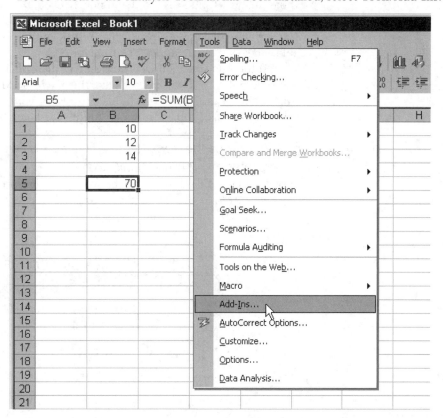

The Add-Ins dialog box will pop up, displaying a list of available add-ins:

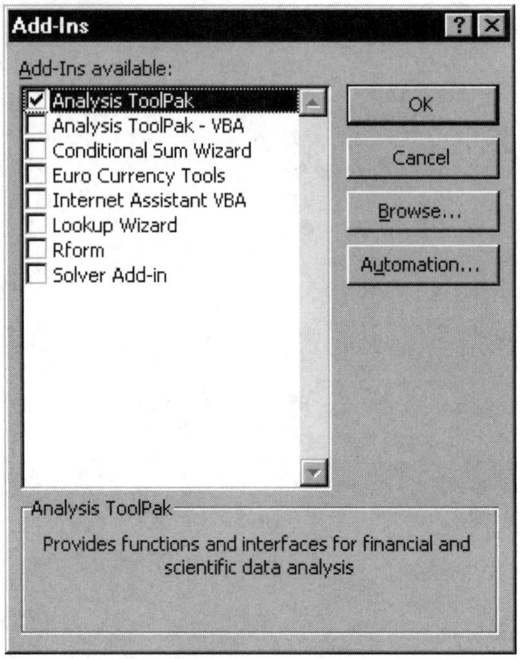

If the Analysis ToolPak is listed, but not checked, the ToolPak has been installed, but not activated. Click the check box next to the Analysis ToolPak to activate the feature. When you leave the dialog box (by clicking the OK button), the Analysis ToolPak will be available and so will the CONVERT() function.

If the Analysis ToolPak is not listed in the Add-Ins list, it has not been installed on your computer. You will need to use the Excel or Microsoft Office CD to install the Analysis ToolPak. Then you will have to restart your computer, restart Excel, and activate the Analysis ToolPak from the Add-ins . . . dialog box.

3.3.2 Once the CONVERT( ) Function Is Available

The CONVERT(value, from_units, to_units) function changes a value in certain units and returns an equivalent value in different units. For example, =CONVERT(1, "in", "cm") returns 2.54:

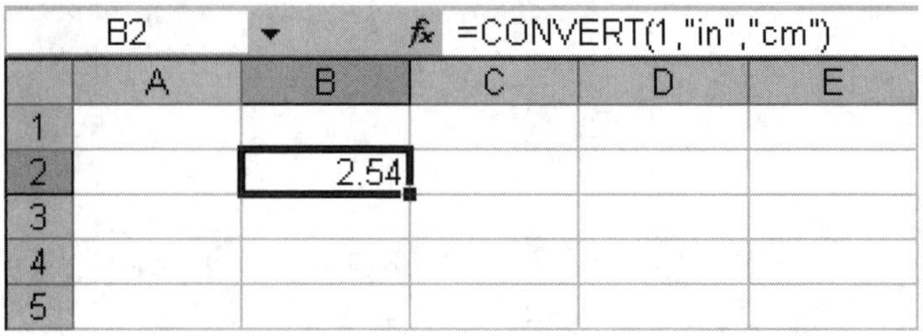

The CONVERT( ) function can be useful for converting a table from one set of units to another:

C4	▼		ƒ×	=CONVERT(B4,"in","cm")		
	A	B	C	D	E	F
1		**Part Sizes**				
2		(inches)	(cm)			
3						
4		1/8	0.318			
5		1/10	0.254			
6		1/16	0.159			
7						

The unit abbreviations used in the CONVERT( ) function are part of the Analysis ToolPak, and only those abbreviations may be used. A complete list of available units is accessible from the help page associated with the CONVERT( ) function. Following is list of commonly used units:

UNIT TYPE	UNIT	USE AS...
Mass	Gram	"g"
	Pound	"lbm"
Length	Meter	"m"
	Mile	"mi"
	Inch	"in"
	Foot	"ft"
Time	Year	"yr"
	Day	"day"
	Hour	"hr"
	Minute	"mn"
	Second	"sec"
Pressure	Pascal	"Pa"
	Atmosphere	"atm"
	mm of Mercury	"mmHg"
Force	Newton	"N"
	Dyne	"dyn"
	Pound force	"lbf"
Energy or Work	Joule	"J"
	Erg	"e"
	Foot-pound	"flb"
	BTU	"BTU"
Power	Horsepower	"HP"
	Watt	"W"
Temperature	Degree Celsius	"C"
	Degree Fahrenheit	"F"
	Degree Kelvin	"K"
Volume	Quart	"qt"
	Gallon	"gal"
	Liter	"l"

For use of the CONVERT() function, it is important to remember the following:

1. The unit names are case sensitive.
2. SI unit prefixes (e.g., "k" for kilo) are available for metric units. In the following example, the "m" prefix for milli has been used with the meter name "m" to report the part sizes in millimeters:

C4	▼	f_x =CONVERT(B4,"in","mm")			
	A	B	C	D	E	F
1		**Part Sizes**				
2		(inches)	(mm)			
3						
4		1/8	3.175			
5		1/10	2.540			
6		1/16	1.588			
7						

A list of prefix names is included in the CONVERT() function help file.

3. Combined units such as gal/mn (gallons per minute) are not supported.

3.4   SIMPLE MATH FUNCTIONS

Most common math operations beyond multiplication and division are implemented as functions in Excel. For example, to take the square root of four, you would use =SQRT(4). SQRT() is Excel's square root function.

3.4.1   Common Math Functions

The following table presents some common mathematical functions in Excel:

OPERATION	FUNCTION NAME
Square Root	SQRT(x)
Absolute Value	ABS(x)
Factorial	FACT(x)
Summation	SUM(range)
Greatest Common Divisor	GCD(x1, x2, . . .)
Least Common Multiple	LCM(x1, x2, . . .)

Many additional mathematical operations are available as built-in functions. To see a list of the functions built into Excel, first click on an empty cell and then click on the Insert Function button on the formula bar:

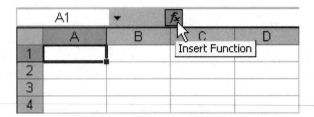

The Insert Function dialog provides access to all of Excel's built-in functions:

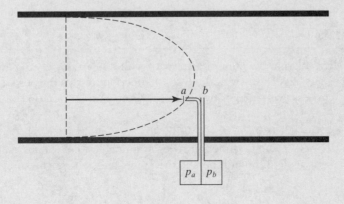

Note: Excel's help files include descriptions of every built-in function, including information about the required and optional arguments.

APPLICATIONS: FLUID MECHANICS

Fluid Velocity From Pitot Tube Data

A pitot tube is a device that measures local velocity. The theory behind the operation of the pitot tube comes from Bernoulli's equation (without the potential energy terms) and relates the change in kinetic energy to the change in fluid pressure:

$$\frac{P_a}{\rho} + \frac{u_a^2}{2} = \frac{p_b}{\rho} + \frac{u_a^2}{2}. \tag{3.1}$$

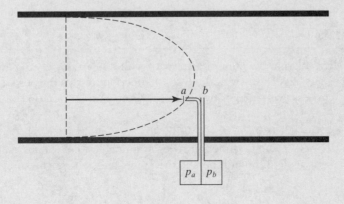

The flow that impacts the tube at point a hits a dead end, and the velocity goes to zero (stagnation). At point b, the flow slides past the pitot tube and does not slow down at all (free stream velocity). The pressure at point b is the free stream pressure. At point a, the energy carried by the fluid has to be conserved, so the kinetic energy of the moving fluid is transformed to pressure energy as the velocity goes to zero. The pitot tube measures a higher pressure at point a than it does at point b, and the pressure difference can be used to compute the free stream velocity at point b.

When the velocity at point a has been set to zero and Bernoulli's equation is rearranged to solve for the local velocity at point b, we get

$$u_b = \sqrt{\frac{2}{\rho}(p_a - p_b)}. \tag{3.2}$$

If the pressure transducer indicates the pressure difference $p_a - p_b = 0.25$ atm, what is the local velocity at point b for a fluid having the specific gravity 0.81?

The formulas used in column C in the following have been displayed in column F:

	C8		f_x =SQRT((2/C6)*C4)				
	A	B	C	D	E	F	G
1	**Pitot Tube**						
2							
3	Pressure Difference:		0.25	atm			
4	Pressure Difference SI:		25331	Pa		=C3*101325	
5	Specific Gravity:		0.81				
6	Density:		810	kg/m³		=C5*1000	
7							
8	Velocity at b:		7.91	m/s		=SQRT((2/C6)*C4)	
9							

PRACTICE!

If the velocity at point b is doubled, what pressure difference would be measured across the pitot tube? [1 atm]

3.5   COMPUTING SUMS

Calculating the sum of a row or column of values is a common operation in a spreadsheet. Excel provides the SUM() function for this purpose, but also tries to simplify the process even further by putting the AutoSum button on the Toolbar.

3.5.1   The SUM( ) Function

The SUM(range) function receives a range of cell values as its argument, computes the sum of all of the numbers in the range, and returns the sum:

B5	▼	f_x =SUM(B1:B3)	
	A	B	C	D
1		10		
2		12		
3		14		
4				
5		36		
6				

Empty cells and text values are ignored:

B5	▼	f_x =SUM(B1:B3)	
	A	B	C	D
1		10		
2		twelve		
3		14		
4				
5		24		
6				

The SUM() function can be used for rows, as in

E2	▼	f_x =SUM(A2:C2)					
	A	B	C	D	E	F	G	H
1							
2	8	9	10		27		
3							
4							

for columns, as in

E2	▼	f_x =SUM(A2:C3)					
	A	B	C	D	E	F	G	H
1							
2	8	9	10		63		
3	11	12	13				
4							
5							

or for odd-shaped collections of cells, as in

E2	▼		f_x	=SUM(A2:C2,A3,B5:C5)			
	A	B	C	D	E	F	G	H
1								
2	8	9	10		63			
3	11							
4								
5		12	13					
6								

3.5.2   The AutoSum Button

The *AutoSum button* on the ToolBar is designed to seek out and find a column or row of values to sum. If you select the cell at the bottom of a column of numbers or to the right of a row of numbers and then press the AutoSum button, Excel will automatically include the entire column or row in the SUM() function in the selected cell.

To sum the values in an expense report, first enter the expenses and select the cell immediately below the list of values. Then click the AutoSum button:

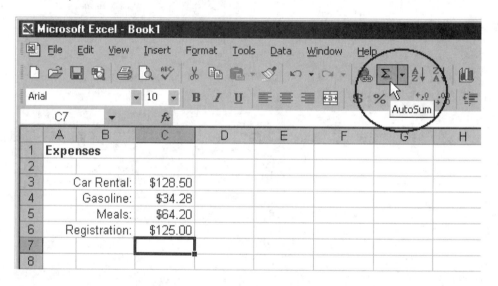

Excel shows you the nearly completed SUM() function so that you can be sure that the correct cells were included:

RATE	▼	✕ ✓	f_x	=SUM(C3:C6)		
	A	B	C	D	E	F
1	**Expenses**					
2						
3		Car Rental:	$128.50			
4		Gasoline:	$34.28			
5		Meals:	$64.20			
6		Registration:	$125.00			
7			=SUM(C3:C6)			
8			SUM(**number1**, [number2], ...)			
9						

If the summation is correct, press [enter] to finish entering the SUM ( ) function:

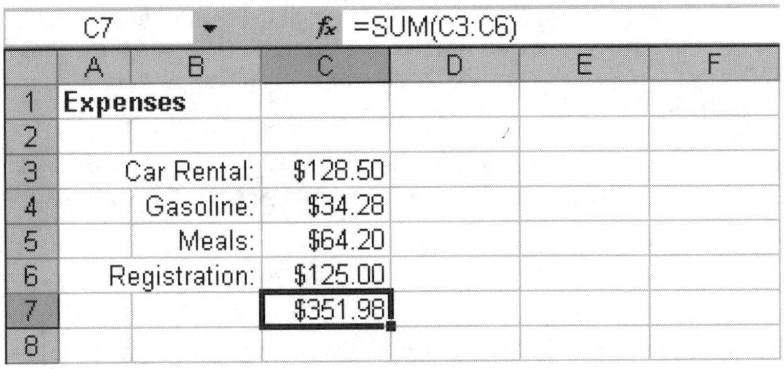

C7	▼	f_x =SUM(C3:C6)			
	A	B	C	D	E	F
1	**Expenses**					
2						
3		Car Rental:	$128.50			
4		Gasoline:	$34.28			
5		Meals:	$64.20			
6		Registration:	$125.00			
7			$351.98			
8						

3.5.3 Logarithm and Exponentiation Functions

Excel provides the following *logarithm* and *exponentiation* functions:

FUNCTION NAME	OPERATION
EXP(x)	Returns *e* raised to the power *x*.
LN(x)	Returns the natural log of *x*.
LOG10(x)	Returns the base-10 log of *x*.
LOG(x, base)	Returns the logarithm of *x* to the specified base.

Excel also provides functions for working with complex numbers:

FUNCTION NAME	OPERATION
IMEXP(x)	Returns the exponential of complex number *x*.
IMLN(x)	Returns the natural log of complex number *x*.
IMLOG10(x)	Returns the base-10 log of complex number *x*.
IMLOG2(x)	Returns the base-2 logarithm of complex number *x*.

PRACTICE!

Try out Excel's functions. First try these easy examples:

    a. =SQRT(4)

    b. =ABS(−7)

    c. =FACT(3)

    d. =LOG 10(100)

Then try these more difficult examples and check the results with a calculator:

    a. =FACT(20)

    b. =LN(2)

    c. =EXP(−0.4)

3.6 TRIGONOMETRIC FUNCTIONS

Excel provides all of the common *trigonometric functions*, such as SIN(x), COS(x), and SINH(x). The x in these functions is an angle, measured in *radians*. If your angles are in *degrees*, you don't have to convert them to radians by hand; Excel provides a RADIANS() function to convert angles in degrees to radians, as in

	B9	▼	*fx*	=RADIANS(A9)	
	A	B	C	D	
1	**Angle**				
2	(Degrees)	(Radians)			
3	0	0.0000			
4	30	0.5236			
5	60	1.0472			
6	90	1.5708			
7	120	2.0944			
8	150	2.6180			
9	180	3.1416			
10					

Similarly, the DEGREES() function takes an angle in radians and returns the same angle in degrees. Also, the PI() function is available whenever π is required in a calculation.

	A5	▼	*fx*	=PI()/2	
	A	B	C	D	
1	**Angle**				
2	(Radians)	(Degrees)			
3	0	0			
4	0.7854	45			
5	1.5708	90			
6	3.1416	180			
7	6.2832	360			
8					

3.6.1 Standard Trigonometric Functions

FUNCTION NAME	OPERATION
SIN(x)	Returns the sine of x.
COS(x)	Returns the cosine of x.
TAN(x)	Returns the tangent of x.

To test these functions, try sin(30°), which should equal 0.5. The 30° may be converted to radians as a preliminary step, as in

C4	▼	f_x =SIN(B4)	
	A	B	C	D
1				
2		x	SIN(x)	
3	Degrees:	30		
4	Radians:	0.5236	0.5	
5				

or the conversion to radians and the calculation of the sine can be combined in a single equation, as in

C3	▼	f_x =SIN(RADIANS(B3))		
	A	B	C	D	E
1					
2		x	SIN(x)		
3	Degrees:	30	0.5		
4					

APPLICATIONS: KINEMATICS

Projectile Motion I

A projectile is launched at the angle 35° from the horizontal with velocity equal to 30 m/s. Neglecting air resistance and assuming a horizontal surface, determine how far away from the launch site will the projectile land.

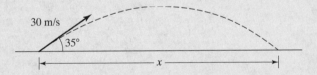

To answer this problem, we will need

1. Excel's trigonometry functions to handle the 35° angle, and
2. Equations relating distance to velocity and acceleration.

When velocity is constant, as in the horizontal motion of our particle (we're neglecting air resistance), the distance traveled is simply the initial horizontal velocity times time:

$$x(t) = v_{ox}t \qquad (3.3)$$

When the acceleration is constant, such as is the acceleration that gravity imposes on the projectile's vertical motion, the distance traveled becomes

$$y(t) = v_{oy}t + \frac{1}{2}g\,t^2 \qquad (3.4)$$

The initial velocity is stated (30 m/s at the angle 35° from the horizontal). We can compute the initial velocity components in the horizontal and vertical directions with Excel's trig. functions:

	A	B	C	D	E
1	**Projectile Motion I**				
2					
3		Angle:	35	degrees	
4		v_o:	30	m/s	
5					
6		v_{oy}:	17.2	m/s	
7		v_{ox}:	24.6	m/s	
8					

Cells C6 and C7 contain the following formulas:

C6: =C4*SIN(RADIANS(C3))
C7: =C4*COS(RADIANS(C3))

The RADIANS() function has been used to convert 35° to radians for compatibility with Excel's trigonometric functions.

The projectile ends up at the same level it started at (a horizontal surface was specified), so $y = 0$. Equation (2) can then be solved for the time of flight:

C13		f_x	=2*C6/C12		
	A	B	C	D	E
1	**Projectile Motion I**				
2					
3		Angle:	35	degrees	
4		v_o:	30	m/s	
5					
6		v_{oy}:	17.2	m/s	
7		v_{ox}:	24.6	m/s	
8					
9	Calculate Time of Flight				
10					
11		y:	0	m	
12		g:	9.8	m/s^2	
13		t:	3.51	s	
14					

The time of flight can then be used in Equation (1) to find the horizontal distance traveled:

	C17		▼		f_x	=C7*C13	
	A	**B**		**C**		**D**	**E**
7		v_{ox}:		24.6		m/s	
8							
9	Calculate Time of Flight						
10							
11		y:		0		m	
12		g:		9.8		m/s²	
13		t:		3.51		s	
14							
15	Calculate Horizontal Travel Distance						
16							
17		x:		86.3		m	
18							

PRACTICE!

If the launch angle is changed to 55°

a. What is the time of flight?

b. How far away will the projectile land?

c. What is the maximum height the projectile will reach? (Without air resistance, the maximum height is attained at half the time of flight.)

3.7 INVERSE TRIGONOMETRIC FUNCTIONS

Excel's *inverse trigonometric functions* return an angle in radians (the DEGREES() function is available if you would rather see the result in degrees):

FUNCTION NAME	OPERATION
ASIN(x)	Returns the angle (between $-\pi/2$ and $\pi/2$) that has a sine value equal to x.
ACOS(x)	Returns the angle (between 0 and π) that has a cosine value equal to x.
ATAN(x)	Returns the angle (between $-\pi/2$ and $\pi/2$) that has a tangent value equal to x.

Resolving Forces

If one person pulls to the right on a rope connected to a hook imbedded in a floor, using the force 400 N at 20° from the horizontal, while another person pulls to the left on the same hook, but using the force 200 N at 45° from the horizontal, what is the net force on the hook? The following diagram describes the situation:

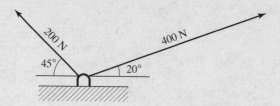

Because both people are pulling up, their vertical contributions combine; but one is pulling left and the other right, so they are (in part) counteracting each other's efforts. To quantify this distribution of forces, we can calculate the horizontal and vertical components of the force being applied by each person. Excel's trigonometric functions are helpful for these calculations.

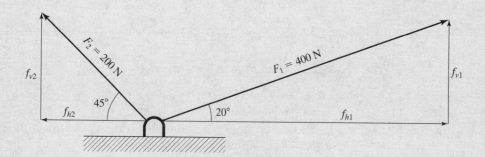

The 400-N force from person 1 resolves into a vertical component, f_{v1}, and a horizontal component, f_{h1}. The magnitudes of these force components can be calculated as follows:

$$f_{v1} = 400\,\sin(20°) = 136.8\text{ N},$$

$$f_{h1} = 400\,\cos(20°) = 375.9\text{ N}.$$

(3.5)

In Excel, these component forces are calculated as follows:

	E5	▼		f_x	=E2*SIN(RADIANS(E3))		
	A	B	C	D	E	F	G
1				**Right Side**			
2				Force	400	N	
3				Angle	20	°	
4							
5				f_{v2}	136.8	N	
6				f_{h2}	375.9	N	
7							

Similarly, the 200-N force from person 2 can be resolved into the following component forces:

$$f_{v2} = 200\sin(45°) = 141.4\text{ N}$$

$$f_{h2} = 200\cos(45°) = 141.4\text{ N} \tag{3.6}$$

We then get

B6	▼		f_x =B2*COS(RADIANS(B3))			
	A	B	C	D	E	F	G
1	**Left Side**			**Right Side**			
2	Force	200	N	Force	400	N	
3	Angle	45	°	Angle	20	°	
4							
5	f_{v2}	141.4	N	f_{v2}	136.8	N	
6	f_{h2}	141.4	N	f_{h2}	375.9	N	
7							

Actually, force component f_{h2} would usually be written as $f_{h2} = -141.421$ N, since it is pointed in the negative x direction. If all angles had been measured from the same position (typically counterclockwise from horizontal), the angle on the 200-N force would have been at 135°, and the signs would have taken care of themselves:

	A	B	C	D	E	F	G
1	**Left Side**			**Right Side**			
2	Force	200	N	Force	400	N	
3	Angle	135	°	Angle	20	°	
4							
5	f_{v2}	141.4	N	f_{v2}	136.8	N	
6	f_{h2}	-141.4	N	f_{h2}	375.9	N	
7							

Once the force components have been computed, the net force in the horizontal and vertical directions can be determined. (Force f_{h2} has a negative value in this calculation.) We obtain

B10	▼		f_x =E6+B6			
	A	B	C	D	E	F	G
1	**Left Side**			**Right Side**			
2	Force	200	N	Force	400	N	
3	Angle	135	°	Angle	20	°	
4							
5	f_{v2}	141.4	N	f_{v2}	136.8	N	
6	f_{h2}	-141.4	N	f_{h2}	375.9	N	
7							
8							
9		**Net Horizontal Force**			**Net Vertical Force**		
10	f_h	234.5	N	f_v	278.2	N	
11							

The net horizontal and vertical components can be recombined to find a combined net force on the hook, F_{net}, at angle θ:

$$F_{net} = \sqrt{f_h^2 + f_v^2} = 363.84 \text{ N},$$

$$\theta = a\tan\left(\frac{f_v}{f_h}\right) = 49.88°. \qquad (3.7)$$

So

B13	▼	f_x	=SQRT(B10^2+E10^2)			
	A	B	C	D	E	F	G
1	**Left Side**			**Right Side**			
2	Force	200	N	Force	400	N	
3	Angle	135	°	Angle	20	°	
4							
5	f_{v2}	141.4	N	f_{v2}	136.8	N	
6	f_{h2}	-141.4	N	f_{h2}	375.9	N	
7							
8							
9		**Net Horizontal Force**			**Net Vertical Force**		
10	f_h	234.5	N	f_v	278.2	N	
11							
12		**Net Force**					
13		363.8	N				
14							

E13	▼	f_x	=DEGREES(ATAN(E10/B10))			
	A	B	C	D	E	F	G
1	**Left Side**			**Right Side**			
2	Force	200	N	Force	400	N	
3	Angle	135	°	Angle	20	°	
4							
5	f_{v2}	141.4	N	f_{v2}	136.8	N	
6	f_{h2}	-141.4	N	f_{h2}	375.9	N	
7							
8							
9		**Net Horizontal Force**			**Net Vertical Force**		
10	f_h	234.5	N	f_v	278.2	N	
11							
12		**Net Force**			**Theta**		
13		363.8	N		49.88	°	
14							

Hyperbolic Trigonometric Functions

Excel provides the common *hyperbolic trigonometric* and *inverse hyperbolic trigonometric functions:*

FUNCTION NAME	OPERATION
SINH (x)	Returns the hyperbolic sine of x.
COSH (x)	Returns the hyperbolic cosine of x.
TANH (x)	Returns the hyperbolic tangent of x.
ASINH (x)	Returns the inverse hyperbolic sine of x.
ACOSH (x)	Returns the inverse hyperbolic cosine of x.
ATANH (x)	Returns the inverse hyperbolic tangent of x.

PRACTICE!

Use Excel's trigonometric functions to evaluate each of the following:

  a. $\sin(\pi/4)$

  b. $\sin(90°)$ *(Don't forget to convert to radians.)*

  c. $\cos(180°)$

  d. $\text{asin}(0)$

  e. $\text{acos}(0)$

3.8 ADVANCED MATH FUNCTIONS

Some of the built-in functions in Excel are pretty specialized. The advanced math functions described here will be very useful to engineers in certain disciplines and of little use to many others.

3.8.1 Logical Functions

Excel provides the following *logical functions:*

FUNCTION NAME	OPERATION
IF(test, Tvalue, Fvalue)	Performs the operation specified by the `test` argument, and then returns `Tvalue` if the test is true, `Fvalue` if the test is false.
TRUE()	Returns the logical value TRUE.
FALSE()	Returns the logical value FALSE.
NOT(test)	Reverses the logic returned by the `test` operation. If test returns TRUE, then NOT (test) returns FALSE.
AND(x1, x2, . . .)	Returns TRUE if all arguments are true, FALSE if any argument is false.
OR (x1, x2, . . .)	Returns TRUE if any argument is true, FALSE if all arguments are false.

A simple example of the use of these functions is

	C5		f_x	=IF(C2<C3,TRUE(),FALSE())		
	A	B	C	D	E	F
1						
2		x_1	3			
3		x_2	4			
4						
5		test:	TRUE			
6						

In this example, Excel tests to see whether 3 is less than 4. Because the test is true, the value TRUE is returned.

The following example shows how a worksheet might monitor the status of a tank being filled:

	B13		f_x =IF(B9>=B8,TRUE(),FALSE())			
	A	B	C	D	E
6					
7	Tank Operating Volume:	1200	liters		
8	Tank Capacity:	1350	liters		
9	Volume in Tank:	800	liters		
10	Tank is Filling:	TRUE			**Column B Contents**
11					
12	Is the tank full?	FALSE			=IF(B9>=B7,TRUE(),FALSE())
13	Is the tank overflowing?	FALSE			=IF(B9>=B8,TRUE(),FALSE())
14					
15	Operator Action:	none required			=IF(B13,"SHUT THE VALVE!","none required")
16					

In this example,

- The IF() function in cell B12 checks to see whether the volume in the tank (B9) has reached or exceeded the operating volume (B7).
- The IF() function in cell B13 checks to see whether the volume in the tank (B9) has reached or exceeded the tank capacity (B7).

If the volume has exceeded the tank capacity (i.e., if the tank is overflowing), the IF() function in cell B15 tells the operator to shut the valve:

	B9		f_x 1350			
	A	B	C	D	E
6					
7	Tank Operating Volume:	1200	liters		
8	Tank Capacity:	1350	liters		
9	Volume in Tank:	1350	liters		
10	Tank is Filling:	TRUE			**Column B Contents**
11					
12	Is the tank full?	TRUE			=IF(B9>=B7,TRUE(),FALSE())
13	Is the tank overflowing?	TRUE			=IF(B9>=B8,TRUE(),FALSE())
14					
15	Operator Action:	SHUT THE VALVE!			=IF(B13,"SHUT THE VALVE!","none required")
16					

PRACTICE!

How would you modify the preceding worksheet to give the operators instructions in the following situations?

1. If the volume in the tank reaches or exceeds the operating volume, tell the operators to shut the valve.

2. If the volume in the tank is less than 200 liters and the tank is not filling, tell the operators to open the valve.

3.9 CONDITIONAL FORMATTING

The IF() function uses a logical test to return one of two possible results. *Conditional formatting*, although not a function at all, uses a logical test to control the way cell contents are displayed. Conditional formatting could be used, for example, to display the "SHUT THE VALVE!" instruction in bold white letters on a black background.

The Conditional Formatting dialog box is accessed from the Format menu by selecting Format/Conditional formatting.... Select the cell(s) to be formatted before opening the Conditional Formatting dialog box.

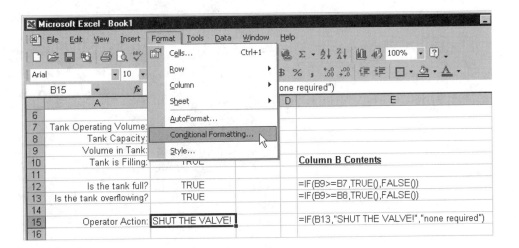

The logical test used to decide whether to apply formatting can be based on the contents of the cell that will be formatted or on a formula. Both approaches are illustrated here.

Note: Up to three conditions (logical tests) can be applied to a cell, so several formats could be used to highlight different conditions.

Using Cell Contents to Set the Display Format If the cell is displaying "SHUT THE VALVE!", then we want to change the format to include bold white letters on a black background. The condition is set as

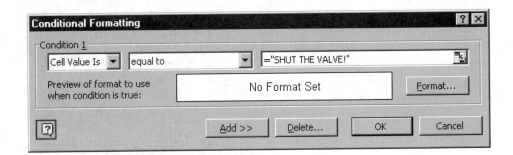

Subsequently the Format... button is used to specify the display format to be used when the condition is met. First, the font characteristics are set to bold and white:

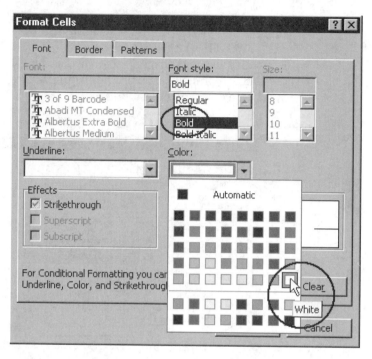

Then, with the Patterns tab, the background color is set to black:

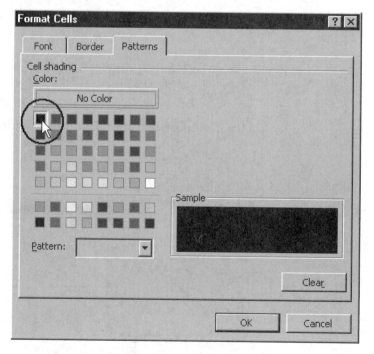

Once the conditional formatting has been set, if cell B15 contains the message "SHUT THE VALVE!" the message will be displayed in bold white letters on a black background, as shown next; but if the tank is not overflowing, the cell format reverts to the original formatting.

	A	B	C	D	E
6					
7	Tank Operating Volume:	1200	liters		
8	Tank Capacity:	1350	liters		
9	Volume in Tank:	1350	liters		
10	Tank is Filling:	TRUE			**Column B Contents**
11					
12	Is the tank full?	TRUE			=IF(B9>=B7,TRUE(),FALSE())
13	Is the tank overflowing?	TRUE			=IF(B9>=B8,TRUE(),FALSE())
14					
15	Operator Action:	**SHUT THE VALVE!**			=IF(B13,"SHUT THE VALVE!","none required")
16					

Using a Formula to Set the Display Format   If cell B13 contains TRUE, the tank is overflowing and the instructions to the operator in cell B15 will be, "SHUT THE VALVE!" We can use a formula to see whether cell B13 contains TRUE and, if so, to change the format used for the cell containing the message (cell B15) to include bold white letters on a black background. When a formula is used, the condition is set as follows:

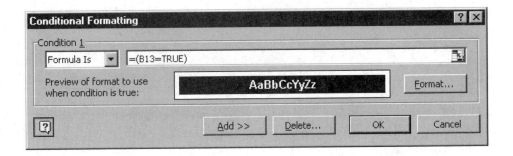

There is an implied "if the formula evaluates to TRUE," so this condition is effectively saying, "if cell B13 contains TRUE, then set the format of the currently selected cell (cell B15)." (The parentheses in the formula are just for readability.)

Both methods, using the contents of cell B15 or a formula to test cell B13, produce the same result; if the tank is overflowing, the operator message is displayed in bold white letters on a black background.

3.10   ERROR FUNCTION

Excel provides two functions for working with *error functions*: ERF() and ERFC(). ERF(x) returns the error function integrated between zero and x, defined as

$$\text{ERF}(x) = \frac{2}{\sqrt{\pi}} \int_0^x e^{-t^2} \, dt. \tag{3.8}$$

Or you can specify two integration limits, as ERF(x1, x2):

$$\text{ERF}(x_1, x_2) = \frac{2}{\sqrt{\pi}} \int_{x_1}^{x_2} e^{-t^2} \, dt. \tag{3.9}$$

The complementary error function, ERFC(x), is also available:

$$\text{ERFC}(x) = 1 - \text{ERF}(x) \tag{3.10}$$

FUNCTION NAME	OPERATION
ERF(x)	Returns the error function integrated between zero and x.
ERF(x1,x2)	Returns the error function integrated between x_1 and x_2.
ERFC(x)	Returns the complementary error function integrated between x and ∞.

Note: The ERF() *and* ERFC() *functions are part of the Analysis Toolpak and are not activated by default. You might have to use Tools/Add Ins ... and then select the Analysis Toolpak from the list of add-ins.*

3.11 BESSEL FUNCTIONS

Bessel functions are commonly needed when integrating differential equations in cylindrical coordinates. These functions are complex and commonly available in tabular form. Bessel and modified Bessel functions are available as built-in functions in Excel:

FUNCTION NAME	OPERATION
BESSELJ(x,n)	Returns the Bessel function, Jn, of x. (N is the order of the Bessel function.)
BESSELY(x,n)	Returns the Bessel function, Yn, of x.
BESSELI(x,n)	Returns the modified Bessel function, In, of x.
BESSELK(x,n)	Returns the modified Bessel function, Kn, of x.

The following spreadsheet uses the BESSELJ(x,n) function to create graphs of the $J0(x)$ and $J1(x)$ Bessel functions:

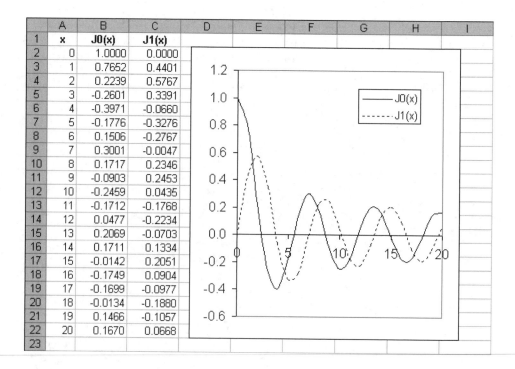

	A	B	C
1	x	J0(x)	J1(x)
2	0	1.0000	0.0000
3	1	0.7652	0.4401
4	2	0.2239	0.5767
5	3	-0.2601	0.3391
6	4	-0.3971	-0.0660
7	5	-0.1776	-0.3276
8	6	0.1506	-0.2767
9	7	0.3001	-0.0047
10	8	0.1717	0.2346
11	9	-0.0903	0.2453
12	10	-0.2459	0.0435
13	11	-0.1712	-0.1768
14	12	0.0477	-0.2234
15	13	0.2069	-0.0703
16	14	0.1711	0.1334
17	15	-0.0142	0.2051
18	16	-0.1749	0.0904
19	17	-0.1699	-0.0977
20	18	-0.0134	-0.1880
21	19	0.1466	-0.1057
22	20	0.1670	0.0668
23			

3.12   WORKING WITH COMPLEX NUMBERS

Excel does provide functions for handling *complex numbers*, although working with complex numbers by using built-in functions is cumbersome at best. Still, if you occasionally need to handle complex numbers, the following functions are available:

FUNCTION NAME	OPERATION
`COMPLEX(real, img, suffix)`	Combines real and imaginary coefficients into a complex number. The `suffix` is an optional text argument allowing you to use a "j" to indicate the imaginary portion. ("i" is used by default.)
`IMAGINARY(x)`	Returns the imaginary coefficient of complex number x.
`IMREAL(x)`	Returns the real coefficient of complex number x.
`IMABS(x)`	Returns the absolute value (modulus) of complex number x.
`IMCOS(x)`	Returns the cosine of complex number x.
`IMSIN(x)`	Returns the sine of complex number x.
`IMLN(x)`	Returns the natural logarithm of complex number x.
`IMLOG10(x)`	Returns the base-10 logarithm of complex number x.
`IMLOG2(x)`	Returns the base-2 logarithm of complex number x.
`IMEXP(x)`	Returns the exponential of complex number x.
`IMPOWER(x,n)`	Returns the value of complex number x raised to the integer power n.
`IMSQRT(x)`	Returns the square root of complex number x.
`IMSUM(x1, x2, ...)`	Adds complex numbers.
`IMSUB(x1, x2)`	Subtracts two complex numbers.
`IMPROD(x1, x2, ...)`	Determines the product of up to 29 complex numbers.
`IMDIV(x1, x2)`	Divides two complex numbers.

3.13   WORKING WITH BINARY, OCTAL, AND HEXADECIMAL VALUES

Excel provides the following functions for converting *binary, octal, decimal,* and *hexadecimal* values:

FUNCTION NAME	OPERATION
`BIN2OCT(number, places)`	Converts a binary number to octal. `Places` can be used to pad leading digits with zeros.
`BIN2DEC(number)`	Converts a binary number to decimal.
`BIN2HEX(number, places)`	Converts a binary number to hexadecimal.
`DEC2BIN(number, places)`	Converts a decimal number to binary.
`DEC2OCT(number, places)`	Converts a decimal number to octal.
`DEC2HEX(number, places)`	Converts a decimal number to hexadecimal.
`OCT2BIN(number, places)`	Converts an octal number to binary.
`OCT2DEC(number)`	Converts an octal number to decimal.
`OCT2HEX(number, places)`	Converts an octal number to hexadecimal.
`HEX2BIN(number, places)`	Converts a hexadecimal number to binary.
`HEX2OCT(number, places)`	Converts a hexadecimal number to octal.
`HEX2DEC(number)`	Converts a hexadecimal number to decimal.

In the next example, the decimal value 43 has been converted to binary, octal, and hexadecimal. In each case, 8 digits were requested, and Excel added leading zeros to the result. The formulas used in column C are listed in column E:

	A	B	C	D	E	F
1						
2		**Binary:**	00101011		=DEC2BIN(C4,8)	
3		**Octal:**	00000053		=DEC2OCT(C4,8)	
4		**Decimal:**	43			
5		**Hexadecimal:**	0000002B		=DEC2HEX(C4,8)	
6						

3.14 MISCELLANEOUS FUNCTIONS

Here are a few built-in functions that, although less commonly used, can be very handy on occasion.

3.14.1 Working with Random Numbers

The RAND() and RANDBETWEEN() functions generate *random numbers*. The RAND() function returns a value greater than or equal to zero and less than 1. The RANDBETWEEN(low, high) function returns a value between the low and high values you specify.

Also, if you select Tools/Data Analysis..., there is a random-number generator that can return a set of random numbers with special properties, such as normally distributed values with a given mean and standard deviation.

Note: Like all of the Add-Ins in the Data Analysis category, this function is not activated by default. You might have to select Tools/Add-Ins ... and then select the Analysis Toolpak from the list of add-ins.

3.14.2 Odd, Even

These functions round away from zero to the next odd or even value:

```
=ODD(2)    rounds to 3
=EVEN(1)   rounds to 2
=ODD(-2)   rounds to -3     (away from zero)
```

The ODD() function rounds zero to 1; EVEN(0) returns a zero.

3.14.3 Date and Time Functions

Excel provides a number of functions that read the calendar and clock on your computer to make date and time information available to your worksheets. The date and time is stored as a *date–time code*, such as 6890.45833. The number to the left of the decimal point represents the number of days after the start date. By default, the start date is January 1, 1900, for PCs, and January 1, 1904, for Macintosh computers. You can ask Excel to use a nondefault date system. For example, to use Macintosh-style date-time codes on a PC, use the following menu commands:

Select Tools/Options/Calculation tab; then check the "1904 Date System" check box

The date-time code 6890.45833 represents a unique moment in history; the eleventh hour ($11/24 = 0.45833$) of the eleventh day of the eleventh month of 1918: the time of the signing of the armistice that ended World War I. November 11, 1918 is 6890 days after the PC start date of 1/1/1900.

You can find the date code for any date after 1900 (or 1904 if the 1904 date system is in use) by using the DATE(year, month, day) function. Using the DATE() function with Armistice Day returns the date value, 6890:

B2	▼		f_x	=DATE(1918,11,11)	
	A	B	C	D	E
1					
2		6890			
3					
4					

Excel tries to make date codes more readable to the user by formatting cells containing date codes in a *date format*. The worksheet in the preceding example actually displayed 11/11/18 in cell B2 until the formatting for that cell was changed from date format to number format.

If you enter a date or time into a cell, Excel converts the value to a date–time code, but formats the cell to display the date or the time:

	A	B	C	D	E	F	G
1		**Date**	**Time**	**Time and Date**			
2		11/11/1918	11:00:00 am	11:00 am 11/11/1918		As entered	
3		6890	0.45833	6890.45833		As converted to date-time codes	
4		11/11/18	11:00 AM	11/11/18 11:00		As displayed (default)	
5							

Date–time codes (also called *serial date values*) are used to allow Excel to perform calculations with dates and times. For example, Excel can calculate the number of days left until Christmas on August 12, 1916 by using the DATE() function (the result is 135 days):

```
=DATE(1916,12,25) - DATE(1916,8,12)
```

Excel provides a number of functions for working with dates and times. Some of these functions are as follows:

FUNCTION NAME	OPERATION
TODAY()	Returns today's date code. Excel displays the date with a date format.
DATE(year, month, day)	Returns the date code for the specified year, month, and day. Excel displays the date code with a date format (shows the date, not the date code). Change the cell format to number or general to see the date code.
DATEVALUE(date)	Returns the date code for the date, which is entered as text such as "12/25/1916" or "December 25, 1916." The quotes are required. Excel displays the date code with a general format (i.e., shows the date code, not the date).
NOW()	Returns the current time code. Excel displays the time with a time format.
TIME(hour, minute, second)	Returns the time code for the specified hour, minute, and second. Excel displays the time code with a time format (shows the time, not the time code). Change the cell format to number or general to see the time code.
TIMEVALUE(time)	Returns the date code for the time, which is entered as text such as "11:00 am." The quotes are required. Excel displays the time code with a general format (i.e., shows the time code, not the time).

Excel also provides a series of functions for pulling particular portions from a date–time code:

FUNCTION NAME	OPERATION
YEAR (date code)	Returns the year referred to by the date represented by the date code. For example, =YEAR (6890) returns 1918.
MONTH (date code)	Returns the month referred to by the date represented by the date code. For example, =MONTH (6890) returns 11.
DAY (date code)	Returns the day referred to by the date represented by the date code. For example, =DAY (6890) returns 11.
HOUR (time code)	Returns the hour referred to by the time represented by the time code or date-time code. For example, =HOUR (6890.45833) returns 11. So does =HOUR (0.45833).
MINUTE (time code)	Returns the minute referred to by the time represented by the time code or date–time code.
SECOND (time code)	Returns the second referred to by the time represented by the time code or date–time code.

When using Excel's date–time codes, remember the following:

1. Date–time codes work only for dates on or after the starting date (1/1/1900 or 1/1/1904).

2. Using four-digit years avoids ambiguity. By default, 00 through 29 is interpreted to mean 2000 through 2029, while 30 through 99 is interpreted to mean 1930 through 1999.

3. Excel tries to be helpful when working with date–time codes. You can enter a date as text when the function requires a date–time code, and Excel will convert the date to a date–time code before sending it to the function:

B2	▼	f_x =YEAR("Nov 11, 1918")		
	A	B	C	D	E
1					
2		1918			
3					

4. Whenever Excel interprets an entry in a cell as a date, it changes the format to date format, and changes the cell contents to a date code. This can be frustrating if the cell contents happen to look like a date, but are not intended to be interpreted as such. For example, if several parts sizes are indicated in inches as 1/8, 1/10, and 1/16, they would appear in a worksheet as follows:

	A	B	C	D
1				
2		**Part Size**		
3		(inches)		
4				
5		8-Jan		
6		10-Jan		
7		16-Jan		
8				

Excel interpreted the 1/8 to be a date (January 8th of the current year), converted the entry to a date code, and changed the cell formatting to display the date. Because Excel has changed the cell contents to a date code, simply fixing the cell format will not give you the desired result. You need to tell Excel that it should interpret the entry as a fraction by setting the display format for the cells to "fraction" before entering the values in the cell. (Or you can type a single quote ['] before the fraction to have Excel interpret the entry as a label. However, with this approach, the part size labels should not be used as values in formulas.)

3.14.4 Text-Handling Functions

Excel provides the following standard text-manipulation functions:

FUNCTION NAME	OPERATION
CHAR (number)	Returns the character specified by the code number.
CONCATENATE (text1, text2)	Joins several text items into one text item. (The & operator can also be used for concatenation.)
EXACT (text1, text2)	Checks to see if two text values are identical. Returns TRUE if the text strings are identical, otherwise FALSE (case-sensitive).
FIND (text_to_find, text_to_search, start_pos)	Finds one text value within another (case-sensitive).
LEFT (text, n)	Returns the leftmost n characters from a text value.
LEN (text)	Returns the number of characters in text string, text.
LOWER (text)	Converts text to lowercase.
MID (text, start_pos, n)	Returns n characters from a text string starting at start_pos.
REPLACE (old_text, start_pos, n, new_text)	Replaces n characters of old_text with new_text starting at start_pos.
RIGHT (text, n)	Returns the rightmost n characters from a text value.
SEARCH (text_to_find, text_to_search, start_pos)	Finds one text value within another (not case-sensitive).
TRIM (text)	Removes spaces from text.
UPPER (text)	Converts text to uppercase.

3.14.5 Lookup and Reference Functions

Because of the tabular nature of a spreadsheet, lists of numbers are common, and Excel provides ways to look up values in tables. Excel's *lookup functions* are summarized here:

FUNCTION NAME	OPERATION
VLOOKUP (value,table,N,matchType)	**Vertical Lookup** Looks up a value in the left column of a table, jumps to column N of the table (the matched column is column 1), and returns the value in that location. If matchType is set to TRUE, the lookup will fail unless the match is exact. If matchType is omitted or set to FALSE, Excel will use the next highest value as the match.
HLOOKUP (value,table,N,matchType)	**Horizontal Lookup** Looks up a value in the top row of a table, jumps down to row N of the same table (same column), and returns the value in that location.

The following spreadsheet example looks up letter grades for students from a table:

C8	▼		f_x	=VLOOKUP(B8,$E$3:$F$14,2)		
	A	B	C	D	E	F	G
1							
2					**Grade Table**		
3	**Student**	**Score**	**Grade**		0	F	
4					59.9	D-	
5	John A.	85	B		62.9	D	
6	Jane B.	93	A		66.9	D+	
7	Jim G.	78	C+		69.9	C-	
8	Pris L.	82	B-		72.9	C	
9					76.9	C+	
10					79.9	B-	
11					82.9	B	
12					86.9	B+	
13					89.9	A-	
14					92.9	A	
15							

KEY TERMS

Argument	Decimal values	Lookup functions
Argument list	Error functions	Name
AutoSum button	Exponentiation functions	Octal values
Bessel functions	Function	Parameter
Binary values	Function name	Radians
Built-in functions	Hexadecimal values	Random numbers
Complex numbers	Hyperbolic trigonometric functions	ScreenTip
Conditional formatting	Inverse hyperbolic trigonometric functions	Serial date value
Date-time code	Logarithm functions	Trigonometric functions
Date format	Logical functions	

SUMMARY

A function is a reusable equation that accepts arguments, uses the argument values in a calculation, and returns the result(s). In this section, the commonly used built-in functions in Excel are collected and displayed in tabular format. Also, descriptions of some other functions (matrix math functions, statistical functions, time-value-of-money functions) are included in the section in an attempt to provide a single source for descriptions of those commonly used functions.

COMMON MATH FUNCTIONS

SQRT (x)	Square Root
ABS (x)	Absolute Value
FACT (x)	Factorial
SUM (range)	Summation
GCD (x1, x2, ...)	Greatest Common Divisor
LCM (x1, x2, ...)	Least Common Multiple

LOG AND EXPONENTIATION	OPERATION
EXP (x)	Returns e raised to the power x.
LN (x)	Returns the natural log of x.
LOG10 (x)	Returns the base-10 log of x.
LOG (x, base)	Returns the logarithm of x to the specified base.

TRIGONOMETRY FUNCTIONS	OPERATION
SIN (x)	Returns the sine of x.
COS (x)	Returns the cosine of x.
TAN (x)	Returns the tangent of x.
RADIANS (x)	Converts x from degrees to radians.
DEGREES (x)	Converts x from radians to degrees.

INVERSE TRIG. FUNCTIONS	OPERATION
ASIN (x)	Returns the angle (between $-\pi/2$ and $\pi/2$) that has a sine value equal to x.
ACOS (x)	Returns the angle (between 0 and π) that has a cosine value equal to x.
ATAN (x)	Returns the angle (between $-\pi/2$ and $\pi/2$) that has a tangent value equal to x.

HYPERBOLIC TRIG. FUNCTIONS	OPERATION
SINH (x)	Returns the hyperbolic sine of x.
COSH (x)	Returns the hyperbolic cosine of x.
TANH (x)	Returns the hyperbolic tangent of x.
ASINH (x)	Returns the inverse hyperbolic sine of x.
ACOSH (x)	Returns the inverse hyperbolic cosine of x.
ATANH (x)	Returns the inverse hyperbolic tangent of x.

LOGICAL FUNCTIONS	OPERATION
IF(test, Tvalue, Fvalue)	Performs the operation specified by the test argument and then returns Tvalue if the test is true, or Fvalue if the test is false.
TRUE ()	Returns the logical value TRUE.
FALSE ()	Returns the logical value FALSE.
NOT (test)	Reverses the logic returned by the test operation. If test returns TRUE, then NOT (test) returns FALSE.
AND (x1, x2, ...)	Returns TRUE if all arguments are true or FALSE if any argument is false.
OR (x1, x2, ...)	Returns TRUE if any argument is true or FALSE if all arguments are false.

FUNCTION NAME	OPERATION
ERF (x)	Returns the error function integrated between zero and x.
ERF (x1, x2)	Returns the error function integrated between x_1 and x_2.
ERFC (x)	Returns the complementary error function integrated between x and ∞.

BESSEL FUNCTIONS	OPERATION
BESSELJ (x,n)	Returns the Bessel function, Jn, of x. (n is the order of the Bessel function.)
BESSELY (x,n)	Returns the Bessel function, Yn, of x.
BESSELI (x,n)	Returns the modified Bessel function, In, of x.
BESSELK (x,n)	Returns the modified Bessel function, Kn, of x.

COMPLEX NUMBER FUNCTIONS	OPERATION
COMPLEX(real, img, suffix)	Combines real and imaginary coefficients into a complex number. The suffix is an optional text argument in case you wish to use a "j" to indicate the imaginary portion. ("i" is used by default.)
IMAGINARY (x)	Returns the imaginary coefficient of complex number x.
IMREAL (x)	Returns the real coefficient of complex number x.
IMABS (x)	Returns the absolute value (modulus) of complex number x.
IMCOS (x)	Returns the cosine of complex number x.
IMSIN (x)	Returns the sine of complex number x.
IMLN (x)	Returns the natural logarithm of complex number x.
IMLOG10 (x)	Returns the base-10 logarithm of complex number x.
IMLOG2 (x)	Returns the base-2 logarithm of complex number x.
IMEXP (x)	Returns the exponential of complex number x.
IMPOWER (x,n)	Returns the value of complex number x raised to the integer power n.
IMSQRT (x)	Returns the square root of complex number x.
IMSUM (x1, x2, . . .)	Adds complex numbers.
IMSUB (x1, x2)	Subtracts two complex numbers.
IMPROD (x1, x2, . . .)	Computes the product of up to 29 complex numbers.
IMDIV (x1, x2)	Divides two complex numbers.

CONVERSION FUNCTIONS	OPERATION
BIN2OCT(number, places)	Converts a binary number to octal. Places can be used to pad leading digits with zeros.
BIN2DEC (number)	Converts a binary number to decimal.
BIN2HEX (number, places)	Converts a binary number to hexadecimal.
DEC2BIN (number, places)	Converts a decimal number to binary.
DEC2OCT (number, places)	Converts a decimal number to octal.
DEC2HEX (number, places)	Converts a decimal number to hexadecimal.
OCT2BIN (number, places)	Converts an octal number to binary.

CONVERSION FUNCTIONS	OPERATION
OCT2DEC (number)	Converts an octal number to decimal.
OCT2HEX (number, places)	Converts an octal number to hexadecimal.
HEX2BIN (number, places)	Converts a hexadecimal number to binary.
HEX2OCT (number, places)	Converts a hexadecimal number to octal.
HEX2DEC (number)	Converts a hexadecimal number to decimal.

DATE AND TIME FUNCTIONS	OPERATION
TODAY()	Returns today's date code. Excel displays the date with a date format.
DATE(year, month, day)	Returns the date code for the specified year, month, and day. Excel displays the date code with a date format (i.e., shows the date, not the date code). Change the cell format to number or general to see the date code.
DATEVALUE (date)	Returns the date code for the date, which is entered as text such as "12/25/1916" or "December 25, 1916." The quotes are required. Excel displays the date code with a general format (i.e., shows the date code, not the date).
NOW ()	Returns the current time code. Excel displays the time with a time format.
TIME(hour, minute, second)	Returns the time code for the specified hour, minute, and second. Excel displays the time code with a time format (i.e., shows the time, not the time code). Change the cell format to number or general to see the time code.
TIMEVALUE (time)	Returns the date code for the time, which is entered as text such as "11:00 am." The quotes are required. Excel displays the time code with a general format (i.e., shows the time code, not the time).
YEAR (date code)	Returns the year referred to by the date represented by the date code. Example: =YEAR (6890) returns 1918.
MONTH (date code)	Returns the month referred to by the date represented by the date code. Example: =MONTH (6890) returns 11.
DAY (date code)	Returns the day referred to by the date represented by the date code. Example: =DAY (6890) returns 11.
HOUR (time code)	Returns the hour referred to by the time represented by the time code or date-time code. Example: =HOUR (6890.45833) returns 11. So does =HOUR (0.45833).
MINUTE (time code)	Returns the minute referred to by the time represented by the time code or date–time code.
SECOND (time code)	Returns the second referred to by the time represented by the time code or date–time code.

TEXT FUNCTIONS	OPERATION
CHAR (number)	Returns the character specified by the code number.
CONCATENATE (text1, text2)	Joins several text items into one text item. (The & operator can also be used for concatenation.)

TEXT FUNCTIONS	OPERATION
EXACT (text1, text2)	Checks to see whether two text values are identical. Returns TRUE if the text strings are identical, FALSE otherwise (case-sensitive).
FIND (text_to_find, text_to_search, start_pos)	Finds one text value within another (case-sensitive).
LEFT (text, n)	Returns the leftmost n characters from a text value.
LEN (text)	Returns the number of characters in text string, text.
LOWER (text)	Converts text to lowercase.
MID (text, start_pos, n)	Returns n characters from a text string starting at start_pos.
REPLACE (old_text, start_pos, n, new_text)	Replaces n characters of old_text with new_text starting at start_pos.
RIGHT (text, n)	Returns the rightmost n characters from a text value.
SEARCH (text_to_find, text_to_search, start_pos)	Finds one text value within another (not case-sensitive).
TRIM (text)	Removes spaces from text.
UPPER (text)	Converts text to uppercase.

FUNCTION NAME	OPERATION
VLOOKUP (value,table,N, matchType)	**Vertical Lookup** Looks up a value in the left column of a table, jumps to column N of the table (to the right of the matched column), and returns the value in that location. If matchType is set to TRUE, the lookup will fail unless the match is exact. If matchType is omitted or set to FALSE, Excel will use the next higher value as the match.
HLOOKUP (value,table,N, matchType)	**Horizontal Lookup** Looks up a value in the top row of a table, jumps down to row N of the same table (same column), and returns the value in that location.

Statistical Functions

FUNCTION NAME	DESCRIPTION
=AVERAGE (range)	Calculates the arithmetic average of the values in the specified range of cells.
=STDEV (range)	Calculates the sample standard deviation of the values in the specified range of cells.
=STDEVP (range)	Calculates the population standard deviation of the values in the specified range of cells.
=VAR (range)	Calculates the sample variance of the values in the specified range of cells.
=VARP (range)	Calculates the population variance of the values in the specified range of cells.

Time-Value-of-Money Functions

FUNCTION NAME	DESCRIPTION
=FV(iP, Nper, Pmt, PV, Type)	Calculates a future value, FV, given a periodic interest rate, i_P, the number of compounding periods, N, a periodic payment, Pmt, an optional present value, PV, and an optional code indicating whether payments are made at the beginning or end of each period, Type: Type = 1 means payments at the beginning of each period. Type = 0 or omitted means payments at the end of each period.
=PV(iP, Nper, Pmt, FV, Type)	Calculates a present value, PV, given a periodic interest rate, i_P, the number of compounding periods, N, a periodic payment, Pmt, an optional future value, FV, and an optional code indicating whether payments are made at the beginning or end of each period, Type.
=PMT(iP, Nper, PV, FV, Type)	Calculates the periodic payment, Pmt, equal to a given present value, PV, and (optionally) future value, FV, given a periodic interest rate, i_P, the number of compounding periods, N. Type is an optional code indicating whether payments are made at the beginning (Type 21) or end (Type 20) of each period.

Internal Rate of Return and Depreciation Functions

FUNCTION NAME	DESCRIPTION
=IRR(Incomes, irrGuess)	Calculates the internal rate of return for the series of incomes and expenses: Incomes is a range of cells containing the incomes (and expenses as negative incomes) (required). IrrGuess is a starting value, or guess value, for the iterative solution required to calculate the internal rate of return, optional.
=NPV (i_P, Inc1, Inc2, . . .)	Calculates the net present value, NPV, of a series of incomes (Inc1, Inc2, . . .) given a periodic interest rate, i_P: Inc1, Inc2, etc., can be values, cell addresses, or cell ranges containing incomes (and expenses as negative incomes).
=VDB (C_init, S, N_SL, Per_start, Per_end, F_DB, NoSwitch)	Calculates depreciation amounts using various methods (straight line, double-declining balance, MACRS). C_{init} — is the initial cost of the asset (required). S — is the salvage value, required (set to zero for MACRS depreciation). N_{SL} — is the service life of the asset, required. Per_{start} — is the start of the period over which the depreciation amount will be calculated, required. Per_{stop} — is the end of the period over which the depreciation amount will be calculated, required. F_{DB} — is declining balance percentage, optional (Excel uses 200% if omitted). NoSwitch — tells Excel whether to switch to straight-line depreciation when the straight-line depreciation factor is larger than the declining-balance depreciation factor, optional (set to FALSE to switch to straight line, TRUE to not switch; the default is FALSE).

Matrix Math Functions

FUNCTION NAME	DESCRIPTION
=MMULT(M1, M2)	Multiplies matrices M1 and M2. Use [Ctrl–Shift–Enter] to tell Excel to enter the matrix function into every cell of the result matrix.
=TRANSPOSE(M)	Transposes matrix M.
=MINVERSE(M)	Calculates the inverse of matrix M (if possible).
=MDETERM(M)	Calculates the determinant of matrix M.

Problems

Trigonometric Functions

1. Devise a test to demonstrate the validity of the following common trigonometric formulas. What values of A and B should be used to test these functions thoroughly?

    a. $\sin(A + B) = \sin(A)\cos(B) + \cos(A)\sin(B)$
    b. $\sin(2A) = 2\sin(A)\cos(A)$
    c. $\sin^2(A) = \frac{1}{2} - \frac{1}{2}\cos(2A)$

    Note: In Excel, $\sin^2(A)$ should be entered as `=(sin(A))^2`. *This causes sin(A) to be evaluated first and that result to be squared.*

Basic Fluid Flow

2. A commonly used rule of thumb is that the average velocity in a pipe should be about 1 m/s or less for "thin" fluids (viscosity about water). If a pipe needs to deliver 6,000 m^3 of water per day, what diameter is required to satisfy the 1-m/s rule?

Projectile Motion II

3. Sports programs' "shot of the day" segments sometimes show across-the-court baskets made just as (or after) the final buzzer sounds. If a basketball player, with three seconds remaining in the game, throws the ball at a 45° angle from 4 feet off the ground, standing 70 feet from the basket, which is 10 feet in the air,

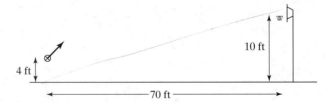

    a.   what initial velocity does the ball need to have in order to reach the basket?
    b.   what is the time of flight? and
    c.   how much time will be left in the game after the shot?

    Ignore air resistance in this problem.

Pulley I

4.   A 200-kg mass is hanging from a hook connected to a pulley, as shown in the accompanying figure. The cord around the pulley is connected to the overhead
support at two points. What is the tension in each cord connected to the support if the angle of the cord from vertical is

a.   0°?

b.   5°?

c.   15°?

Force Components and Tension in Wires I

5.   A 150-kg mass is suspended by wires from two hooks, as shown in the accompanying figure. The lengths of the wires have been adjusted so that the wires
are each 50° from horizontal. Assume that the mass of the wires is negligible.

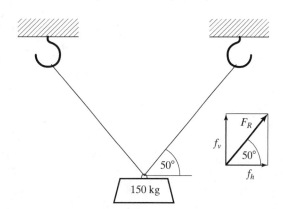

a.   Two hooks support the mass equally, so the vertical component of force exerted by either hook will be equal to the force resulting from 75 kg being
acted on by gravity. Calculate this vertical component of force, f_v on the
right hook. Express your result in newtons.

b.   Compute the horizontal component of force, f_h, by using the result obtained in part (a) and trigonometry.

c.   Determine the force exerted on the mass in the direction of the wire F_R
(equal to the tension in the wire).

d.   If you moved the hooks farther apart to reduce the angle from 50° to 30°
would the tension in the wires increase or decrease? Why?

Force Components and Tension in Wires II

6.   If two 150-kg masses are suspended on a wire as shown in the accompanying figure, such that the section between the loads (wire B) is horizontal, then wire B is under tension, but is doing no lifting. The entire weight of the 150-kg mass on the right is being held up by the vertical component of the force in wire C. In the same way, the mass on the left is being supported entirely by the vertical component of the force in wire A. What is the tension on wire B?

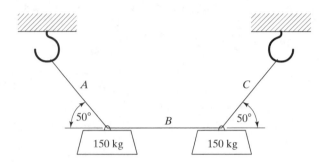

Force Components and Tension in Wires III

7.   If the hooks shown in the previous problem are pulled farther apart, the tension in wire B will change, and the angle of wires A and C with respect to the horizontal will also change.

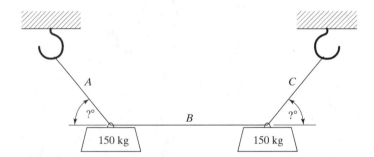

If the hooks are pulled apart until the tension in wire B is 2000 N, compute

a.   the angle between the horizontal and wire C.

b.   the tension in wire C.

How does the angle in part (a) change if the tension in wire B is increased to 3000 N?

Finding the Volume of a Storage Bin I

8.   A fairly common shape for a dry-solids storage bin is a cylindrical silo with a conical collecting section at the base where the product is removed. (See the accompanying figure.) To calculate the volume of the contents, you use the formula for

a cone, as long as the height of product, h, is less than the height of the conical section, h_{cone}:

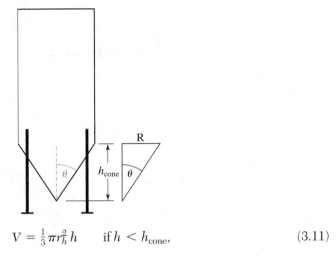

$$V = \tfrac{1}{3}\pi r_h^2 h \qquad \text{if } h < h_{cone},\qquad\qquad (3.11)$$

Here, r_h is the radius at height h and can be calculated from h by using trigonometry:

$$r_h = h_{cone}\tan(\theta).\qquad\qquad (3.12)$$

If the height of the stored product is greater than the height of the conical section, the equation for a cylinder must be added to the volume of the cone:

$$V = \tfrac{1}{3}\pi R^2 h_{cone} + \pi R^2(h - h_{cone}) \qquad \text{if } h > h_{cone}.\qquad (3.13)$$

If the height of the conical section is 3 meters, the radius of the cylindrical section is 2 m, and the total height of the storage bin is 10 meters, what is the maximum volume of material that can be stored?

Finding the Volume of a Storage Bin II

9. Consider the storage bin described in the previous problem.

    a. Calculate the angle θ as shown in the diagram.
    b. For a series of h values between 0 to 10 meters, calculate r_h values and bin volumes.
    c. Plot the volume-vs.-height data.

Pulley Problem II

10. A 200-kg mass is attached to a pulley, as shown here:

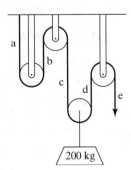

    a.   What force must be exerted on cord e to keep the mass from moving?

    b.   When the mass is stationary, what is the tension in cords a through e?

    c.   Which, if any, of the solid supports connecting the pulleys to the overhead support is in compression?

Nonideal Gas Equation

11.   The Soave–Redlich–Kwong (SRK) equation is a commonly used equation of state that relates the absolute temperature T, the absolute pressure P, and the molar volume $\hat{V}$ of a gas under conditions in which the behavior of the gas cannot be considered ideal (e.g., moderate temperature and high pressure). The SRK equation[1] is

$$P = \frac{RT}{(\hat{V} - b)} - \frac{\alpha a}{\hat{V}(\hat{V} + b)} \tag{3.14}$$

where R is the ideal gas constant and α, a, and b are parameters specific to the gas, calculated as follows:

$$a = 0.42747 \frac{(RT_C)^2}{P_C},$$

$$b = 0.08664 \frac{RT_C}{P_C},$$

$$m = 0.48508 + 1.55171\omega - 0.1561\omega^2,$$

$$T_r = \frac{T}{T_C},$$

$$\alpha = \left[1 + m\left(1 - \sqrt{T_r}\right)\right]^2. \tag{3.15}$$

Here,

    T_C is the critical temperature of the gas,
    P_C is the critical pressure of the gas, and
    ω is the Pitzer acentric factor.

Each of these is readily available for many gases.

Calculate the pressure exerted by 20 gram-moles of ammonia at 300 K in a 5-liter container, using

    a.   the ideal gas equation,

    b.   the SRK equation.

The required data for ammonia are tabulated as follows:

T_C	405.5K
P_C	111.3 atm
ω	0.25

(3.16)

Projectile Motion III

12.   Major-league baseball outfield fences are typically about 350 feet from home plate. Home run balls need to be at least 12 feet off the ground to clear the fence. Calculate the minimum required initial velocity (in miles per hour or km/hr) for a home run, for baseballs hit at the angles $10°$, $20°$, $30°$, and $40°$ from the horizontal.

---

[1]From *Elementary Principles of Chemical Processes*, 3d ed., R. M. Feldar and R. W. Rousseau, New York: Wiley, 2000.

4

Matrix Operations in Excel

4.1 INTRODUCTION

Matrix manipulations are a natural for spreadsheets: The spreadsheet grid provides a natural home for the columns and rows of a matrix. All standard spreadsheets provide the standard matrix operations. Excel goes a step further and allows many matrix operations to be performed by using array functions rather than menu commands. This makes the matrix operations "live"; they will automatically be recalculated if any of the data is changed.

4.2 MATRICES, VECTORS, AND ARRAYS

Excel uses the term *array* to refer to a collection of values organized in rows and columns that should be kept together. For example, mathematical operations performed on any element of the array are performed on each element of the array, and Excel will not allow you to delete a portion of an array; the entire array must be deleted or converted to values using Edit/Paste Special/Values. Functions designed to operate on or return arrays are called *array functions*.

Standard mathematics nomenclature calls a collection of related values organized in rows and columns a *matrix*. A matrix with a single row or column is called a *vector*. In Excel, both would be considered arrays.

Defining and Naming Arrays

You define an array in Excel simply by filling a range of cells with the contents of the array. For example, the 3×2 matrix

$$A = \begin{bmatrix} 1 & 3 \\ 7 & 2 \\ 8 & 11 \end{bmatrix}$$

can be entered into a 3×2 range of cells:

OBJECTIVES

After reading this chapter, you will know

- How Excel handles matrix math
- How to use basic cell arithmetic to multiply matrices by a scalar
- How to understand matrix addition
- How to use Excel's array function MMULT() to multiply matrices
- How to transpose matrices, using two methods: menu commands Edit/Paste Special/Transpose, and array function TRANSPOSE()
- How to invert a matrix, using array function MINVERSE()
- How to find the determinant of a matrix, using array function MDETERM()
- How to solve systems of simultaneous linear equations, using matrix math

	A	B	C	D
1				
2				
3				
4	[A], 3x2	1	3	
5		7	2	
6		8	1	
7				
8				
9				

Naming a range of cells allows you to use the name (e.g., A) in place of the cell range (e.g., B4:C6) in array formulas. Excel's matrix math functions do not require named arrays, but naming the cell ranges that contain the arrays is commonly done. Using named arrays not only expedites entering array formulas, it also makes your spreadsheets easier to read and understand.

To give a name to the range of cells that hold an array, first select the array:

B4	▼	f_x 1		
	A	B	C	D
1		**Name box shows**		
2				
3		**top-left cell address**		
4	[A], 3x2	1	3	
5		7	2	
6		8	1	
7				
8				
9				

Then, enter the desired name in the Name box at the left side of the Formula Bar. Here, the array has been called "A".

A	▼	f_x 1		
	A	B	C	D
1		**Selected range has**		
2				
3		**been named "A"**		
4	[A], 3x2	1	3	
5		7	2	
6		8	1	
7				
8				
9				

Alternatively, a selected range can be assigned a name by using Insert/Name/Define.

4.3 HOW EXCEL HANDLES MATRIX MATH

Excel provides mechanisms for performing each of the standard matrix operations, but they are accessed in differing ways:

- Addition and *scalar* multiplication are handled through either basic cell arithmetic or array math operations.
- Matrix transposition, multiplication, and inversion are handled by array functions.

The various spreadsheet programs handle matrix operations, such as transposition, multiplication, and inversion, in different ways. Some programs treat a matrix as a range of values, and matrix operations (from menus, not functions) produce a new range of values. Excel provides many standard matrix operations using menu functions, but also provides array functions for matrix operations such as transposition, multiplication, and inversion. There are pros and cons to each approach. Excel's array functions require more effort on the user's part to perform matrix transposition, multiplication, and inversion, but, by using array functions, Excel can recalculate the resulting matrices if any of the input data change. When spreadsheet programs do not use functions for matrix transposition, multiplication, and inversion, they cannot automatically recalculate your matrix results when your spreadsheet changes.

4.4 BASIC MATRIX OPERATIONS

4.4.1 Adding Two Matrices

The two matrices to be added must be the same size. Begin by entering the two matrices to be added:

	A	B	C	D
1	**ADDITION**			
2	Matrices must be the same size			
3				
4	[A], 3x2	1	3	
5		7	2	
6		8	11	
7				
8	[B], 3x2	4	8	
9		6	1	
10		0	5	
11				

The matrices are added element by element. For example, the top left cell of the resultant matrix will hold the formula required to add 1 + 4 (cell B4 plus cell B8). The formula would be written as =B4+B8. When this formula is placed in cell B12, the result is displayed:

B12	▼	f_x =B4+B8	
	A	B	C	D
1	ADDITION			
2	Matrices must be the same size			
3				
4	[A], 3x2	1	3	
5		7	2	
6		8	11	
7				
8	[B], 3x2	4	8	
9		6	1	
10		0	5	
11				
12	[A] + [B]	5		
13				

If the contents of cell B12 are now copied to cells B12:C14, Excel will add the [A] matrix and the [B] matrix element by element:

C13	▼	f_x =C5+C9	
	A	B	C	D
1	ADDITION			
2	Matrices must be the same size			
3				
4	[A], 3x2	1	3	
5		7	2	
6		8	11	
7				
8	[B], 3x2	4	8	
9		6	1	
10		0	5	
11				
12	[A] + [B]	5	11	
13		13	3	
14		8	16	
15				

Matrix Addition Using Array Math

If the cell ranges holding matrices [A] and [B] are named, then the array names can be used to perform the addition. Here, the name "A" was applied to cell range B4:C6, and cells B8:C10 were named "B".

In Excel, any time array math is used, the size of the resulting array must be indicated before the entering of the array formula. The result of adding matrices [A] and [B] will be a 3×2 matrix, so a 3×2 region of cells (B16:C18) is selected:

	A	B	C	D	E	F	G
1	**ADDITION**						
2	Matrices must be the same size.						
3							
4	[A], 3x2	1	3				
5		7	2				
6		8	1				
7							
8	[B], 3x2	4	8				
9		6	1				
10		0	5				
11							
12	[A] + [B]	5	11		<< Using cell arithmetic		
13		13	3				
14		8	16				
15							
16	{[A] + [B]}				<< Using array math		
17							
18							
19							

Then the formula =A + B is entered in the top left cell of the selected range:

WEEKDAY	▼ ✗ ✓ *fx*	=A+B					
	A	B	C	D	E	F	G
1	**ADDITION**						
2	Matrices must be the same size.						
3							
4	[A], 3x2	1	3				
5		7	2				
6		8	11				
7							
8	[B], 3x2	4	8				
9		6	1				
10		0	5				
11							
12	[A] + [B]	5	11		<< Using cell arithmetic		
13		13	3				
14		8	16				
15							
16	{[A] + [B]}	=A+B			<< Using array math		
17							
18							
19							

Notice that, when the named arrays were entered in the formula, Excel put a box around the named arrays. This allows you to quickly see that the correct matrices are being added.

Excel requires a special character sequence when entering array formulas: [Ctrl–Shift–Enter], not just the [Enter] key. The [Ctrl–Shift–Enter] key combination tells Excel to fill the entire array (i.e., the selected region) with the results of the formula, not just one cell. After the pressing of [Ctrl–Shift–Enter], the spreadsheet will look like this:

	A	B	C	D	E	F	G
	B16 ▾		*fx* {=A+B}				
1	ADDITION						
2	Matrices must be the same size.						
3					**Each cell in the new**		
4	[A], 3x2	1	3		**array contains {=A+B}**		
5		7	2				
6		8	11				
7							
8	[B], 3x2	4	8				
9		6	1				
10		0	5				
11							
12	[A] + [B]	5	11		<< Using cell arithmetic		
13		13	3				
14		8	16				
15							
16	{[A] + [B]}	5	11		<< Using array math		
17		13	3				
18		8	16				
19							

Each cell in the new array (B16:C18) contains the same formula, {=A + B}. The braces { } indicate that array math was used and that the result is an array. This means that the six cells in B16:C18 are considered a collection, and individual elements of the new array cannot be edited or deleted. However, if either matrix [A] or matrix [B] is changed, the new array will automatically be updated.

Note: Named arrays are not required for array math in Excel; the same result could have been obtained by selecting the B16:C18 range, then typing in the array formula =B4:C6 + B8:C10 and pressing [Ctrl–Shift–Enter]. The array formula {=B4:C6 + B8:C10} would have been entered into every cell of the new array, and matrices [A] and [B] would have been added.

4.4.2 Multiplying a Matrix by a Scalar

Begin by entering the *scalar* (constant value) and the matrix that it is to multiply:

	A	B	C	D
1	**SCALAR MULTIPLICATION**			
2				
3	Scalar:	10		
4				
5	[A], 3x2	1	3	
6		7	2	
7		8	11	
8				

Multiplying a matrix by a scalar simply requires you to multiply each element of the matrix by the scalar:

B9   ▼ f_x =$B$3*B5

	A	B	C	D
1	**SCALAR MULTIPLICATION**			
2				
3	Scalar:	10		
4				
5	[A], 3x2	1	3	
6		7	2	
7		8	11	
8				
9	10 [A]	10		
10				

By placing the formula =$B$3*B5 in cell B9, you instruct Excel to multiply the contents of cell B3 (the scalar) and the contents of cell B5 (the top left element of the matrix). The dollar signs before the 'B' and '3' in $B$3 are in preparation for copying the formula to other cells. When copied, the new formulas will continue to reference the scalar in cell B3; the dollar signs tell Excel not to increment either the column letter ($B) or row number ($3) during the copy process. You can type the dollar signs as you enter the formula, or just type the B3 and press [F4] to insert the dollar signs:

C11   ▼ f_x =$B$3*C7

	A	B	C	D	E
1	**SCALAR MULTIPLICATION**				
2					
3	Scalar:	10			
4					
5	[A], 3x2	1	3		
6		7	2		
7		8	11		
8					
9	10 [A]	10	30		
10		70	20	**Fill Handle**	
11		80	110		
12					

Copying the formula in cell B9 to the range B9:C11 completes the process of multiplying the matrix [A] by the scalar, 10. The results are shown in cells B9:C11. The *fill handle* (small square at bottom right corner of indicator box) is pointed out in the figure. If, after entering the first formula in cell B9, you grab the fill handle with the mouse and drag the cell indicator down to cell B11, the contents of cell B9 will automatically be copied to cells B10:B11. Then, you can use the fill handle again to copy the contents of cells B9:B11 to C9:C11. (Alternatively, you can copy and paste.)

Scalar Multiplication Using Array Math

The same result can be obtained using array math and a named array. If the cells holding the [A] matrix (B5:B7 in this example) are given the name "A", then the array name can be used to perform the scalar multiplication.

First, the size of the result matrix is indicated by selecting a 3×2 range of cells (B13:C15), and the array formula =B3 * A is entered in the top-left cell of the new array:

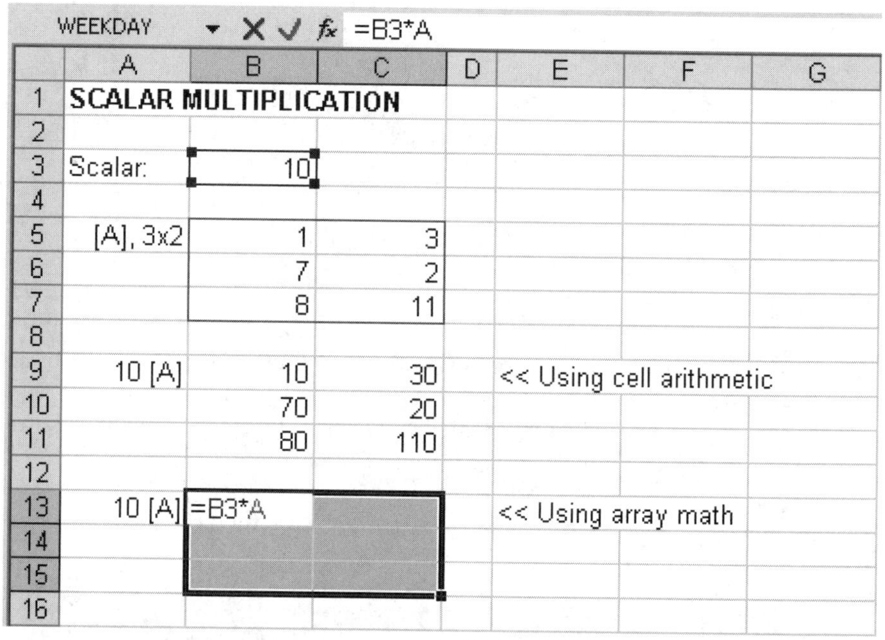

WEEKDAY	▾ ✗ ✓ ƒ× =B3*A						
	A	B	C	D	E	F	G
1	SCALAR MULTIPLICATION						
2							
3	Scalar:	10					
4							
5	[A], 3x2	1	3				
6		7	2				
7		8	11				
8							
9	10 [A]	10	30		<< Using cell arithmetic		
10		70	20				
11		80	110				
12							
13	10 [A]	=B3*A			<< Using array math		
14							
15							
16							

The formula is concluded by pressing [Ctrl–Shift–Enter] to tell Excel to fill the entire selected region with the array formula. Excel places the array formula {=B3 ° A} in each cell in the result array:

	B13	▼	*fx* {=B3*A}				
	A	B	C	D	E	F	G
1	SCALAR MULTIPLICATION						
2							
3	Scalar:	10					
4							
5	[A], 3x2	1	3				
6		7	2				
7		8	11				
8							
9	10 [A]	10	30		<< Using cell arithmetic		
10		70	20				
11		80	110				
12							
13	10 [A]	10	30		<< Using array math		
14		70	20				
15		80	110				
16							

4.4.3 Multiplying Two Matrices

In order to multiply two matrices, the number of columns in the first matrix must equal the number of rows in the second matrix. Again, begin by entering the two matrices to be multiplied:

	A	B	C	D	E	F
1	MULTIPLICATION					
2	Inside dimensions must match (2, 2)					
3	Dimensions of resulting matrix from outside dimensions (3 x 1)					
4						
5	[A], 3x2	1	3			
6		7	2			
7		8	11			
8						
9	[e], 2x1	4				
10		8				
11						

Begin the matrix multiplication process by indicating where the product matrix will go.

Note: You should determine the actual size of the product matrix (from the outside dimensions of the matrices being multiplied) and indicate the correct number of cells—but too big is better than too small if you want to guess.

In this example the product matrix will be 3 rows by 1 column, as shown here:

	A	B	C	D	E	F
1	**MULTIPLICATION**					
2	Inside dimensions must match (2, 2)					
3	Dimensions of resulting matrix from outside dimensions (3 x 1)					
4						
5	[A], 3x2	1	3			
6		7	2			
7		8	11			
8						
9	[e], 2x1	4				
10		8				
11						
12	[A][e], 3x1					
13						
14						
15						

Once you have indicated where the new matrix is to go, begin entering the matrix multiplication array function, MMULT(first matrix, second matrix). After entering the equal sign, function name, and the opening parenthesis, use the mouse to indicate the cells that contain the first matrix:

4				
5	[A], 3x2	1	3	
6		7	2	
7		8		
8				
9	[e], 2x1	4		
10		8		
11				
12	[A][e], 3x1	=mmult(B5:C7		
13		MMULT(**array1**, array2)		
14				
15				

After the first matrix has been indicated, enter comma, and indicate the cells containing the second matrix:

4				
5	[A], 3x2	1	3	
6		7	2	
7		8	11	
8				
9	[e], 2x1	4		
10				
11				
12	[A][e], 3x1	=mmult(B5:C7,B9:B10		
13		MMULT(array1, **array2**)		
14				
15				

Finish entering the matrix multiplication formula by entering the final parenthesis and pressing [Ctrl-Shift-Enter], not just the [Enter] key. The MMULT() function is an array function, and the [Ctrl-Shift-Enter] is used to enter the function into *all* of the cells from B12 to B14:

4					
5	[A], 3x2	1	3		
6		7	2		
7		8	11		
8					
9	[e], 2x1	4			
10		8			
11					
12	[A][e], 3x1	=mmult(B5:C7,B9:B10)			
13			**then [Ctrl-Shift-Enter]**		
14					
15					

Pressing [Enter] alone would enter the formula only in cell B12, and only one element of the product matrix would be displayed.

If the indicated size of the product matrix is incorrect, one of the following things will happen:

- If you indicated too few cells to hold the entire product matrix, only part of your result matrix will be displayed.
- If you indicated more cells than are needed to hold the product matrix, Excel will put "#N/A" in the extra cells, indicating that no formula was placed in the extra cells.

After pressing [Ctrl-Shift-Enter], you will find the result displayed in the indicated cells:

B12	▼		f_x {=MMULT(B5:C7,B9:B10)}		
	A	**B**	**C**	**D**	**E**	**F**
1	**MULTIPLICATION**					
2	Inside dimensions must match (2, 2)					
3	Dimensions of resulting matrix from outside dimensions (3 x 1)					
4						
5	[A], 3x2	1	3			
6		7	2			
7		8	11			
8						
9	[e], 2x1	4				
10		8				
11						
12	[A][e], 3x1	28				
13		44				
14		120				
15						

Note: Excel places the formulas for multiplying the matrices in the result cells, not just the final values. Because of this, any changes in the first two matrices will automatically cause the product matrix to be updated.

PRACTICE!

Can the following matrices be multiplied, and, if so, what size will the resulting matrices be?

$$[A]_{3\times2}[B]_{2\times2} \qquad [C]_{3\times3}[D]_{1\times3} \qquad [D]_{1\times3}[C]_{3\times3} \qquad (4.1)$$

EXAMPLE 4.1

Multiplication of $[A]_{3\times2}$ and $[G]_{2\times4}$ matrix produces matrix $[H]_{3\times4}$. Skipping a lot of the details, we see that the multiplication process is summarized as

1. Enter the [A] and [G] matrices.
2. Indicate the size (3 × 4) and location of the result matrix.
3. Begin entering the MMULT() array function, and indicate the cells containing the first matrix, [A]: B36:C38
4. Indicate the cells containing the second matrix, [G]: B40:E41, and add the final parenthesis.

	A	B	C	D	E	F
1	MULTIPLICATION					
2	Inside dimensions must match (2, 2)					
3	Dimensions of resulting matrix from outside dimensions (3 x 4)					
4						
5	[A], 3x2	1	3			
6		7	2			
7		8	11			
8						
9	[G], 2x4	1	2	3	4	
10		5	6	7	8	
11						
12	[H], 3x4	=mmult(B5:C7,B9:E10)				
13						
14						
15						

5. Press [Ctrl-Shift-Enter] to create the product matrix, [H]: B43. The result of the multiplication is the 3 × 4 matrix called [H]:

	B12	▼		*fx* {=MMULT(B5:C7,B9:E10)}		
	A	B	C	D	E	F
1	**MULTIPLICATION**					
2	Inside dimensions must match (2, 2)					
3	Dimensions of resulting matrix from outside dimensions (3 x 4)					
4						
5	[A], 3x2	1	3			
6		7	2			
7		8	11			
8						
9	[G], 2x4	1	2	3	4	
10		5	6	7	8	
11						
12	[H], 3x4	16	20	24	28	
13		17	26	35	44	
14		63	82	101	120	
15						

4.4.4 Transposing a Matrix

Any matrix can be transposed. To *transpose* a matrix, simply interchange the rows and columns. You can transpose a matrix in two ways: as values or by using array function TRANSPOSE(). Using values is simpler, but the result will not be automatically recalculated if the input data change. Both methods are described here.

Using PASTE SPECIAL to Transpose as Values The process in Excel can be summarized as follows:

1. Enter the original matrix.
2. Begin the transposition process by selecting the cells containing the matrix and copying the matrix to the Windows clipboard by selecting Edit/Copy.
3. Indicate the cell that will contain the top left corner of the result matrix.
4. Select Edit/Paste Special.
5. The Paste Special Dialog box is displayed.
6. Select Values from the Paste section of the dialog box, and check the Transpose check box at the bottom of the dialog box.
7. Click on OK to close the dialog box and create the transposed matrix (as values, not array functions).

Using a Matrix Function to Transpose a Matrix Using the TRANSPOSE() array function, the process is as follows:

1. Enter the original matrix.
2. Indicate where the result should be placed, showing the exact size of the transposed matrix:

	A	B	C	D	E
1	**TRANSPOSE**				
2	Interchange rows and columns				
3	Any size matrix				
4					
5	[A], 3x2	1	3		
6		7	2		
7		8	11		
8					
9	[A-trans], 2x3				
10					
11					

3. Enter the first portion of the TRANSPOSE($matrix$) array function, and use the mouse to indicate the cells containing the matrix to be transposed:

	A	B	C	D	E
1	**TRANSPOSE**				
2	Interchange rows and columns				
3	Any size matrix				
4					
5	[A], 3x2	1	3		
6		7	2		
7		8	11		
8					
9	[A-trans], 2x3	=transpose(B5:C7			
10		TRANSPOSE(**array**)			
11					

4. Finish typing the function by adding the final parenthesis and press [Ctrl-Shift-Enter] to transpose the matrix:

B9	▼	f_x {=TRANSPOSE(B5:C7)}		

	A	B	C	D	E
1	**TRANSPOSE**				
2	Interchange rows and columns				
3	Any size matrix				
4					
5	[A], 3x2	1	3		
6		7	2		
7		8	11		
8					
9	[A-trans], 2x3	1	7	8	
10		3	2	11	
11					

PRACTICE!

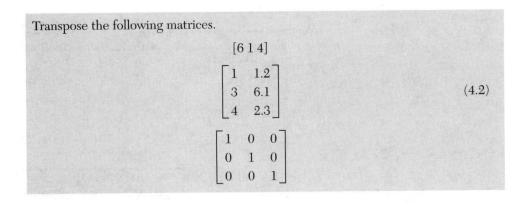

Transpose the following matrices.

$$[6\ 1\ 4]$$

$$\begin{bmatrix} 1 & 1.2 \\ 3 & 6.1 \\ 4 & 2.3 \end{bmatrix} \qquad (4.2)$$

$$\begin{bmatrix} 1 & 0 & 0 \\ 0 & 1 & 0 \\ 0 & 0 & 1 \end{bmatrix}$$

4.4.5   Inverting a Matrix

Only square matrices (number of rows equal to number of columns) can possibly be inverted, and not even all square matrices can actually be inverted. (They must be *nonsingular* to be inverted.)

The procedure to *invert* a nonsingular, square matrix in Excel, using the MINVERSE() array function, is as follows:

1. Enter the matrix to be inverted.
2. Indicate where the inverted matrix should be placed and the correct size (same size as original matrix).
3. Enter the first portion of the MINVERSE(`matrix`) array function, and use the mouse to indicate the cells containing the matrix to be inverted. Finish the function by typing the final parenthesis:

	A	B	C	D	E
1	**INVERT**				
2	Matrix must be square and non-singular				
3					
4	[J], 3x3	2	3	5	
5		7	2	4	
6		8	11	6	
7					
8	[J-inv], 3x3	=minverse(B4:D6)			
9					
10					
11					

4. Press [Ctrl-Shift-Enter] to enter the array function in all the cells making up the result matrix:

B8		f_x {=MINVERSE(B4:D6)}			
	A	B	C	D	E
1	**INVERT**				
2	Matrix must be square and non-singular				
3					
4	[J], 3x3	2	3	5	
5		7	2	4	
6		8	11	6	
7					
8	[J-inv], 3x3	-0.1517	0.1754	0.0095	
9		-0.0474	-0.1327	0.1280	
10		0.2891	0.0095	-0.0806	
11					

4.4.6   Matrix Determinant

The *determinant* of a matrix is a single value, calculated from a matrix, that is often used in solving systems of equations. One of the most straightforward ways to use the determinant is to see whether a matrix can be inverted. If the determinant is zero, the matrix is *singular* and cannot be inverted. You can calculate a determinant only for square matrices.

Note: Calculating a determinant for a large matrix requires a lot of calculations and can result in round-off errors on digital computers. A very small, nonzero determinant (e.g., 1×10^{-14}) is probably a round-off error, and the matrix most likely cannot be inverted.

Excel's MDETERM(*matrix*) array function is used to compute determinant values. In the previous section, the [J] matrix was inverted, so it must have had a nonzero determinant.

B8		f_x =MDETERM(B4:D6)			
	A	B	C	D	E
1	**DETERMINANT**				
2	Matrix must be square				
3					
4	[J], 3x3	2	3	5	
5		7	2	4	
6		8	11	6	
7					
8	Det(J)	211			
9					

If the matrix is singular (and so cannot be inverted), the determinant will be zero. In the [K] matrix below, the second row is the same as the first row. Whenever a matrix contains two identical rows the matrix cannot be inverted, and the determinant will be zero:

B8	▼		f_x =MDETERM(B4:D6)		
	A	B	C	D	E
1	DETERMINANT				
2	Matrix must be square				
3					
4	[K], 3x3	2	3	5	
5		2	3	5	
6		8	11	6	
7					
8	Det(K)	0			
9					

A matrix is singular (and cannot be inverted) if

- any row (or column) contains all zeros;
- any two rows (or columns) are identical;
- any row (or column) is equal to a linear combination of other rows (or columns). (When two or more rows are multiplied by constants and then added, the result is a *linear combination* of the rows. When a linear combination of two or more rows is equal to another row in the matrix, the matrix is singular.)

4.5   SOLVING SYSTEMS OF LINEAR EQUATIONS

One of the most common uses of matrix operations is to solve *systems of linear algebraic equations*. The process of solving simultaneous equations by using matrices works as follows:

1. Write the equations in matrix form (*coefficient matrix* multiplying an *unknown vector*, equal to a *right-hand-side vector*).
2. Invert the coefficient matrix.
3. Multiply both sides of the equation by the inverted coefficient matrix.

The result of step 3 is a solution matrix containing the answers to the problem.

In order for inverting the coefficient matrix to be possible, it must be nonsingular. In terms of solving simultaneous equations, this means that you will be able to invert the coefficient matrix only if there is a solution to the set of equations. If there is no solution, the coefficient matrix will be singular.

Consider the following three equations in three unknowns:

$$3x_1 + 2x_2 + 4x_3 = 5,$$
$$2x_1 + 5x_2 + 3x_3 = 17,$$
$$7x_1 + 2x_2 + 2x_2 = 11. \tag{4.3}$$

Step 1. Write the Equations in Matrix Form

The unknowns are x_1, x_2, and x_3, which can be written as the vector of unknown, $[x]$:

$$[x] = \begin{bmatrix} x_1 \\ x_2 \\ x_3 \end{bmatrix}.$$

The coefficients multiplying the various x's can be collected in a coefficient matrix [C]:

$$[C] = \begin{bmatrix} 3 & 2 & 4 \\ 2 & 5 & 3 \\ 7 & 2 & 2 \end{bmatrix}.$$

PRACTICE!

Try multiplying [C] times [x] (symbolically) to see that you do indeed get back the left side of the preceding equation.

The constants on the right side of the equations can be written as a right-hand-side vector [r]:

$$[r] = \begin{bmatrix} 5 \\ 17 \\ 11 \end{bmatrix}.$$

The three equations in three unknowns can now be written as

$$[C][x] = [r].$$

In a spreadsheet, they might resemble this:

	A	B	C	D	E	F	G	H
1	**Simultaneous Equations**							
2								
3	**[C]**	3	2	4		**[r]**	5	
4		2	5	3			17	
5		7	2	2			11	
6								

Step 2. Invert the Coefficient Matrix

Use the array function MINVERSE() to invert the [C] matrix:

B7		f_x {=MINVERSE(B3:D5)}						
	A	B	C	D	E	F	G	H
1	**Simultaneous Equations**							
2								
3	**[C]**	3	2	4		**[r]**	5	
4		2	5	3			17	
5		7	2	2			11	
6								
7	**[C-inv]**	-0.05	-0.05	0.18				
8		-0.22	0.28	0.01				
9		0.40	-0.10	-0.14				
10								

Step 3. Multiply Both Sides of the Equation by the Inverted [C] Matrix:

$$[C_{inv}][C][x] = [C_{inv}][r]$$
$$[I][x] = [C_{inv}][r]$$
$$[x] = [C_{inv}][r].\tag{4.4}$$

Multiplying the inverted [C] matrix and the original [C] matrix returns an identity matrix. (Try it!)

Multiplying the [x] vector by an identity matrix returns the [x] vector unchanged.

So multiplying the inverted [C] matrix and the [r] vector returns the [x] values and gives the solution:

	G7	▼		f_x {=MMULT(B7:D9,G3:G5)}				
	A	B	C	D	E	F	G	H
1	**Simultaneous Equations**							
2								
3	**[C]**	3	2	4		**[r]**	5	
4		2	5	3			17	
5		7	2	2			11	
6								
7	**[C-inv]**	-0.05	-0.05	0.18	**[x] = [C-inv][r]**		0.846	
8		-0.22	0.28	0.01			3.846	
9		0.40	-0.10	-0.14			-1.31	
10								

The solutions to the original three equations in three unknowns are

$$x_1 = 0.846,$$
$$x_2 = 3.846,$$
$$x_3 = -1.31.$$

APPLICATIONS: MULTILOOP CIRCUITS I

Multiloop circuits are analyzed by using Kirchhoff's laws of voltage and current:

Kirchhoff's voltage law says that

for a closed circuit (a loop), the algebraic sum of all changes in voltage must be zero.

Kirchhoff's current law says that

at any junction in a circuit, the input current(s) must equal the output current(s).

We will use these laws, with the following circuit, to compute the three unknown currents, i_1 through i_3:

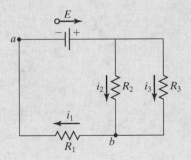

The voltage from the battery and the three resistances are as follows:

Known Values	
E	12 volts
R_1	30 ohms
R_2	40 ohms
R_3	50 ohms

Applying the current law at point b gives one equation:

$$i_1 = i_2 + i_3 \tag{4.5}$$

Applying the voltage law to the left loop and the overall loop provides two more equations:

$$E - V_2 - V_1 = 0,$$
$$E - V_3 - V_1 = 0.$$

In terms of current and resistance, these are

$$E - i_2 R_2 - i_1 R_1 = 0,$$
$$E - i_3 R_3 - i_1 R_1 = 0.$$

We now have three equations for i_1, i_2, and i_3. Writing them in matrix form with constants on the right side of the equal sign yields

$$1i_1 - 1i_2 - 1i_3 = 0,$$
$$R_1 i_1 + R_2 i_2 + 0i_3 = E,$$
$$R_1 i_1 + 0i_2 + R_3 i_3 = E.$$

The coefficients one and zero have been included in the equations as a reminder to include them in the coefficient matrix:

$$C = \begin{bmatrix} 1 & -1 & -1 \\ R_1 & R_2 & 0 \\ R_1 & 0 & R_3 \end{bmatrix} \qquad r = \begin{bmatrix} 0 \\ E \\ E \end{bmatrix}.$$

Substituting the known values, we get

$$C = \begin{bmatrix} 1 & -1 & -1 \\ 30 & 40 & 0 \\ 30 & 0 & 50 \end{bmatrix} \qquad r = \begin{bmatrix} 0 \\ 12 \\ 12 \end{bmatrix}.$$

In Excel, the coefficient matrix and right-hand-side vector are entered as

	A	B	C	D	E	F	G	H	I
1	Multiloop Circuits I								
2									
3		[C]	1	-1	-1		[r]	0	
4			30	40	0			12	
5			30	0	50			12	
6									

A quick check with Excel's MDETERM() array function shows that a solution is possible:

C7	▼	f_x	=MDETERM(C3:E5)					
	A	B	C	D	E	F	G	H	I

	A	B	C	D	E	F	G	H	I
1	**Multiloop Circuits I**								
2									
3		**[C]**	1	-1	-1		**[r]**	0	
4			30	40	0			12	
5			30	0	50			12	
6									
7		**Determinant:**	4700						
8									

So the coefficient matrix is inverted by using the MINVERSE() array function:

C9	▼	f_x	{=MINVERSE(C3:E5)}					

	A	B	C	D	E	F	G	H	I
1	**Multiloop Circuits I**								
2									
3		**[C]**	1	-1	-1		**[r]**	0	
4			30	40	0			12	
5			30	0	50			12	
6									
7		**Determinant:**	4700						
8									
9		**[C$_{inv}$]**	0.43	0.01	0.01				
10			-0.32	0.02	-0.01				
11			-0.26	-0.01	0.01				
12									

The currents are found by multiplying the inverted [C] matrix and the [r] vector, using Excel's MMULT() array function:

H9	▼	f_x	{=MMULT(C9:E11,H3:H5)}					

	A	B	C	D	E	F	G	H	I
1	**Multiloop Circuits I**								
2									
3		**[C]**	1	-1	-1		**[r]**	0	
4			30	40	0			12	
5			30	0	50			12	
6									
7		**Determinant:**	4700						
8									
9		**[C$_{inv}$]**	0.43	0.01	0.01		**[i]**	0.23	amp
10			-0.32	0.02	-0.01			0.13	
11			-0.26	-0.01	0.01			0.10	
12									

The unknown currents have been found to be $i_1 = 0.23$, $i_2 = 0.13$, and $i_3 = 0.10$ amp, respectively.

Wheatstone Bridge I

A fairly commonly used circuit in instrumentation applications is the *Wheatstone Bridge*, shown in the accompanying figure. The bridge contains two known resistances, R_1 and R_2, and an adjustable resistance (i.e., a potentiometer). In use, the setting on the potentiometer is adjusted until points a and b are at the same voltage. Then the known resistances can be used to calculate an unknown resistance, shown as R_3 in the diagram:

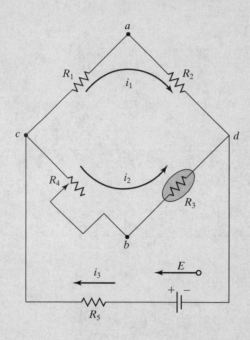

R_1 and the potentiometer, R_4, are connected at point c, and the circuit has been adjusted to have the same potential at points a and b, so the voltage drops across R_1 and R_4 must be equal:

$$V_1 = V_4. \tag{4.6}$$

But the voltages can be written in terms of current and resistance:

$$i_1 R_1 = i_2 R_4. \tag{4.7}$$

Similarly, the voltage drops across R_2 and R_3 must be equal, so

$$V_2 = V_3, \tag{4.8}$$

or

$$i_1 R_2 = i_2 R_3. \tag{4.9}$$

Solving one equation for i_1 and substituting into the other yields an equation for the unknown resistance, R_3, in terms of the known resistance values:

$$R_3 = R_4 \frac{R_2}{R_1}. \tag{4.10}$$

a. Given the known resistances and battery potential listed in the accompanying table, what is the resistance of R_3 when the potentiometer (R_4) has been adjusted to 24 ohms to balance the bridge?

b. Use Kirchhoff's laws to find currents i_1 through i_3.

Known Values	
E	12 volts
R_1	20 ohms
R_2	10 ohms
R_5	100 ohms

First, the known values are entered into the spreadsheet:

	A	B	C	D	E
1	Wheatstone Bridge I				
2					
3		E:	12	volts	
4		R_1:	20	ohms	
5		R_2:	10	ohms	
6		R_4:	24	ohms	
7		R_5:	100	ohms	
8					

Then the unknown resistance, R_3, is computed:

C9			f_x	=C6*(C5/C4)	
	A	B	C	D	E
1	Wheatstone Bridge I				
2					
3		E:	12	volts	
4		R_1:	20	ohms	
5		R_2:	10	ohms	
6		R_4:	24	ohms	
7		R_5:	100	ohms	
8					
9		R_3:	12	ohms	
10					

To find the currents, we need three equations to solve for the three unknowns. First, Kirchhoff's current law can be applied at point d:

$$i_1 + i_2 = i_3.$$

Then, the voltage law can be applied to the outer loop so that

$$E - V_5 - V_1 - V_2 = 0,$$

or, in terms of current and voltage,

$$E - i_3 R_5 - i_1 R_1 - i_1 R_2 = 0.$$

Finally, the voltage law can be applied to the inner loop, so

$$E - V_5 - V_4 - V_3 = 0,$$

or

$$E - i_3 R_5 - i_2 R_4 - i_2 R_3 = 0.$$

In matrix form, these equations become

$$C = \begin{bmatrix} 1 & 1 & -1 \\ R_1 + R_2 & 0 & R_5 \\ 0 & R_4 + R_3 & R_5 \end{bmatrix} \quad r = \begin{bmatrix} 0 \\ E \\ E \end{bmatrix}. \tag{4.11}$$

In Excel, the coefficient matrix and right-hand-side vector are

	B13	▼		f_x	=C4+C5		
	A	B	C	D	E	F	G	H
11								
12	**[C]**	1	1	-1		**[r]**	0	
13		30	0	100			12	
14		0	36	100			12	
15								

The coefficient matrix is then inverted:

	B16	▼		f_x	{=MINVERSE(B12:D14)}		
	A	B	C	D	E	F	G	H
11								
12	**[C]**	1	1	-1		**[r]**	0	
13		30	0	100			12	
14		0	36	100			12	
15								
16	**[C$_{inv}$]**	0.469	0.018	-0.013				
17		0.391	-0.013	0.017				
18		-0.141	0.005	0.004				
19								

It is next multiplied by the right-hand-side vector to find the currents:

	G16	▼		f_x	{=MMULT(B16:D18,G12:G14)}			
	A	B	C	D	E	F	G	H
11								
12	[C]	1	1	-1		[r]	0	
13		30	0	100			12	
14		0	36	100			12	
15								
16	[C$_{inv}$]	0.469	0.018	-0.013		[i]	0.056	amp
17		0.391	-0.013	0.017			0.047	
18		-0.141	0.005	0.004			0.103	
19								

The results are

$$i_1 = 0.056 \text{ amp,}$$
$$i_2 = 0.047 \text{ amp,}$$
$$i_3 = 0.103 \text{ amp.}$$

(4.12)

APPLICATION: CONDUCTION HEAT TRANSFER

The differential equation

$$\frac{\partial T}{\partial t} = \frac{k}{\rho \, C_p} \left[\frac{\partial^2 T}{\partial x^2} + \frac{\partial^2 T}{\partial y^2} \right]$$

describing energy conduction through a two-dimensional region can be applied to a surface exposed to various boundary temperatures. We might want to know the temperature distribution in a 50-cm- × -40-cm metal plate exposed to boiling water (100°C) along two edges, ice water (0°C) on one edge, and room temperature (25°C) along another, as represented in the following diagram:

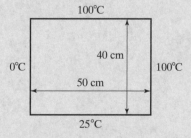

At steady state, $\frac{\partial T}{\partial t} = 0$, the equation simplifies considerably, to a form known as *Laplace's equation*:

$$0 = \frac{\partial^2 T}{\partial x^2} + \frac{\partial^2 T}{\partial y^2}$$

The partial derivatives can be approximated by using finite differences to produce the algebraic equation

$$0 = \left[\frac{T_{i+1,j} - 2T_{i,j} + T_{i-1,j}}{(\Delta x)^2}\right] + \left[\frac{T_{i,j+1} - 2T_{i,j} + T_{i,j-1}}{(\Delta y)^2}\right],$$

where subscript i,j represents a point on the plate at which the temperature is $T_{i,j}$; $T_{i-1,j}$ represents the temperature at a point to the left of i,j; and $T_{i,j-1}$ represents the temperature at a point above i,j, as indicated in the following diagram:

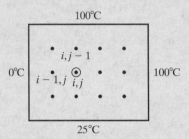

And, if we choose to make $\Delta x = \Delta y$, we get a particularly simple result:

$$0 = \lfloor T_{i+1,j} - 2T_{i,j} + T_{i-1,j} \rfloor + \lfloor T_{i,j+1} - 2T_{i,j} + T_{i,j-1} \rfloor,$$

or

$$4\,T_{i,j} = \lfloor T_{i+1,j} + T_{i-1,j} \rfloor + \lfloor T_{i,j+1} + T_{i,j-1} \rfloor \qquad \text{(general equation)}.$$

This equation says that the sum of the temperatures at the four points around any central point (i.e., any i,j) is equal to four times the temperature at the central point. (Remember, this is true only at steady state and only when $\Delta x = \Delta y$.) We can apply this equation at each interior point to develop a system of equations that, when solved simultaneously, will yield the temperatures at each point.

To help see how this is done, let's assign each interior point a letter designation and show the four points surrounding point A with a circle:

Applying the general equation at point A yields

$$4\,T_A = 0 + 100 + T_B + T_E$$

Then, we move the circle to point B:

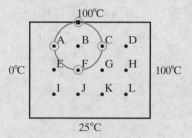

Applying the general equation at point B yields

$$4\,T_B = T_A + 100 + T_C + T_F$$

By continuing to apply the general equation at each interior point (points A through L), we generate 12 equations, one for each interior point:

$$4\,T_A = 0 + 100 + T_B + T_E$$
$$4\,T_B = T_A + 100 + T_C + T_F$$
$$4\,T_C = T_B + 100 + T_D + T_G$$
$$4\,T_D = T_C + 100 + 100 + T_H$$
$$4\,T_E = 0 + T_A + T_F + T_I$$
$$4\,T_F = T_E + T_B + T_G + T_J$$
$$4\,T_G = T_F + T_C + T_H + T_K$$
$$4\,T_H = T_G + T_D + 100 + T_L$$
$$4\,T_I = 0 + T_E + T_J + 25$$
$$4\,T_J = T_I + T_F + T_K + 25$$
$$4\,T_K = T_J + T_G + T_L + 25$$
$$4\,T_L = T_K + T_H + 100 + 25$$

In matrix form, these equations can be written as

$$[C][T] = [r]$$

where

$$[C] = \begin{bmatrix} -4 & 1 & 0 & 0 & 1 & 0 & 0 & 0 & 0 & 0 & 0 & 0 \\ 1 & -4 & 1 & 0 & 0 & 1 & 0 & 0 & 0 & 0 & 0 & 0 \\ 0 & 1 & -4 & 1 & 0 & 0 & 1 & 0 & 0 & 0 & 0 & 0 \\ 0 & 0 & 1 & -4 & 0 & 0 & 0 & 1 & 0 & 0 & 0 & 0 \\ 1 & 0 & 0 & 0 & -4 & 1 & 0 & 0 & 1 & 0 & 0 & 0 \\ 0 & 1 & 0 & 0 & 1 & -4 & 1 & 0 & 0 & 1 & 0 & 0 \\ 0 & 0 & 1 & 0 & 0 & 1 & -4 & 1 & 0 & 0 & 1 & 0 \\ 0 & 0 & 0 & 1 & 0 & 0 & 1 & -4 & 0 & 0 & 0 & 1 \\ 0 & 0 & 0 & 0 & 1 & 0 & 0 & 0 & -4 & 1 & 0 & 0 \\ 0 & 0 & 0 & 0 & 0 & 1 & 0 & 0 & 1 & -4 & 1 & 0 \\ 0 & 0 & 0 & 0 & 0 & 0 & 1 & 0 & 0 & 1 & -4 & 1 \\ 0 & 0 & 0 & 0 & 0 & 0 & 0 & 1 & 0 & 0 & 1 & -4 \end{bmatrix} \quad [T] = \begin{bmatrix} T_A \\ T_B \\ T_C \\ T_D \\ T_E \\ T_F \\ T_G \\ T_H \\ T_I \\ T_J \\ T_K \\ T_L \end{bmatrix} \quad [r] = \begin{bmatrix} -100 \\ -100 \\ -100 \\ -200 \\ 0 \\ 0 \\ 0 \\ -100 \\ -25 \\ -25 \\ -25 \\ -125 \end{bmatrix}$$

The coefficient and right-hand-side (constant-value) matrices can be entered into Excel:

	A	B	C	D	E	F	G	H	I	J	K	L	M	N	O	P	Q	R
1	Heat Conduction																	
2																		
3			A	B	C	D	E	F	G	H	I	J	K	L				
4	[C]	A	-4	1	0	0	1	0	0	0	0	0	0	0		[r]	-100	
5		B	1	-4	1	0	0	1	0	0	0	0	0	0			-100	
6		C	0	1	-4	1	0	0	1	0	0	0	0	0			-100	
7		D	0	0	1	-4	0	0	0	1	0	0	0	0			-200	
8		E	1	0	0	0	-4	1	0	0	1	0	0	0			0	
9		F	0	1	0	0	1	-4	1	0	0	1	0	0			0	
10		G	0	0	1	0	0	1	-4	1	0	0	1	0			0	
11		H	0	0	0	1	0	0	1	-4	0	0	0	1			-100	
12		I	0	0	0	0	1	0	0	0	-4	1	0	0			-25	
13		J	0	0	0	0	0	1	0	0	1	-4	1	0			-25	
14		K	0	0	0	0	0	0	1	0	0	1	-4	1			-25	
15		L	0	0	0	0	0	0	0	1	0	0	1	-4			-125	
16																		

Now the coefficient matrix can be inverted and then multiplied by the [r] vector to find the temperatures at the interior points:

17																	
18	[Cinv]	-0.3	-0.1	-0	-0	-0.1	-0.1	-0	-0	-0	-0	-0	-0		[T]	50.4	=T_A
19		-0.1	-0.3	-0.1	-0	-0.1	-0.1	-0.1	-0	-0	-0	-0	-0		(°C)	70.6	=T_B
20		-0	-0.1	-0.3	-0.1	-0	-0.1	-0.1	-0.1	-0	-0	-0	-0			81.4	=T_C
21		-0	-0	-0.1	-0.3	-0	-0	-0.1	-0.1	-0	-0	-0	-0			90.2	=T_D
22		-0.1	-0.1	-0	-0	-0.3	-0.1	-0.1	-0	-0.1	-0.1	-0	-0			31.1	=T_E
23		-0.1	-0.1	-0.1	-0	-0.1	-0.4	-0.1	-0.1	-0.1	-0.1	-0.1	-0			50.8	=T_F
24		-0	-0.1	-0.1	-0.1	-0.1	-0.1	-0.4	-0.1	-0	-0.1	-0.1	-0.1			64.7	=T_G
25		-0	-0	-0.1	-0.1	-0	-0.1	-0.1	-0.3	-0	-0	-0.1	-0.1			79.5	=T_H
26		-0	-0	-0	-0	-0.1	-0.1	-0	-0	-0.3	-0.1	-0	-0			23.2	=T_I
27		-0	-0	-0	-0	-0.1	-0.1	-0.1	-0	-0.1	-0.3	-0.1	-0			36.6	=T_J
28		-0	-0	-0	-0	-0	-0.1	-0.1	-0.1	-0	-0.1	-0.3	-0.1			47.3	=T_K
29		-0	-0	-0	-0	-0	-0	-0.1	-0.1	-0	-0	-0.1	-0.3			62.9	=T_L
30																	

KEY TERMS

Array
Array function
Coefficient matrix
Determinant
Fill handle
Invert

Laplace's equation
Linear combination
Matrix
Nonsingular
Right-hand-side vector
Scalar

Singular
System of linear algebraic
    equations
Transpose
Unknown vector
Vector

SUMMARY

Basic Matrix Math Operations

Matrix Multiplication by a Scalar	Use basic cell arithmetic
Adding Two Matrices	Use basic cell arithmetic
Multiplying Two Matrices	Use array function MMULT ( )

Transposing a Matrix, two methods:

- Use Edit/Copy and then Edit/Paste Special/Transpose
- Use array function TRANSPOSE ( )

Invert a Matrix	Use array function MINVERSE ( )
Matrix Determinant	Use array function MDETERM ( )
Completing an Array Function	Use [Ctrl-Shift-Enter] to tell Excel to enter the matrix function into every cell of the result matrix.

Solving Simultaneous Linear Equations

1. Write the equations in the matrix form $[C][x] = [r]$:

$$3x_1 + 2x_2 = 7,$$
$$4x_1 + 6x_2 = 9, \tag{4.13}$$

becomes

$$\begin{bmatrix} 3 & 2 \\ 4 & 6 \end{bmatrix} \begin{bmatrix} x_1 \\ x_2 \end{bmatrix} = \begin{bmatrix} 7 \\ 9 \end{bmatrix} \tag{4.14}$$

2. Invert the coefficient matrix $[C]$, using array function `MINVERSE()` to get the inverted coefficient matrix:

$$[C_{inv}] \tag{4.15}$$

3. Multiply the inverted coefficient matrix $[C_{inv}]$ and the right-hand-side vector $[r]$ to get the solution vector

$$[x]. \tag{4.16}$$

That is,

$$[C_{inv}][C][x] = [C_{inv}][r],$$
$$[x] = [C_{inv}][r]. \tag{4.17}$$

Problems

Simultaneous Equations I

1. Use Excel's `MDETERM()` array function to see whether there is a solution to the simultaneous equations represented by each of the coefficient and right-hand-side matrices shown.

$$\text{a.} \quad C = \begin{bmatrix} 3 & 0 & 5 \\ 8 & 7 & 8 \\ 0 & 3 & 7 \end{bmatrix} \quad r = \begin{bmatrix} 3 \\ 8 \\ 2 \end{bmatrix},$$

$$\text{b.} \quad C = \begin{bmatrix} 1 & 2 & 0 \\ 0 & 1 & 1 \\ 2 & 5 & 1 \end{bmatrix} \quad r = \begin{bmatrix} 4 \\ 3 \\ 3 \end{bmatrix},$$

$$\text{c.} \quad C = \begin{bmatrix} 4 & 3 & 3 \\ 3 & 8 & 4 \\ 3 & 2 & 8 \end{bmatrix} \quad r = \begin{bmatrix} 7 \\ 3 \\ 2 \end{bmatrix}.$$

If the solution exists, solve the equations by using matrix methods.

Simultaneous Equations II

2. Write the following sets of simultaneous equations in matrix form, and check the determinant to see whether there is a solution:

a. $0x_1 + 7x_2 + 1x_3 = 3,$
$3x_1 + 6x_2 + 3x_3 = 8,$
$-3x_1 + 8x_2 - 1x_3 = 2.$

b. $1x_1 + 8x_2 + 4x_3 = 0,$
$-1x_1 + 1x_2 + 7x_3 = 7,$
$6x_1 + 7x_2 - 2x_3 = 3.$

c. $6x_1 + 5x_2 - 1x_3 + 6x_4 = 0,$
$-2x_1 + 2x_2 + 2x_3 + 2x_4 = 1,$
$1x_1 - 1x_2 + 1x_3 - 2x_4 = 1,$
$7x_1 - 3x_2 + 8x_3 + 4x_4 = 4.$

If a solutions exists, solve the equations using matrix methods.

Simultaneous Equations III

3. Write the following sets of simultaneous equations in matrix form and solve the new equations (if possible):

a. $3x_1 + 1x_2 + 5x_3 = 20,$
$2x_1 + 3x_2 - 1x_3 = 5,$
$-1x_1 + 4x_2 \quad = 7.$

b. $6x_1 + 2x_2 + 8x_3 = 14,$
$x_1 + 3x_2 + 4x_3 = 5,$
$5x_1 + 6x_2 + 2x_3 = 7.$

c. $4y_1 + 2y_2 + y_3 + 5y_4 = 52.9,$
$3y_1 + y_2 + 4y_3 + 7y_4 = 74.2,$
$2y_1 + 3y_2 + y_3 + 6y_4 = 58.3,$
$3y_1 + y_2 + y_3 + 3y_4 = 34.2.$

Multiloop Circuits II

4. Find the currents i_1 through i_3 in the following circuit (the battery potential and the resistances are tabulated):

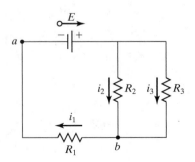

KNOWN VALUES	
E	12 volts
R_1	10 ohms
R_2	20 ohms
R_5	50 ohms

Material Balances on a Gas Absorber

5.   The following equations are material balances for CO_2, SO_2, and N_2 around the gas absorber shown in the accompanying figure.

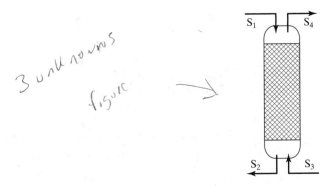

Stream S_1 is known to contain 99 mole % monoethanolamine (MEA) and 1 mole % CO_2. The flow rate in S_1 is 100 moles per minute. The compositions (mole fractions) used in the material balances are tabulated as

	S_1	S_2	S_3	S_4
CO_2	0.01000	0.07522	0.08000	0.00880
SO_2	0	0.01651	0.02000	0.00220
N_2	0	0	0.90000	0.98900
MEA	0.99000	0.90800	0	0

CO_2 Balance:	CO_2 in S_1 + CO_2 in S_3 = CO_2 in S_2 + CO_2 in S_4
	$1 \text{ mole} + 0.08000 \cdot S_3 = 0.07522 \cdot S_2 + 0.00880 \cdot S_4$
SO_2 Balance:	SO_2 in S_1 + SO_2 in S_3 = SO_2 in S_2 + SO_2 in S_4
	$0 + 0.02000 \cdot S_3 = 0.01651 \cdot S_2 + 0.00220 \cdot S_4$
N_2 Balance:	N_2 in S_1 + N_2 in S_3 = N_2 in S_2 + N_2 in S_4
	$0 + 0.90000 \cdot S_3 = 0 + 0.98900 \cdot S_4$

Solve the material balances for the unknown flow rates, S_2 through S_4.

Material Balances on an Extractor

6.   This problem focuses on a low-cost, high-performance, chemical extraction unit: a drip coffee maker. The ingredients are water, coffee solubles (CS), and coffee grounds (CG). Stream S_1 is water only, and the coffee maker is designed to hold 1 liter of it. Stream S_2 is the dry coffee placed in the filter and contains 99% grounds and 1% soluble ingredients. The product coffee (S_3) contains 0.4% CS and 99.6% water. Finally, the waste product (S_4) contains 80% CG, 19.6% water, and 0.4% CS. (All percentages are on a volume basis.) The following is a diagram of the unit:

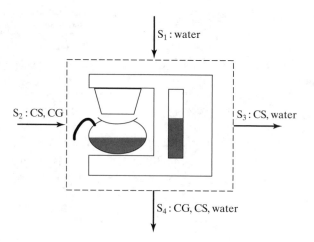

Write material balances on water, CS, and CG. Then solve the material balances for the volumes S_2 through S_4.

Flash Distillation

7. When a hot, pressurized liquid is pumped into a tank (flash unit) at a lower pressure, the liquid boils rapidly. This rapid boiling is called a flash. If the liquid contains a mixture of chemicals, the vapor and liquid leaving the flash unit will have different compositions, and the flash unit can be used as a separator. The physical principle involved is vapor–liquid equilibrium; the vapor and liquid leaving the flash unit are in equilibrium. This allows the composition of the outlet streams to be determined from the operating temperature and pressure of the flash unit. Multiple flash units can be used together to separate multicomponent mixtures.

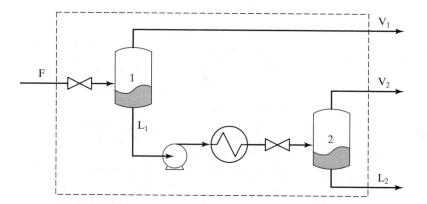

A mixture of methanol, butanol, and ethylene glycol is fed to a flash unit (Unit 1) operating at 165°C and 7 atm. The liquid from the first flash unit is recompressed, reheated, and sent to a second flash unit (Unit 2) operating at 105°C and 1 atm. The mixture is fed to the process at a rate of 10,000 kg/h. Write material balances for each chemical, and solve for the mass flow rate of each product stream (V_1, V_2, L_2). A material balance is simply a mathematical statement that all of the methanol (for example) going into the process has to come out again. (This assumes a steady state and no chemical reactions.) The following methanol balance is shown as an example:

$$\text{methanol in } F = \text{methanol in } V_1 + \text{methanol in } V_2 + \text{methanol in } L_2$$
$$0.300 \cdot (10{,}000 \text{ kg/h}) = 0.716 \cdot V_1 + 0.533 \cdot V_2 + 0.086 \cdot L_2$$

The compositions of the feed stream F and of the three product streams are listed here:

COMPONENT	MASS FRACTION IN STREAM			
	F	V_1	V_2	L_2
Methanol	0.300	0.716	0.533	0.086
Butanol	0.400	0.268	0.443	0.388
Ethylene Glycol	0.300	0.016	0.024	0.526

Wheatstone Bridge II

8. Because the resistances of metals vary with temperature, measuring the resistance of a metal sensor inserted in a material is one way to determine the temperature of the material. Resistance temperature detectors (RTDs) are devices commonly used to make temperature measurements.

    To compute the resistance of the RTD, it can be built into a Wheatstone bridge as the unknown resistance—R_3 in the following figure:

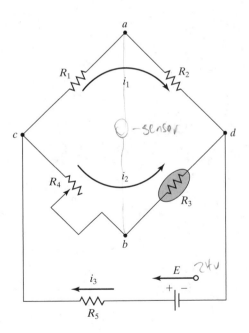

Once the bridge has been balanced, the unknown resistance can be computed from the known resistances, R_1 and R_2, and the setting on the potentiometer, R_4:

$$R_3 = R_4 \frac{R_2}{R_1} \qquad (4.18)$$

The temperature can be determined from the resistance as

$$R_T = R_0[1 + \alpha T],$$ (4.19)

where

R_T is the resistance at the unknown temperature, T;

R_0 is the resistance at 0°C (known, one of the RTD specifications); and

α is the linear temperature coefficient (known, one of the RTD specifications).

a. For the known resistances and battery potential listed below, what is the resistance of the RTD when the potentiometer (R_4) has been adjusted to 24 ohms to balance the bridge?

b. Use Kirchhoff's laws to find currents i_1 through i_3.

KNOWN VALUES	
E	24 volts
R_1	15 ohms
R_2	15 ohms
R_5	40 ohms

5

Linear Regression in Excel

5.1 INTRODUCTION

When most people think of *linear regression*, they think about finding the slope and intercept of the best-fit straight line through some data points. While that is linear regression, it's only the beginning. The "linear" in linear regression means that the equation used to fit the data points must be linear in the coefficients; it does not imply that the curve through the data points must be a straight line or that the equation can have only two coefficients (i.e., slope and intercept).

Excel provides several methods for performing linear regressions making possible both very simple and quick analyses and highly detailed advanced regression models.

5.2 LINEAR REGRESSION BY USING EXCEL FUNCTIONS

When all you need is the *slope* and *intercept* of the best-fit straight line through your data points, Excel's SLOPE() and INTERCEPT() functions are very handy. Include the RSQ() function to calculate the *coefficient of determination*, or R^2 value, and a lot of simple data analysis problems are covered.

5.2.1 A Simple Example

The following temperature and time values are a simple data set that we can use to investigate the basics of regression analysis in Excel:

OBJECTIVES

After reading this chapter, you will know

- How to use regression functions to calculate slopes, intercepts, and R^2 values for data sets
- How to perform simple regression analyses directly from a graph of your data, using trendlines
- A general approach that can handle any linear-regression model

	A	B	C	D	E
1	**Linear Regression Using Excel Functions**				
2					
3	**Time**	**Temp.**			
4	(min.)	(K)			
5					
6	0	298			
7	1	299			
8	2	301			
9	3	304			
10	4	306			
11	5	309			
12	6	312			
13	7	316			
14	8	319			
15	9	322			
16					

The SLOPE() function can be used to determine the slope of the best-fit straight line through these data. The SLOPE() function takes two arguments: the cell range containing the *y*-values (*dependent variable* values) and the cell range containing the *x*-values (*independent variable* values)—in that order.

Here, temperature depends on time, not the other way around, so temperature (cells B6:B15) is the dependent variable, and time (A6:A15) is the independent variable:

D6			f_x =SLOPE(B6:B15,A6:A15)		
	A	B	C	D	E
1	**Linear Regression Using Excel Functions**				
2					
3	**Time**	**Temp.**			
4	(min.)	(K)			
5					
6	0	298	**Slope:**	2.78	
7	1	299	**Intercept:**		
8	2	301	**R^2:**		
9	3	304			
10	4	306			
11	5	309			
12	6	312			
13	7	316			
14	8	319			
15	9	322			
16					

Similarly, the intercept can be obtained by using the INTERCEPT() function, with the same arguments:

	D7	▼	*fx*	=INTERCEPT(B6:B15,A6:A15)	
	A	B	C	D	E
1	**Linear Regression Using Excel Functions**				
2					
3	**Time**	**Temp.**			
4	(min.)	(K)			
5					
6	0	298	**Slope:**	2.78	
7	1	299	**Intercept:**	296.1	
8	2	301	**R²:**		
9	3	304			
10	4	306			
11	5	309			
12	6	312			
13	7	316			
14	8	319			
15	9	322			
16					

Next, the coefficient of determination (R^2) can be computed by using the RSQ() function with the same arguments:

	D8	▼	*fx*	=RSQ(B6:B15,A6:A15)	
	A	B	C	D	E
1	**Linear Regression Using Excel Functions**				
2					
3	**Time**	**Temp.**			
4	(min.)	(K)			
5					
6	0	298	**Slope:**	2.78	
7	1	299	**Intercept:**	296.1	
8	2	301	**R²:**	0.9864	
9	3	304			
10	4	306			
11	5	309			
12	6	312			
13	7	316			
14	8	319			
15	9	322			
16					

This tells us that the best-fit line through the data has the slope (b_1) 2.7758 K/min (the units were inferred from the data) and the intercept (b_0) 296.1 K. The R^2 value is 0.9864.

Thus, we have a slope, an intercept, and an R^2 value. Is the straight line a good fit to the data? It is always a good idea to plot the data and the regression line to verify the fit visually; but, if you are going to graph your data, Excel provides an even easier way to perform a linear regression: the trendline.

5.3 LINEAR REGRESSION BY USING EXCEL'S TRENDLINE CAPABILITY

Performing a linear-regression analysis in Excel is incredibly simple. Once the data have been graphed, regression takes just a few mouse clicks. This ease of use creates a situation in which people sometimes use linear regression because it is easy, even when the relationship between their independent and dependent variables might be nonlinear. A rule of thumb for fitting curves to data: *always graph the fitted curve against the original data to inspect the quality of the fit visually*. When the regression is done with a *trendline*, the fitted curve is automatically added to the graph of the original data.

5.3.1 Simple Slope–Intercept Linear Regression

The process of performing a linear regression for a slope and intercept requires the computation of various sums involving the x-(independent) and y-(dependent) values in your data set. With these sums, you could use the following equations to calculate the slope b_1 and intercept b_0 of the straight line that best fits your data, *but Excel will do it for you*:

$$b_1 = \frac{\sum_i x_i y_i - \dfrac{1}{N_{\text{data}}} \sum_i x_i \sum_i y_i}{\sum_i (x_i^2) - \dfrac{1}{N_{\text{data}}} \left(\sum_i x_i\right)^2},$$

$$b_0 = \frac{\sum_i y_i - b_1 \sum_i x_i}{N_{\text{data}}}. \tag{5.1}$$

In these equations, $\sum_i$ implies summation over all data points, $i = 1$ to N_{data}.

When you add a trendline to a graph, Excel calculates all of the required summations, then calculates the slope b_1, the intercept b_0, and the coefficient of determination R^2. The trendline is added to the graph so that you can see how well the fitted equation matches the data points. By default, the equation of the trendline and the R^2 value are not shown on the graph. You have to request them by using the Options panel on the Trendline dialog box.

To see how to use trendlines, we will again use the temperature-vs.-time data. The first step in using a trendline to obtain a regression equation is to plot the data, using an *XY (Scatter) plot* with the data points indicated by data markers.

Note: It is customary to plot data values with markers and fitted results with curves, so that it is easy to see whether the fitted line actually goes through the data points.

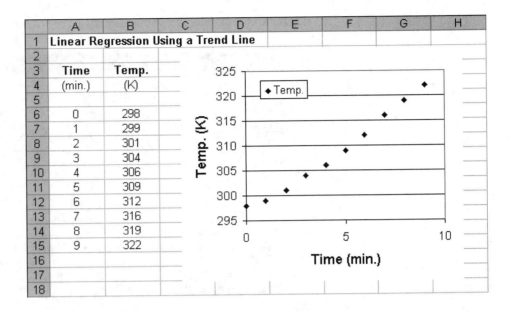

Then right click on any data point on the graph and select Add Trendline . . . from the pop-up menu:

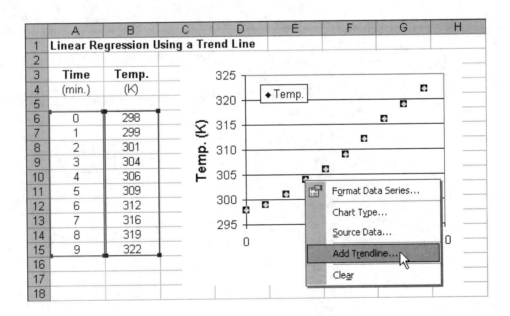

In the Add Trendline dialog box, select Linear from the Type panel:

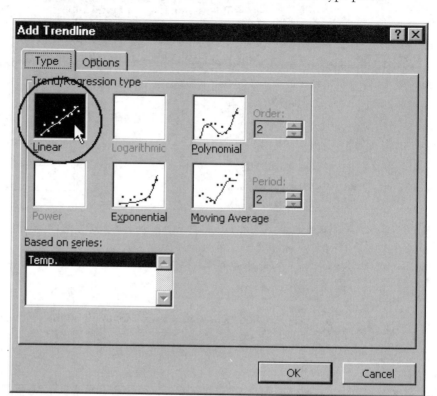

Press OK to close the Add Trendline dialog box and add the trendline to the plot:

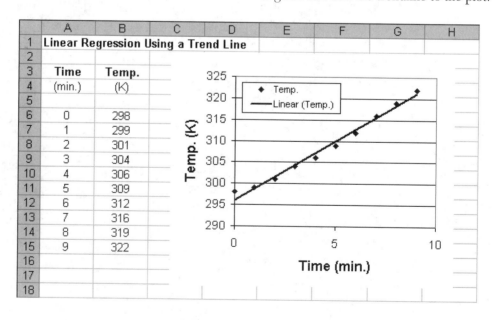

Excel has added the trendline to the graph. To create the line, Excel performed a linear regression on the graphed data. The equation of the regression line is available; you need only ask Excel to display it. Double click on the trendline to bring up the Format Trendline dialog box.

In the Options panel, check "Display equation on chart" and "Display R-squared value on chart" to have these items included on the graph. Also, you can see where the trendline text in the plot's legend could be defined in the Custom field on the Trendline name box:

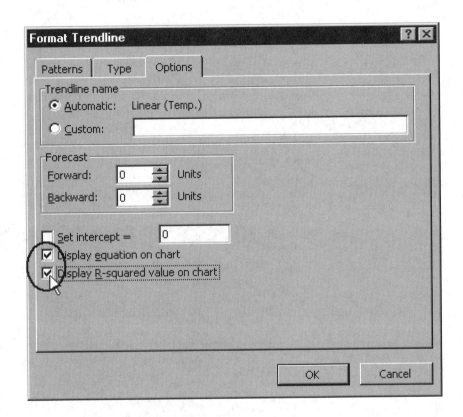

When you press OK, Excel will add the regression equation and the R^2 value to the plot:

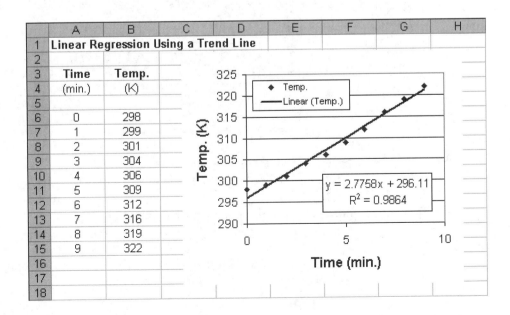

From the graph, we see that the regression equation is

$$y = 2.7758x + 296.11, \tag{5.2}$$

with the R^2 value 0.9864. (These are the same results we obtained by using the SLOPE(), INTERCEPT(), and RSQ() functions.) This tells us that the best-fit line through the data has the slope (b_1) 2.7758 K/min (the units were inferred from the data), the intercept (b_0) 296.1 K, and R^2 value 0.9864. The R^2 value 1 means a perfect fit, so the value 0.9864 indicates that this is a less-than-perfect fit. The plotted trendline allows you to verify visually that the best-fit line really does (or doesn't) fit the data. In this case, the straight trendline is not fitting the data well. We need to try to find a regression equation that allows for some curvature.

5.3.2 Forcing the Regression Line through the Origin (0,0)

If you do not want Excel to compute an intercept (i.e., if you want to force the curve to go through $y = 0$ when $x = 0$), there is a check box you can select on the Options panel of the Format Trendline (or Add Trendline) dialog box.

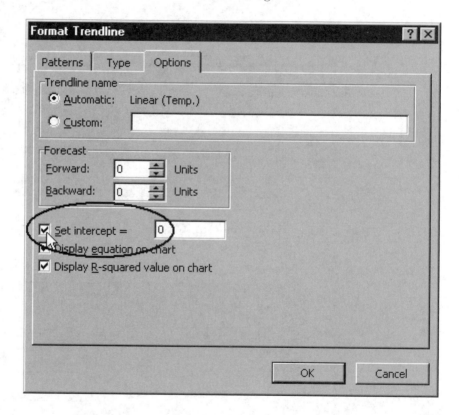

The temperature in our data set does not go to zero at time zero, so there is no reason to force the intercept through the origin for this data.

5.4 OTHER TWO-COEFFICIENT LINEAR-REGRESSION MODELS

Any equation relating x-values to y-values that is linear in the coefficients can be used in regression analysis, but there are a number of two-coefficient models that are commonly used for linear regression. These include the following:

NAME	EQUATION	LINEAR FORM	DATA MANIPULATION
Exponential Fit	$y = k_0 e^{k_1 x}$	$\ln(y) = \ln(k_0) + k_1 x$ $= b_0 + b_1 x$ $b_0 = \ln(k_0)$ $b_1 = k_1$	Take the natural log of all y-values prior to performing regression. Regression returns b_0 and b_1 which can be related to k_0 and k_1 through the equations on the left.
Logarithmic Fit	$y = k_0 + k_1 \ln(x)$	$y = k_0 + k_1 \ln(x)$ $b_0 = k_0$ $b_1 = k_1$	Take the natural log of all x-values prior to performing regression.
Power Fit	$y = k_0 x^{k_1}$	$\ln(y) = \ln(k_0) + k_1 \ln(x)$ $= b_0 + b_1 \ln(x)$ $b_0 = \ln(k_0)$ $b_1 = k_1$	Take the natural log of all x and y-values prior to performing regression.

Note: Excel will manipulate the data as needed to create these trendlines. You simply need to select the type of trendline you want to use.

Each of these is available as a regressed trendline in Excel. You select the type of regression you want Excel to perform, using the Type panel of the Format Trendline (or Add Trendline) dialog box.

The slight upward bending visible in the data suggests an exponential fit might work. It's easy to give it a try; just double click on the existing trendline to bring up the Format Trendline dialog box:

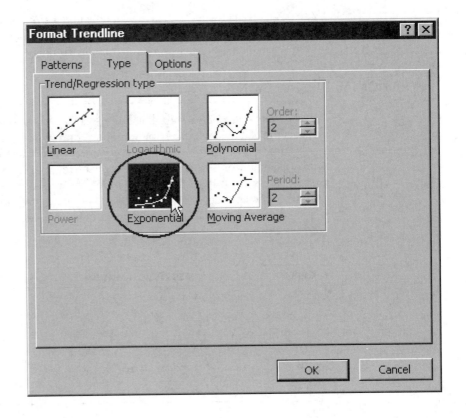

Then select Exponential from the Type panel. When you click the OK button, Excel will perform the regression, using an exponential fit.

Note: The Logarithmic and Power types are not available for this data set, because it contains x = 0 *(i.e., time = 0). Both of those regression equations take the natural log of* x. *Because ln(0) is not defined, these regression equations cannot be used with this data set.*

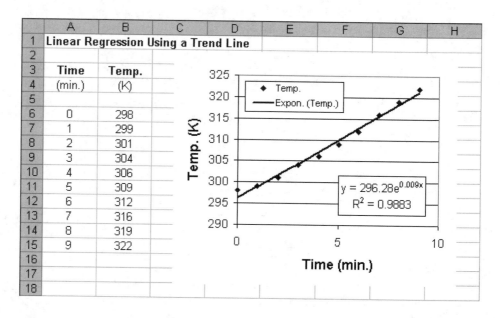

The result from using an exponential fit doesn't look much better, and the R^2 value is about the same as what we obtained with the linear model. (Excel offers one more type of regression on the Format Trendline (or Add Trendline) dialog: polynomial regression.)

5.5   POLYNOMIAL REGRESSION

Polynomial regression is still a linear regression, because the regression polynomials are linear in the coefficients (the *b*-values). Since, during a regression analysis, all of the *x* values are known, each of the polynomial equations listed next is a linear equation, so regression using any of these equations is a linear regression. It differs from the other regression models in that there are more than two coefficients. Technically, simple slope–intercept regression is also a polynomial regression, but the term is usually reserved for polynomials of order two or higher. Excel's polynomial trendlines can be of order 2 through 6:

ORDER	REGRESSION MODEL	NUMBER OF COEFFICIENTS
2	$y_p = b_0 + b_1x + b_2x^2$	3
3	$y_p = b_0 + b_1x + b_2x^2 + b_3x^3$	4
4	$y_p = b_0 + b_1x + b_2x^2 + b_3x^3 + b_4x^4$	5
5	$y_p = b_0 + b_1x + b_2x^2 + b_3x^3 + b_4x^4 + b_5x^5$	6
6	$y_p = b_0 + b_1x + b_2x^2 + b_3x^3 + b_4x^4 + b_5x^5 + b_6x^6$	7

Note: It is fairly standard nomenclature to write polynomials with the higher powers to the right. However, Excel reports polynomial trendline equations with the high powers first. It's not a big deal, but read the equations carefully.

To request a second-order polynomial regression, bring up the Format Trendline dialog box by double clicking on the existing trendline:

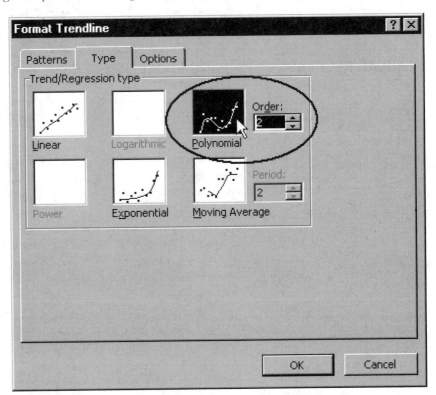

Select Polynomial type, and set the order to 2. Click the OK button to have Excel perform the regression:

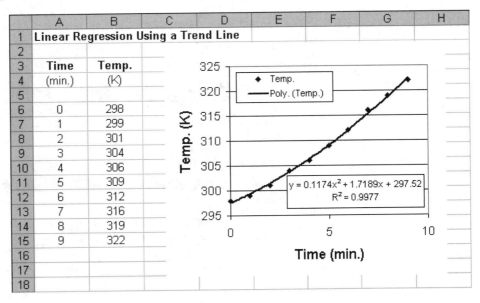

The second-order polynomial trendline appears to fit the data well, and the R^2 value is much closer to unity.

Polynomial regression using trendlines is as easy and simple as slope–intercept regression. But what happens if you want a 7th-order fit, and Excel's trendline capability can't be used? You can still use Excel's regression-analysis package, which is available from the main menu under Tools/Data Analysis/Regression. You also might want to use the regression-analysis package if you want more details about the regression results than are available with the use of trendlines.

5.6 LINEAR REGRESSION BY USING EXCEL'S REGRESSION-ANALYSIS PACKAGE

Excel's regression analysis package is fairly easy to use, but is more involved than simply asking for a trendline on a graph. There are two reasons why you might want to use the more complex approach:

1. You want to use a regression model that is not available as a trendline. The regression-analysis package can handle any linear-regression model.
2. You want more details about the regression process than are provided by using trendlines.

5.6.1 A Simple Linear Regression for Slope and Intercept

As a first example of working with the regression-analysis package, we will regress the temperature-vs.-time data to calculate the slope and intercept:

Step 1. Open the Data Analysis list box.

Regression is one of the data-analysis tools available on the Tools menu. Select Tools/Data Analysis . . . to bring up the Data Analysis list box.

Note: By default, the data-analysis package is installed, but not an activated part of Excel. If the Data Analysis menu option does not appear under the Tools menu, it has not been activated. Select Tools/Add-Ins . . . , and activate the Analysis option. (This need be done only once.)

Step 2. Select Regression from the Data Analysis list box.

Click Regression, and then OK.

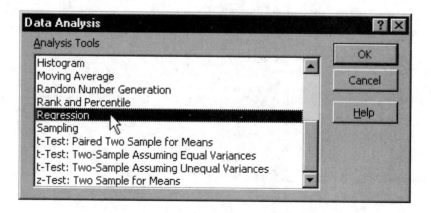

This will open the Regression dialog box:

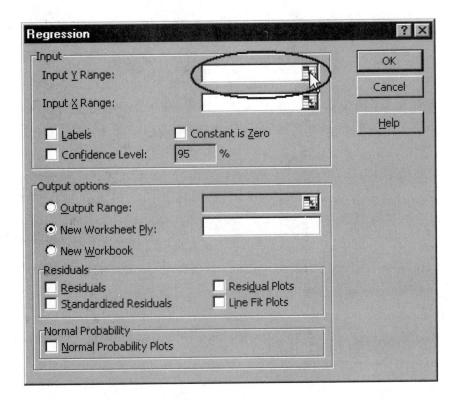

Step 3(a). Tell Excel where to find the *y*-values for the regression: Use the "to the spreadsheet" button.

The Regression dialog box has a field labeled "Input Y Range" as shown before. At the right side of the input field, there is a small button that will take you to the spreadsheet (so you can select the cells containing *y*-values). Click the "to the spreadsheet" button.

Note: Common linear regression assumes that all of the uncertainty in the data is in the y-values and that the x-values are known precisely. It is important, therefore, to call the values that are imprecise the y-values for regression analysis.

Step 3(b). Select the cells containing the *y*-values.

When the spreadsheet is displayed, drag the mouse to indicate the cells containing the *y*-values. When the *y*-values have been selected, click on the "return to dialog" button (circled in the figure below) or press [Enter] to return to the Regression dialog box.

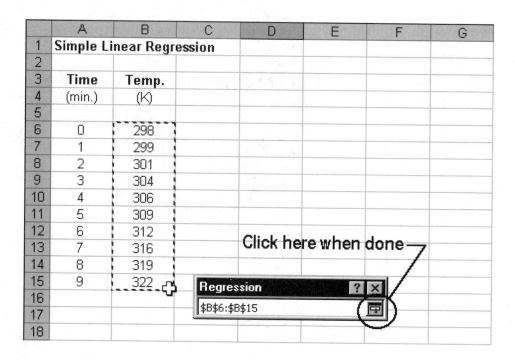

Step 4(a). Tell Excel where to find the *x*-values: Use the "to the spreadsheet" button.

Similarly, indicate the *x*-values:

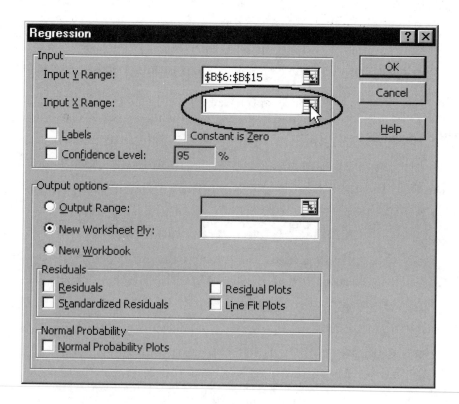

Step 4(b). Select the cells containing the *x*-values.

Use the mouse to indicate which cells contain the *x*-values, then return to the Regression dialog box:

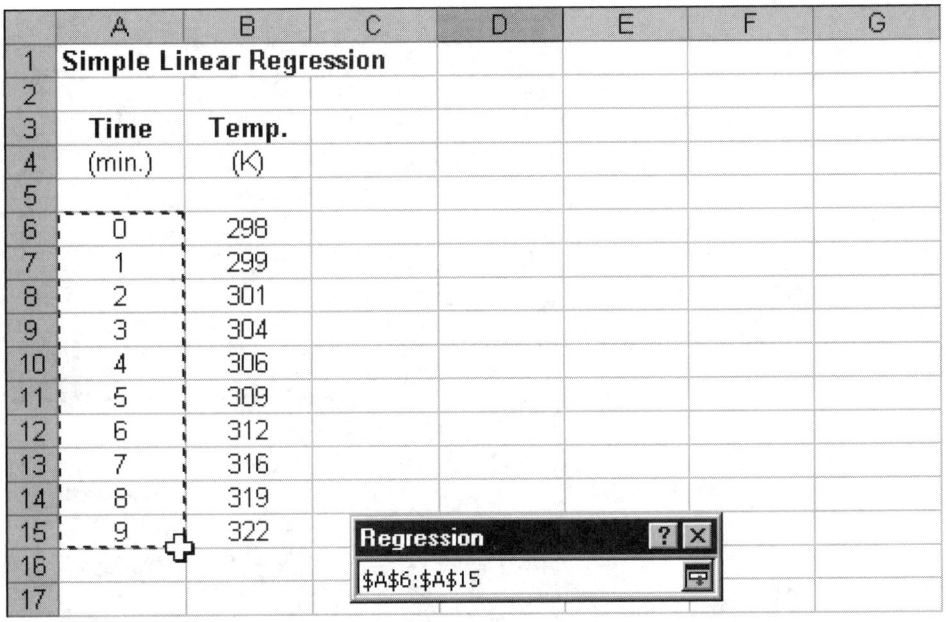

	A	B	C	D	E	F	G
1	**Simple Linear Regression**						
2							
3	**Time**	**Temp.**					
4	(min.)	(K)					
5							
6	0	298					
7	1	299					
8	2	301					
9	3	304					
10	4	306					
11	5	309					
12	6	312					
13	7	316					
14	8	319					
15	9	322					
16				Regression ? X			
17				$A$6:$A$15			

Step 5. Choose a location for the results of the regression analysis.

Because the results take up quite a bit of space, putting the output on a new worksheet ply (a new spreadsheet) is the most common. Do this by clicking the button next to "New Worksheet Ply:"

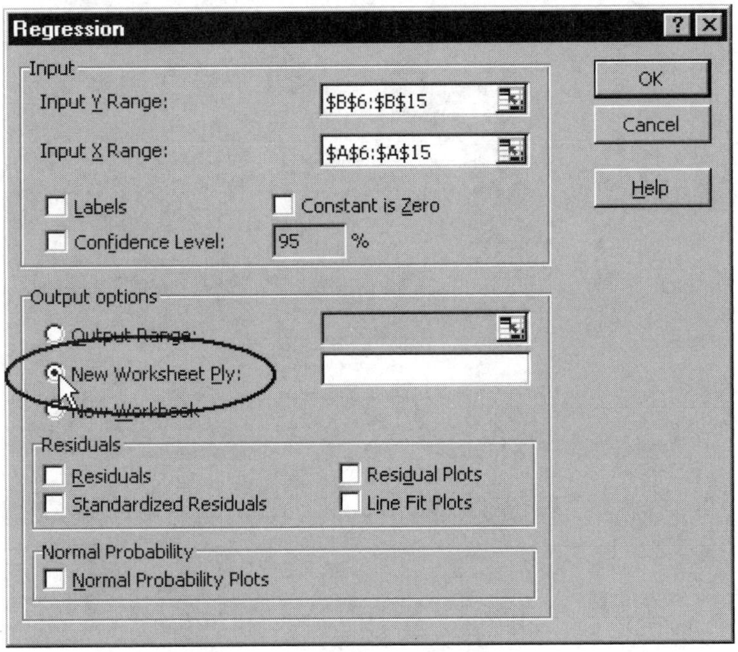

Regression   ? X

Input

Input Y Range: $B$6:$B$15

Input X Range: $A$6:$A$15

☐ Labels   ☐ Constant is Zero
☐ Confidence Level: 95   %

OK
Cancel
Help

Output options
○ Output Range:
◉ New Worksheet Ply:
○ New Workbook

Residuals
☐ Residuals   ☐ Residual Plots
☐ Standardized Residuals   ☐ Line Fit Plots

Normal Probability
☐ Normal Probability Plots

Step 6. Indicate whether you want the results to be graphed.

It is always a good idea to look at the residual plot and the line-fit plot to check visually on whether the regression line actually fits the data:

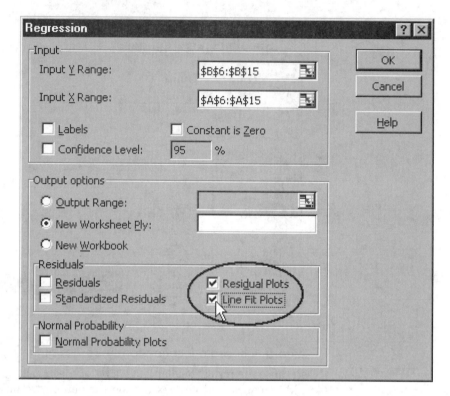

Step 7. Perform the regression.

Click the [OK] button to perform the regression. The results are presented as tables and graphs (if you requested them.) Only a portion of the output is shown here:

	A	B	C	D	E	F
1	SUMMARY OUTPUT					
2						
3	*Regression Statistics*					
4	Multiple R	0.99318635				
5	R Square	0.98641913				
6	Adjusted R Square	0.98472152				
7	Standard Error	1.04591558				
8	Observations	10				
9						
10	ANOVA					
11		*df*	*SS*	*MS*	*F*	*Significance F*
12	Regression	1	635.6484848	635.6485	581.0637	9.35281E-09
13	Residual	8	8.751515152	1.093939		
14	Total	9	644.4			
15						
16		*Coefficients*	*Standard Error*	*t Stat*	*P-value*	*Lower 95%*
17	Intercept	296.109091	0.614740869	481.6812	3.86E-19	294.691495
18	X Variable 1	2.77575758	0.115151515	24.10526	9.35E-09	2.510217534
19						

The output page tells us that the best-fit line through the data has slope (b_1) 2.775 K/min (the units were inferred from the data) and intercept (b_0) 296.1 K. The R^2 value is 0.9864. These are the same results we obtained with the other methods.

The plots allow you to verify visually that the best-fit line really does (or doesn't) fit the data. The line-fit plot was created by Excel and placed on the output page:

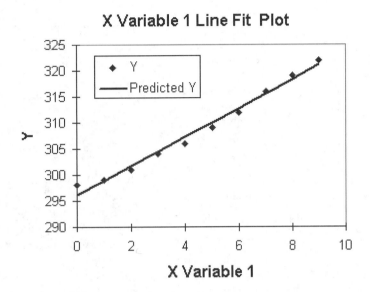

This is essentially the same as the plot created by using the linear trendline. Poor fits become even more apparent when you look at the residual plot:

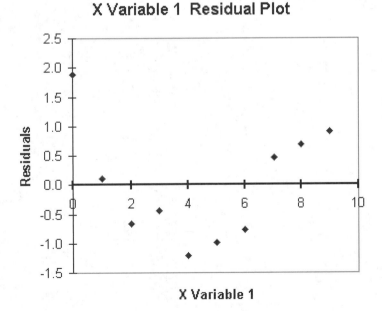

The *residual* is the difference between the data point *y*-value and the regression line *y*-value at each *x*-value. A *residual plot* highlights poor fits and makes them easy to spot. Strong patterns in a residual plot, such as the "U" shape shown here, indicate that there is something happening in your data that the model (the straight line) is not accounting for. It's a strong hint that you should look for a better regression model.

This example was included as a reminder that, although you can perform a simple (straight-line) linear regression on any data set, it is not always a good idea. Nonlinear data require a more sophisticated curve-fitting approach.

5.6.2   Polynomial Regression by Using the Regression Analysis Package

The best fit we obtained to the temperature–time data by using trendlines was obtained by using the second-order polynomial

$$y_p = b_0 + b_1 x + b_2 x^2 \tag{5.3}$$

or, in terms of temperature, T, and time, t,

$$T_p = b_0 + b_1 t + b_2 t^2. \tag{5.4}$$

We will use this polynomial to demonstrate how to solve for more than two coefficients when using Excel's regression-analysis package.

The regression-analysis package in Excel allows you only one column of y-values (dependent variables), but you can have multiple columns of x-values (independent variables). These can be completely independent variables [e.g., enthalpy (dependent) as a function of temperature (independent) and pressure (independent)], or they can be the same variable in multiple forms [e.g., x and x^2 or z and $\ln(z)$]

For this second-order polynomial, we'll need two columns of independent values, containing t and t^2. The dependent values (T-values) have been moved to column A to allow the t- and t^2-values to be placed next to each other in columns B and C:

	C9	▼	f_x =B9^2	
	A	B	C	D
1	**Generalized Linear Regression**			
2				
3	**Temp.**	**Time**	**Time^2**	
4	(K)	(min.)	(min.^{2})	
5				
6	298	0	0	
7	299	1	1	
8	301	2	4	
9	304	3	9	
10	306	4	16	
11	309	5	25	
12	312	6	36	
13	316	7	49	
14	319	8	64	
15	322	9	81	
16				

The procedure for getting Excel to regress this data is almost the same as that used in the previous example, except in step 4 when you tell Excel where to find the *x*-values. In this case, you need to indicate *both* column B and column C:

	A	B	C	D	E	F	G
1	**Generalized Linear Regression**						
2							
3	**Temp.**	**Time**	**Time²**				
4	(K)	(min.)	(min.²)				
5					Regression ? ✕		
6	298	0	0		$B$6:$C$15		
7	299	1	1				
8	301	2	4				
9	304	3	9				
10	306	4	16				
11	309	5	25				
12	312	6	36				
13	316	7	49				
14	319	8	64				
15	322	9	81				
16							

The rest of the regression process is unchanged. The regression output page shows the results (again, only part of it is shown here):

	A	B	C	D	E	F
1	SUMMARY OUTPUT					
2						
3	*Regression Statistics*					
4	Multiple R	0.99885781				
5	R Square	0.99771693				
6	Adjusted R Square	0.99706462				
7	Standard Error	0.45844646				
8	Observations	10				
9						
10	ANOVA					
11		*df*	*SS*	*MS*	*F*	*Significance F*
12	Regression	2	642.9287879	321.4644	1529.522	5.68616E-10
13	Residual	7	1.471212121	0.210173		
14	Total	9	644.4			
15						
16		*Coefficients*	*Standard Error*	*t Stat*	*P-value*	*Lower 95%*
17	Intercept	297.518182	0.36045142	825.4044	1.01E-18	296.6658503
18	X Variable 1	1.71893939	0.186520848	9.215803	3.66E-05	1.277887988
19	X Variable 2	0.11742424	0.019951321	5.885537	0.000608	0.070246898
20						

Again, the results obtained by using the regression-analysis package are the same as those obtained by using the polynomial trendline. One piece of information that is available only from the regression-analysis package is the residual plot:

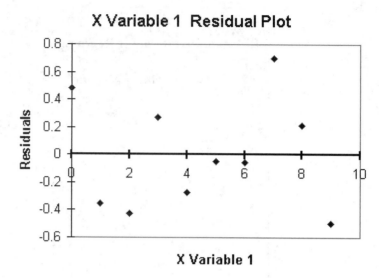

The residual plot shows no obvious trends. This suggests that the model is doing about as good a job of fitting these data as can be done.

5.6.3   Other Linear Models

The models used in the preceding examples,

$$T_p = b + b_1 t \text{ and}$$

$$T_p = b_0 + b_1 t + b_2 t^2, \tag{5.5}$$

are both linear models (linear in the coefficients, not in time). Excel's regression-analysis package works with any linear model, so you could try fitting equations such as

$$T_p = b_0 + b_1 \sinh(t) + b_2 \operatorname{atan}(t^2) \tag{5.6}$$

or

$$T_p = b_0 \exp(t^{0.5}) + b_1 \ln(t^3). \tag{5.7}$$

There is no reason to suspect that either of these last two models would be a good fit to the temperature-vs.-time data, but both equations are linear in the coefficients (the b's) and are compatible with generalized regression analysis. (There is one problem: The natural logarithm in the last model won't work with the $t = 0$ that appears in the data set.) In general, you choose a linear model either from some theory that suggests a relationship between your variables or from looking at a plot of the data set.

You could also use a regression equation that has multiple independent variables, such as

$$V_p = b_0 + b_1 P + b_2 T, \tag{5.8}$$

which says that the volume of a gas depends on pressure and temperature.

5.6.4 Forcing the Regression Line through the Origin (0,0)

If you do not want Excel to compute an intercept (i.e., if you want to force the curve to go through $y = 0$ when $x = 0$ by setting $b_0 = 0$), there is a check box you can select on the Regression dialog box:

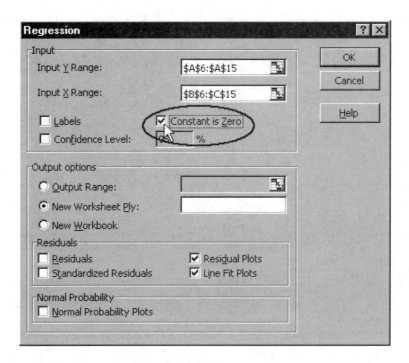

There are times when theory predicts that the curve should go through (0,0). In such a situation, you force the regression line through the origin by setting the Constant is Zero check box.

PRACTICE!

In the accompanying table, two sets of y-values are shown. The noisy data were calculated from the clean data by adding random values. Try linear regression on these data sets. What is the impact of noisy data on the calculated slope, intercept, and R^2-value?

X	Y$_{CLEAN}$	Y$_{NOISY}$
0	1	0.85
1	2	1.91
2	3	3.03
3	4	3.96
4	5	5.10
5	6	5.90

EXAMPLE 5.1

Thermocouple output voltages are commonly considered to be directly proportional to temperature. But, over large temperature ranges, a power fit can do a better job. Compare the simple linear fit and the exponential fit for copper–constantan output voltages (reference junction at 0°C) between 10 and 400°C.[1]

First, create a simple linear regression for a slope and intercept:

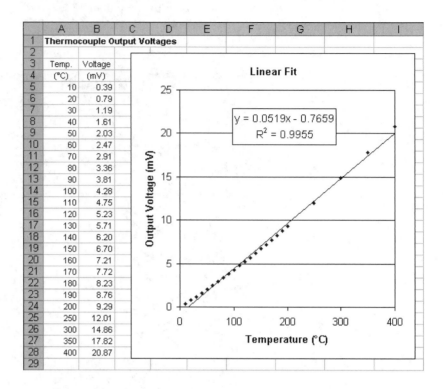

	A	B
1	**Thermocouple Output Voltages**	
3	Temp.	Voltage
4	(°C)	(mV)
5	10	0.39
6	20	0.79
7	30	1.19
8	40	1.61
9	50	2.03
10	60	2.47
11	70	2.91
12	80	3.36
13	90	3.81
14	100	4.28
15	110	4.75
16	120	5.23
17	130	5.71
18	140	6.20
19	150	6.70
20	160	7.21
21	170	7.72
22	180	8.23
23	190	8.76
24	200	9.29
25	250	12.01
26	300	14.86
27	350	17.82
28	400	20.87

The R^2 value, 0.9955, looks OK, but there is some difference between the data values and the regression line.

Now try the power fit:

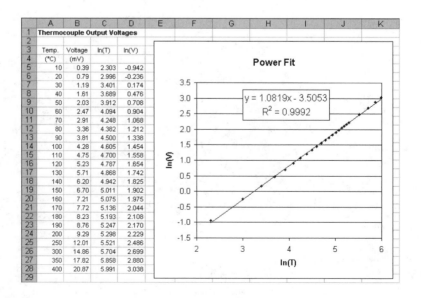

	A	B	C	D
1	**Thermocouple Output Voltages**			
3	Temp.	Voltage	ln(T)	ln(V)
4	(°C)	(mV)		
5	10	0.39	2.303	-0.942
6	20	0.79	2.996	-0.236
7	30	1.19	3.401	0.174
8	40	1.61	3.689	0.476
9	50	2.03	3.912	0.708
10	60	2.47	4.094	0.904
11	70	2.91	4.248	1.068
12	80	3.36	4.382	1.212
13	90	3.81	4.500	1.338
14	100	4.28	4.605	1.454
15	110	4.75	4.700	1.558
16	120	5.23	4.787	1.654
17	130	5.71	4.868	1.742
18	140	6.20	4.942	1.825
19	150	6.70	5.011	1.902
20	160	7.21	5.075	1.975
21	170	7.72	5.136	2.044
22	180	8.23	5.193	2.108
23	190	8.76	5.247	2.170
24	200	9.29	5.298	2.229
25	250	12.01	5.521	2.486
26	300	14.86	5.704	2.699
27	350	17.82	5.858	2.880
28	400	20.87	5.991	3.038

[1]Data from *Transport Phenomena Data Companion* by L.P.B.M. Janssen and M.M.C.G. Warmoeskerken, Arnold D.U.M. (1987). Copper–constantan was selected for this example because its performance is the least linear of the commonly used thermocouple types.

The R^2 value is closer to unity, although there is still some difference between the plotted data points [$\ln(T)$ and $\ln(V)$]. Let's see how the power model does at predicting voltages:

$$b_0 = -3.5053, \qquad \text{so,} \qquad k_0 = e^{b_0} = 0.03004,$$
$$b_1 = 1.0819, \qquad \text{so,} \qquad k_1 = b_1 = 1.0819.$$

Thus, the power model gives $V = k_0 T^{k_1} = 0.03004 T^{1.0819}$.

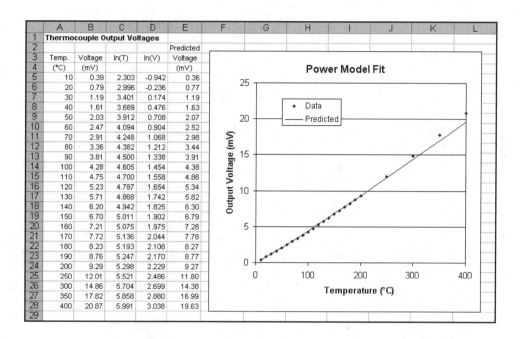

The power model fits the data better at lower temperatures, but it misses at the higher end.

KEY TERMS

Coefficient of
  Determination (R^2)
Dependent variable (y-axis)
Independent variable (x-axis)

Intercept
Linear regression
Polynomial regression
Residual

Residual plot
Slope
Trendline
XY Scatter plot

SUMMARY

Using Excel's Regression Functions

For simple slope–intercept calculations, Excel's built-in regression functions are useful. The drawback is they don't show you the regression line superimposed on the data values, and verifying the fit visually is always a good idea. If you plot the data, it is faster to use a trendline to find the slope and intercept, but, although the trendline displays the values, it does not make them available for use in the rest of the spreadsheet.

SLOPE(y, x)	Returns the slope of the straight line through the x- and y-values.
INTERCEPT(y, x)	Returns the intercept of the straight line through the x- and y-values.
RSQ(y, x)	Returns the coefficient of determination (R^2) for the straight line through the x- and y-values.

APPLICATIONS: RECALIBRATING A FLOW METER

Flow meters in industrial situations come with calibration charts for standard fluids (typically air or water), but, for critical applications, the calibration must be checked periodically to see whether the instrument still provides a reliable reading.

When purchased, the calibration sheet for a turbine flow meter provided the following equation relating the meter output (frequency, Hz) to water flow velocity (m/s):

$$v = 0.0023 + 0.0674f$$

After the meter had been in use for a period of one year, it was removed from service for recalibration. During a preliminary test, the following data were obtained from a test system:

Velocity	Frequency
(m/s)	(Hz)
0.05	0.8
0.27	4.2
0.53	8.2
0.71	10.9
0.86	13.1
1.10	16.8
1.34	20.5
1.50	23.0
1.74	26.6
1.85	28.4
2.15	32.9
2.33	35.6
2.52	38.5
2.75	42.1

Does the meter need to be recalibrated? If so, what is the new calibration equation?

First, we can use the original calibration equation to calculate predicted velocity values and plot the results to see whether the original calibration equation is still working.

C6	▼	f_x =0.0023+0.0674*A6	

	A	B	C	D
1	**Flow Meter Calibration Test**			
2		**Experimental**	**Predicted**	
3	**Frequency**	**Velocity**	**Velocity**	
4	(Hz)	(m/s)	(m/s)	
5				
6	0.8	0.05	0.1	
7	4.2	0.27	0.3	
8	8.2	0.53	0.6	
9	10.9	0.71	0.7	
10	13.1	0.86	0.9	
11	16.8	1.10	1.1	
12	20.5	1.34	1.4	
13	23.0	1.50	1.6	
14	26.6	1.74	1.8	
15	28.4	1.85	1.9	
16	32.9	2.15	2.2	
17	35.6	2.33	2.4	
18	38.5	2.52	2.6	
19	42.1	2.75	2.8	
20				

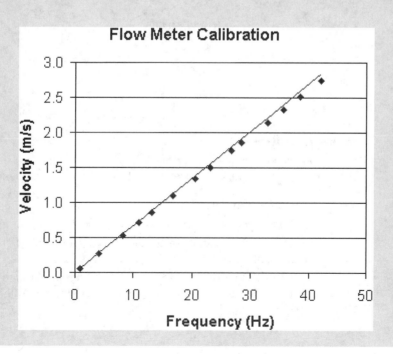

Flow Meter Calibration

The data values do not seem to agree with the original calibration line at higher velocities, so it is time to recalibrate. To do so, simply add a linear trendline to the graph and ask Excel to display the equation of the line. The result (without the original calibration line) is

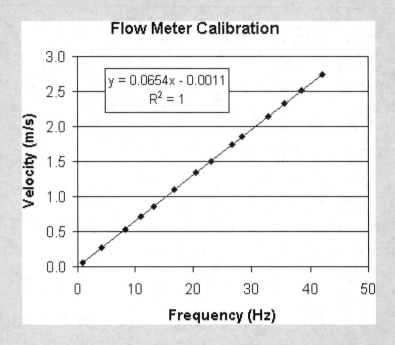

The new calibration equation is

$$v = -0.0011 + 0.0654\,f$$

Using TrendLines

If you don't need a lot of information about the regression results—just the equation and the R^2 value—then trendlines are fast and easy. To add a trendline, do the following:

1. Graph your data.
2. Right click on any data point, and select Trendline from the pop-up menu.
3. Use the Type tab to select the type of trendline. (Set the order of the polynomial, if needed.)
4. Use the Options tab to force the intercept through the origin or to have Excel display the equation of the line and the R^2 value, if desired.

Using the Regression-Analysis Package
General Procedure

1. Choose a linear regression model: The model must be linear in the coefficients.
2. Set up the required columns of x- and y-values in the spreadsheet:

   - You may have only one column of y-values, but multiple columns of x-values (e.g., x, x^2, x^3, etc., as required for the regression model you want to fit).

3. Have Excel perform the regression analysis.
4. Find your regression results in the output table created by Excel. As a minimum, check for

   - The coefficients are listed as Intercept (if you asked Excel to compute one) and the Coefficients for X Variable 1, X Variable 2, and so on (as many as are needed for your model).
   - The R^2 value; the value 1.0 indicates a perfect fit.

5. Check the line-fit plot and residual plots (if you requested them) to verify visually that your model does (or does not) fit the data.

Using the Regression-Analysis Package

1. Open the Data Analysis List Box by selecting Tools/Data Analysis . . . .
2. Select Regression from the list.
3. Show Excel where to find the y-values.

   a. Click on the "go to spreadsheet button" at the right side of the Input Y Range field.
   b. When the spreadsheet is displayed, drag the mouse over the cells containing the y-values to select those cells.
   c. Click the "go to dialog" button on the small Regression box to return to the Regression dialog box.

4. Show Excel where to find the x-values.
5. Choose a location for the regression results; on a new worksheet ply is the most common location.
6. Indicate whether you want the line-fit plot and residual plot prepared (generally recommended).
7. Click [OK] on the Regression dialog box to have Excel perform the regression and produce the output summary tables, and any graphs you requested.

Problems

Graphing Functions

1. Graph the following common regression functions, using the specified coefficients and ranges:

   a. $y = a + bx$, $a = 2$, $b = 0.3, 0 \leq x \leq 5$.

b. $y = a + b/x$ $a = 2,\ \ b = 0.3, 1 \leq x \leq 5.$

c. $y = ae^{bx},$ $a = 2,\ \ b = 0.3, 0 \leq x \leq 5.$

d. $y = ae^{-bx},$ $a = 2,\ \ b = 0.3, 0 \leq x \leq 5.$

Compare the general shape of the curves produced by the regression functions with the following plot:

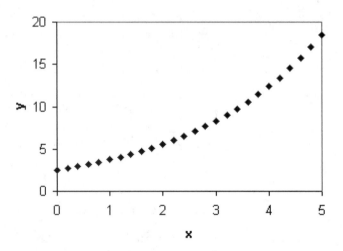

Which regression function is most likely to provide a good fit to the data?

Simple Linear Regression

2.   Plot each of the three data sets presented in the accompanying table to see whether a straight line through each set of points seems reasonable. If so, add a linear trendline to the graph and have Excel display the equation of the line and the R^2 value on the graph.

The same x-data have been used in each example, to minimize typing.

X	Y_1	Y_2	Y_3
0	2	0.4	10.2
1	5	3.6	4.2
2	8	10.0	12.6
3	11	9.5	11.7
4	14	12.0	28.5
5	17	17.1	42.3
6	20	20.4	73.6
7	23	21.7	112.1

Thermocouple Calibration Curve

3.   Thermocouples are made by joining two dissimilar metal wires. The contact of the two metals results in a small, but measurable, voltage drop across the junction. This voltage drop changes as the temperature of the junction changes; thus, the thermocouple can be used to measure temperature if you know the

relationship between temperature and voltage. Equations for common types of thermocouples are available, or you can simply take a few data points and prepare a calibration curve. This is especially easy for thermocouples, because, for small temperature ranges, the relationship between temperature and voltage is nearly linear.

Use a linear trendline to find the coefficients of a straight line through the data shown here:

T(°C)	V (mV)
10	0.397
20	0.798
30	1.204
40	1.612
50	2.023
60	2.436
70	2.851
80	3.267
90	3.682

Note: The thermocouple voltage changes because the temperature changes—that is, the voltage depends on the temperature. For regression, the independent variable (temperature) should always be on the x-axis, and the dependent variable (voltage) should be on the y-axis.

Conduction Heat Transfer

4. Thermal conductivity is a property of a material related to the material's ability to transfer energy by conduction. Materials with high thermal conductivities, such as copper and aluminum, are good conductors. Materials with low thermal conductivities are used as insulating materials.

    The next figure depicts a device that could be used to measure the thermal conductivity of a material. A rod of the material to be tested is placed between a resistance heater (on the right) and a block containing a cooling coil (on the left). Five thermocouples are inserted into the rod at evenly spaced intervals. The entire apparatus is placed under a bell jar, and the space around the rod is evacuated to reduce energy losses.

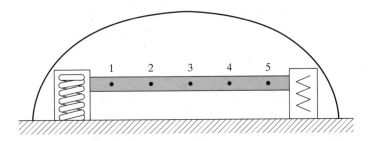

To run the experiment, a known amount of power is sent to the heater, and the system is allowed to reach steady state. Once the temperatures are steady, the power level and the temperatures are recorded. A data sheet from an experiment with the device is reproduced next.

Thermal Conductivity Experiment	
Rod Diameter: 2 cm	
Thermocouple Spacing: 5 cm	
Power: 100 watts	
THERMOCOUPLE	**TEMPERATURE (K)**
1	348
2	387
3	425
4	464
5	503

The thermal conductivity can be determined by using Fourier's law,

$$\frac{q}{A} = -k\frac{dT}{dx}, \tag{5.9}$$

where

q is the power applied to the heater
A is the cross-sectional area of the rod

The q/A is the energy flux and is a vector quantity, having both magnitude and direction. With the energy source on the right, the energy moves to the left or in the negative x-direction, so the flux in this problem is negative.

a. From the information supplied on the data sheet, prepare a table of position, x, and temperature T, values, in a spreadsheet, with the position of thermocouple 1 set as $x = 0$.

b. Graph the temperature-vs-position data, and add a linear trendline to the data. Ask Excel to display the equation of the line and the R^2 value.

c. Use Fourier's law and the slope of the regression line to compute the material's thermal conductivity.

Thermal conductivity is usually a function of temperature. Does your graph indicate that the thermal conductivity of this material changes significantly over the temperature range in this problem? Explain your reasoning.

Calculating Heat Capacity

5. The heat capacity at constant pressure is defined as

$$C_p = \left(\frac{\partial \hat{H}}{\partial T}\right)_p, \tag{5.10}$$

where

$\hat{H}$ is specific enthalpy
T is absolute temperature

C = Heat Capacity
P = Constant pressure

If enthalpy data are available as a function of temperature at constant pressure, the heat capacity can be computed. For steam, these data are readily available.[2] For example, the following table shows the specific enthalpy for various absolute temperatures at a pressure of 5 bars:

T	$\hat{H}$
(°C)	(kJ/kg)
200	2855
250	2961
300	3065
350	3168
400	3272
450	3379
500	3484
550	3592
600	3702
650	3813
700	3926
750	4040

a. Plot the specific enthalpy on the y-axis against absolute temperature on the x-axis.

b. Add a linear trendline to the data, and ask Excel to display the equation of the line and the R^2 value.

c. Compute the heat capacity of steam from the slope of the trendline.

d. The ideal gas (assumed) heat capacity for steam is approximately 2.10 kJ/kg K at 450°C. How does your result compare with this value?

e. Does it appear that the heat capacity of steam is constant over this temperature range? Why or why not?

Vapor–Liquid Equilibrium

6. When a liquid mixture is boiled, the vapor that leaves the vessel is enriched in the more volatile component of the mixture. The vapor and liquid in a boiling vessel are in equilibrium, and vapor–liquid equilibrium (VLE) data are available for many mixtures. The data are usually presented in tabular form, as in the table shown next, which represents VLE data for mixtures of methanol (MeOH) and ethanol (EtOH) boiling at 1 atm. From the graph of the VLE data, we see that, if a 50:50 liquid mixture of the alcohols is boiled, the vapor will contain about 60% methanol:

---

[2]The data are from a steam table in *Elementary Principles of Chemical Engineering*, 3d ed., by R. M. Felder and R. W. Rousseau, New York: Wiley, 2000. These data are also available at *http://www.coe.montana.edu/che/Excel*.

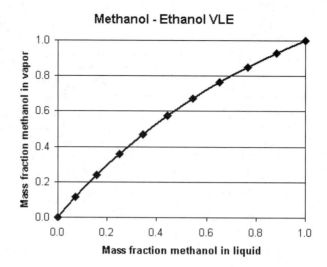

VLE data are commonly used in designing distillation columns, but an equation relating vapor mass fraction to liquid mass fraction is a lot more handy than tabulated values.

Use trendlines on the tabulated VLE data to obtain an equation relating the mass fraction of methanol in the vapor (y) to the mass fraction of vapor in the liquid (x). Test several different linear models (e.g., polynomials) to see which model gives a good fit to the experimental data.

Data The VLE values shown next were generated by using Excel, with the assumption that these similar alcohols form an ideal solution. The x-column represents the mass fraction of methanol in the boiling mixture. (Mass fraction of ethanol in the liquid is calculated as $1 - x$ for any mixture.) The mass fraction of methanol in the vapor leaving the solution is shown in the y-column.

These data are available in electronic form at the text's website *http:// www.coe.montana.edu/che/Excel*:

X_{MeOH}	Y_{MeOH}
1.000	1.000
0.882	0.929
0.765	0.849
0.653	0.764
0.545	0.673
0.443	0.575
0.344	0.471
0.250	0.359
0.159	0.241
0.072	0.114
0.000	0.000

Calculating Latent Heat of Vaporization

7. The Clausius–Clapeyron equation can be used to determine the latent heat (or enthalpy change) of vaporization, $\Delta \hat{H}_v$. The equation is

$$\ln(p_{\text{vapor}}) = -\frac{\Delta \hat{H}_v}{R\,T} + k, \tag{5.11}$$

where

p_{vapor} is the vapor pressure of the material,
R is the ideal gas constant,
T is the absolute pressure,
k is a constant of integration.

There are a few assumptions built into this equation:

1. The molar volume of the liquid must be much smaller than the molar volume of the gas (not true at high pressures).
2. The gas behaves as an ideal gas.
3. The latent heat of vaporization is not a function of temperature.

Water vapor data from the website of an Honors Chemistry class[3] is reported in the table shown next. A plot of the natural log of vapor pressure vs. T^{-1} should, if the assumptions are valid, produce a straight line on the graph.

a. Create a plot of natural log of vapor pressure of water on the y-axis, against T^{-1} on the x-axis.
b. Add a linear trendline, and have Excel show the equation of the line and the R^2 value.
c. What is the latent heat of vaporization of water, as computed from the slope of the trendline?
d. Does it look as if the assumptions built into the Clausius–Clapeyron equation are valid for this data set? Why, or why not?

T (°C)	p_{vapor}(mm Hg)	$\ln(p_{\text{vapor}})$
90	525.8	6.265
92	567.0	6.340
94	610.9	6.415
96	657.6	6.489
98	707.3	6.561
100	760.0	6.633
102	815.9	6.704
104	875.1	6.774
106	937.9	6.844
108	1004.4	6.912
110	1074.6	6.980

[3]Dr. Tom Bitterwolf's Honors Chemistry class at the University of Idaho. Data used with the permission of Dr. Bitterwolf.

Note: What do you do with the units on vapor pressure inside that natural logarithm? In this problem, it doesn't matter; the units you choose for vapor pressure change the value of k, but not the slope of the trendline.

Orifice Meter Calibration

8. Orifice meters are commonly used to measure flow rates, but they are highly nonlinear devices. Because of this, special care must be taken when preparing calibration curves for these meters. The equation relating volumetric flow rate, V, to the measured pressure drop, ΔP, across the orifice is

$$\dot{V} = \frac{A_O C_O}{\sqrt{1 - \beta^4}} \sqrt{\frac{2 g_c \Delta P}{\rho}}. \tag{5.12}$$

For purposes of creating a calibration curve, the details of the equation are unimportant (as long as the other terms stay constant). It is necessary that we see the theoretical relationship between flow rate and pressure drop:

$$\dot{V} \propto \sqrt{\Delta P}. \tag{5.13}$$

Also, the pressure drop across the orifice plate depends on the flow rate, not the other way around. So, the $\sqrt{\Delta P}$ should be regressed as the dependent variable (y-values) and the volumetric flow rate as the independent variable (x-values):

$\dot{V}$ (ft³ · MIN)	ΔP (psi)
3.9	0.13
7.9	0.52
11.8	1.18
15.7	2.09
19.6	3.27
23.6	4.71
27.5	6.41
31.4	8.37
35.3	10.59
39.3	13.08

a. Calculate $\sqrt{\Delta P}$ values at each flow rate from the tabulated data.
b. Regress $\dot{V}$ and $\sqrt{\Delta P}$, using Excel's regression package (found by selecting Tools/Data Analysis . . .) to create a calibration curve for this orifice meter.
c. Check the line-fit and residual plots to make sure your calibration curve really fits the data.

6

Iterative Solutions Using Excel

6.1 INTRODUCTION

Some equations are easy to rearrange and solve directly. For example,

$$PV = nRT \qquad (6.1)$$

is easily rearranged to solve for the volume of an ideal gas:

$$V = \frac{nRT}{P}. \qquad (6.2)$$

Sometimes, things are a little tougher, such as when you need to solve for the superficial velocity ($\overline{V}_0$) or porosity (ε) in the Ergun equation[1]:

$$\frac{\Delta p}{L} = \frac{150\,\overline{V}_0\mu}{\Phi_s^2 D_p^2}\cdot\frac{(1-\varepsilon)^2}{\varepsilon^3} + \frac{1.75\,\rho\overline{V}_0^2}{\Phi_s D_p}\cdot\frac{1-\varepsilon}{\varepsilon^3}, \qquad (6.3)$$

Here,

$\dfrac{\Delta p}{L}$ is the pressure drop per unit length of packing in a packed bed,

$\overline{V}_0$ is the superficial velocity of the fluid through the bed,

Φ_s is the sphericity of the particles in the bed,

D_P is the particle diameter,

OBJECTIVES

After reading this chapter, you will know

- How to find solutions, or roots, of equations by using several methods, including both graphing and iterative methods, such as direct substitution
- How to use in-cell iteration to find roots
- How to use Excel's Solver to solve for roots of equations
- How to include constraints with Excel's Solver
- How to solve optimization problems by using Excel's Solver
- How to solve a nonlinear regression problem by using Excel's Solver
- How to set up and solve linear-programming problems by using Excel's Solve

---

[1]McCabe, W. L., J. C. Smith, and P. Harriott, *Unit Operations of Chemical Engineering*, 5th ed., New York: McGraw-Hill, 1993, p. 154.

ε is the porosity of the packing,

μ is the absolute viscosity of the fluid flowing through the bed, and

ρ is the density of the fluid flowing through the bed.

There are times when it is easier to find a solution, or *root,* by using iterative methods, and sometimes they are the only way. Iterative techniques can be used to solve a variety of complex equations.

6.2   ITERATIVE SOLUTIONS

A common way of describing an *iterative solution* is to say it is a "guess and check" method. Nearly all iterative solution techniques require an *initial guess* to be provided by the user. Then the equation is solved, using the *guessed value*, and a result is calculated. A test is performed to see whether the guess is "close enough" to the correct answer. If not, a new guess value is used; the repeating process is called *iteration.* The key to efficient iterative solutions is a solver that is good at coming up with guess values. We will look at several methods:

1.   Using a Plot to Search for Roots.
2.   Guessing and Checking.
3.   Direct Substitution Method.
4.   In-Cell Iteration.
5.   Excel's Solver.

Excel Solver uses the Generalized Reduced Gradient (GRG2) nonlinear optimization code developed by Leon Lasdon, University of Texas at Austin, and Allan Waren, Cleveland State University in Ohio.[2] This technique is a good, general–purpose iterative solver with many applications.

6.2.1   Standard Forms

The equation to be solved should be put into a *standard form* if it is to be solved by iterative methods. Two standard forms will be used in this chapter. For example, consider $x^3 + 12 = 17x$:

FORM	EXAMPLE	CONVENIENT FOR...
(1) Set equation equal to zero	$x^3 - 17x + 12 = 0$	Plot method, Excel's Solver
(2) Get an x by itself on the left side	$x = \dfrac{x^3 + 12}{17}$	Direct-substitution method and in-cell iteration

---

[2]Excel 2000 help files, topic: Algorithm and methods used by Solver.

The direct-substitution method and *in-cell iteration* in Excel require form 2. Form 1 is usually more convenient for finding solutions when using graphs, and a version of form 1 is required when using the Solver. (The right-hand side may be any constant, not just zero.) Either form can be used if you just "guess and check" by hand until you find a solution.

6.2.2 Using a Plot to Search for Roots

If you have a fair idea that the root or roots of the equation should lie within a finite range of values, you can run a series of guesses in the range through your equation, solve for the computed values, plot the results, and look for the roots. The first standard form, with all terms on one side of the equation, works well when using a plot to search for roots. The equation listed previously can be written as

$$x^3 - 17x + 12 = 0 \tag{6.4}$$

in form 1. However, when we try guess values of x in the equation, the result will usually not equal zero (unless we have found a root). The equation might be written as

$$f(x) = x^3 - 17x + 12. \tag{6.5}$$

To search for roots, we will try various values of x (i.e. various guesses), solve for $f(x)$, and graph $f(x)$ vs. x. When $f(x) = 0$, we have found a root. First, we create a column of x-values in the range $0 \leq x \leq 5$:

	A	B	C	D	E
1	**Using a Plot to Search For Roots**				
2					
3		**x**			
4		0.0			
5		0.4			
6		0.8			
7		1.2			
8		1.6			
9		2.0			
10		2.4			
11		2.8			
12		3.2			
13		3.6			
14		4.0			
15		4.4			
16		4.8			
17		5.0			
18					

Then we compute $f(x)$ at each x-value:

	C4	▼	f_x =B4^3-17*B4+12		
	A	**B**	**C**	**D**	**E**
1	**Using a Plot to Search For Roots**				
2					
3		**x**	**f(x)**		
4		0.0	12.00		
5		0.4	5.26		
6		0.8	-1.09		
7		1.2	-6.67		
8		1.6	-11.10		
9		2.0	-14.00		
10		2.4	-14.98		
11		2.8	-13.65		
12		3.2	-9.63		
13		3.6	-2.54		
14		4.0	8.00		
15		4.4	22.38		
16		4.8	40.99		
17		5.0	52.00		
18					

Finally, we graph the $f(x)$- and x-values, using an XY Scatter plot:

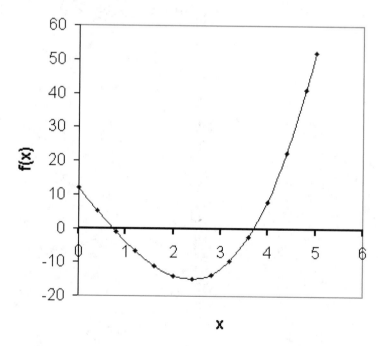

From the graph, it looks as if there are roots near $x = 0.8$ and $x = 3.7$. To get more accurate results, narrow the range of x values used. For example, we might use 20 x values between 0.6 and 1.0 to try to find out the value of the root near $x = 0.8$ more accurately.

The x^3 in the original equation suggests that there are three roots, but this graph shows only two. We should expand the range of x-values to search for the third root:

	A	B	C	D	E	F	G	H	I	J
1	Using a Plot to Search For Roots									
2										
3		x	f(x)							
4		-5.0	-28.00							
5		-4.5	-2.63							
6		-4.0	16.00							
7		-3.5	28.63							
8		-3.0	36.00							
9		-2.5	38.88							
10		-2.0	38.00							
11		-1.5	34.13							
12		-1.0	28.00							
13		-0.5	20.38							
14		0.0	12.00							
15		0.5	3.63							
16		1.0	-4.00							
17		1.5	-10.13							
18		2.0	-14.00							
19		2.5	-14.88							
20		3.0	-12.00							
21		3.5	-4.63							
22		4.0	8.00							
23										

From the new plot, it looks as if the additional root is near $x = -4.5$.

A plot is a fairly easy way to get approximate values for the roots, but, if you want more precise values, then numerical iteration techniques can be useful. These numerical iteration techniques require initial guesses, and a plot is a great way to get a good initial guess.

6.2.3 Simple "Guess and Check" Iteration

If you know that you have a root near $x = 0.8$, one of the easiest ways to find the root is simply to create a spreadsheet with a place to enter guess values and a formula that evaluates $f(x)$:

B4	▼	f_x	=B3^3-17*B3+12	
	A	B	C	D
1	Guess and Check			
2				
3	Guess:	0.8		
4	f(x):	-1.088		
5				

Then keep entering guesses until the $f(x)$-value is nearly zero (because the equation was written in form 1):

	A	B	C
1	Guess and Check		
2			
3	Guess:	0.728	
4	f(x):	0.009828	
5			

In the foregoing spreadsheet, $f(x)$ is getting close to, but is not, zero. Iterative methods can go on forever if you try to get $f(x)$ exactly to zero. Instead, you search for a root value that is close enough for your needs. Later, we will see how a *tolerance* value is used to judge whether the value is close enough. For now, realize that the guess value shown here is good to two decimal places; there is still inaccuracy in the third decimal place, the digit that is still being "guessed."

Sometimes, it is easier to keep track of the iteration process if you leave a trail of previous guesses and calculated values:

	A	B	C	D
1	**Guess and Check**			
2				
3		**Guess**	**f(x)**	
4		0.8	-1.088	
5		0.7	0.443	
6		0.75	-0.32813	
7		0.72	0.133248	
8		0.73	-0.02098	
9		0.725	0.056078	
10		0.727	0.025241	
11		0.728	0.009828	
12				

6.2.4 Writing Your Equation in Standard Form 2 ("Guess and Check" Form)

When your equation is in standard form 1, you keep trying guess values of x until $f(x)$ is close enough to zero. The alternative is to put your equation into standard form 2, sometimes called "guess and check" form. This simply requires getting one of the variables to be solved for by itself on the left side of the equation. The example equation can be rewritten in form 2 as

$$x = \frac{x^3 + 12}{17} \tag{6.6}$$

or

$$x = \sqrt[3]{17x - 12}. \tag{6.7}$$

Either version will work. We will use the first version.

The isolated variable is called the *computed value* and is renamed x_C. The x-value in the expression on the right side of the equation is called the guess value and is renamed x_G:

$$x_C = \frac{x_G^3 + 12}{17}. \tag{6.8}$$

During the solution process, x_G will be assigned values, and x_C will be computed. When x_C and x_G are equal, you have found a root. This equation actually has three roots ($x = -4.439$, 0.729, and 3.710), but only one can be found at a time. You can try to find all roots by starting the iteration process with different guess values, but not all iterative solution methods can find all roots for all equations.

With your equation in form 2, you keep trying x_G-values until x_C is equal (actually, "close enough") to x_G. The following spreadsheet shows the initial guess $x_G = 0.8$ and the initial computed value $x_C = 0.736$:

		B4	▼	f_x	=(B3^3+12)/17	
		A	B	C	D	
1	Guess and Check - Form 2				
2					
3		x_G:	0.8		
4		x_C:	0.736		
5					

After a few iterations, we are getting close to a root value:

	A	B	C	D
1	Guess and Check - Form 2			
2				
3	x_G:	0.728		
4	x_C:	0.728578		
5				

We can quantify how close we are getting to a root by calculating the difference between the guessed and computed values:

$$\Delta x = x_C - x_G. \tag{6.9}$$

Often a *precision setting*, or *tolerance* (e.g., TOL = 0.001) is used with the calculated difference to specify stopping criteria such as "Stop when $\Delta x <$ TOL":

		B6	▼	f_x	=B4-B3	
		A	B	C	D	
1	Guess and Check - Form 2				
2					
3		x_G:	0.728		
4		x_C:	0.728578		
5					
6		Δx:	0.000578		
7					

When you get tired of guessing values yourself, you might want to come up with an automatic guessing mechanism—that's what the remaining methods are all about.

6.2.5 Direct Substitution

The simplest automatic guessing mechanism is to use the previous computed value as the next guess value. Standard form 2 must be used to make this possible. This is called *direct substitution* of the computed value. Using the direct-substitution method, all you need to provide is the initial guess; the method takes it from there. The chief virtue of this method is simplicity, but there are roots that this method simply cannot find.

When the direct-substitution method is used in Excel, the original guess is still entered by hand. Here, it has been entered in cell A4:

B4			f_x =(A4^3+12)/17		
	A	B	C	D	E
1	**Direct Substitution Method**				
2					
3	x_G	x_C			
4	0.8	0.7360			
5					

The value in cell B4 is computed from the guess value in cell A4.

The key to the direct substitution method is using the previous computed value as the next guess value:

A5			f_x =B4		
	A	B	C	D	E
1	**Direct Substitution Method**				
2					
3	x_G	x_C			
4	0.8	0.7360			
5	0.7360				
6					

The "formula" for the new guess value in cell A5 is just =B4. To complete row 5, the formula in cell B4 is copied to cell B5 to calculate the new computed value:

B5			f_x =(A5^3+12)/17		
	A	B	C	D	E
1	**Direct Substitution Method**				
2					
3	x_G	x_C			
4	0.8	0.7360			
5	0.7360	0.7293			
6					

For more iterations, simply copy the cells in row 5 down the spreadsheet as far as necessary:

	A	B	C	D	E
1	**Direct Substitution Method**				
2					
3	x_G	x_C			
4	0.8	0.7360			
5	0.7360	0.7293			
6	0.7293	0.7287			
7	0.7287	0.7286			
8	0.7286	0.7286			
9	0.7286	0.7286			
10	0.7286	0.7286			
11					

As you can see in the preceding spreadsheet, the direct substitution method did find a root: $x = 0.7286$. Adding the difference calculation allows us to see how quickly this method found this root:

D10		f_x	=B10-A10	

	A	B	C	D	E
1	**Direct Substitution Method**				
2					
3	x_G	x_C		Δx	
4	0.8	0.7360		-0.06400	
5	0.7360	0.7293		-0.00667	
6	0.7293	0.7287		-0.00063	
7	0.7287	0.7286		-0.00006	
8	0.7286	0.7286		-0.00001	
9	0.7286	0.7286		0.00000	
10	0.7286	0.7286		0.00000	
11					

It took just a few steps to find this solution.

To try to find the root near $x = 3.7$, let's try an initial guess slightly larger than the expected root, say, $x_G = 3.8$:

	A	B	C	D	E
1	**Direct Substitution Method**				
2					
3	x_G	x_C		Δx	
4	3.8	3.9336		0.13365	
5	3.9336	4.2863		0.35268	
6	4.2863	5.3383		1.05197	
7	5.3383	9.6546		4.31628	
8	9.6546	53.6419		43.98736	
9	53.6419	########		########	
10	########	########		########	
11					

The pound symbols indicate that Excel cannot fit the number into the format specified for the cell. In this case, the numbers are getting too large: The direct substitution method is *diverging*. To try to get around this, try an initial guess on the other side of the root, say, $x_G = 3.6$:

	A	B	C	D	E
1	**Direct Substitution Method**				
2					
3	x_G	x_C		Δx	
4	3.6	3.4504		-0.14965	
5	3.4504	3.1221		-0.32822	
6	3.1221	2.4961		-0.62604	
7	2.4961	1.6207		-0.87540	
8	1.6207	0.9563		-0.66440	
9	0.9563	0.7573		-0.19897	
10	0.7573	0.7314		-0.02589	
11	0.7314	0.7289		-0.00253	
12	0.7289	0.7287		-0.00024	
13	0.7287	0.7286		-0.00002	
14	0.7286	0.7286		0.00000	
15					

This time the direct substitution method did find a root ($x = 0.7286$), but not the root near $x = 3.7$. Direct substitution simply cannot find the root at 3.71 from this equation (unless you are lucky enough to enter 3.71 as the initial guess).

6.2.6   In-Cell Iteration

In the preceding method, we caused Excel to repeatedly calculate (iterate) by copying row 5 down the spreadsheet. Excel can perform this iteration within a cell. To tell Excel to iterate within a cell, you use the value displayed in a cell in a formula in the *same* cell. Excel calls this a *circular reference*. Generally, you try to avoid circular references in spreadsheets, but if you want Excel to iterate within a cell, you must intentionally create a circular reference.

In-cell iteration requires your equation to be written in standard form 2, the guess-and-check form. Excel displays the result of the calculation (i.e., the computed value x_C). If we put the right side of our equation, $(x_G^3 + 12)/17$, into cell A5 and refer to this cell wherever the equation calls for x_G, we create the circular reference:

A5	▼		f_x	=(A5^3+12)/17		
	A	B	C	D	E	F
1	**In-Cell Iteration**					
2						
3	Cell A5 contains the formula and displays the final result.					
4						
5	0.0000					
6						

If you try this in Excel, you will probably get an error message saying there is a circular reference and Excel cannot solve the formula. In-cell iteration is not activated by default in Excel, because iteration slows down spreadsheet performance, and, most of the time, a circular reference indicates a data entry error, not an intentional request for an iterative solution. To activate iteration, select Tools/Options…

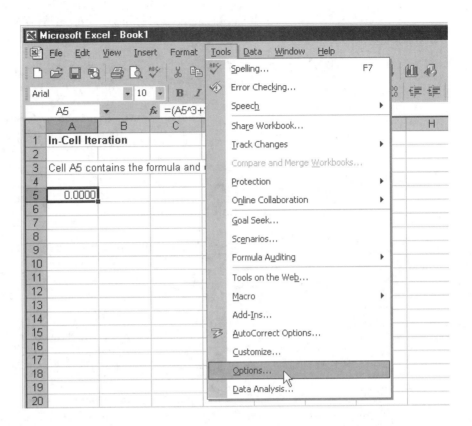

Then click on the Calculation tab on the Options dialog box:

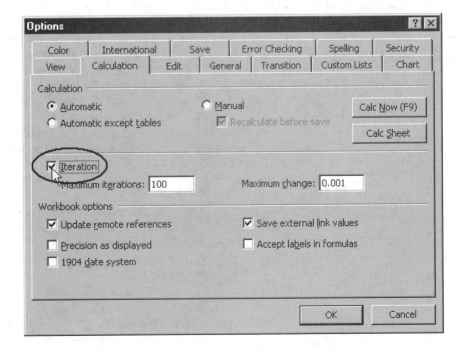

Activate iteration by checking the box labeled "Iteration". By default, Excel will iterate for no more than 100 cycles and will stop iterating when the computed value differs from the guess by less than 0.001. You can change these values, and solutions using this direct-substitution method could well require more than 100 iteration cycles.

Note: If the root is not found within the specified number of iterations, this method provides no feedback, but simply leaves the last result displayed in the cell. You should always check the result to make sure the displayed value actually satisfies your equation.

When you press the OK button to leave the Options dialog box, Excel iterates the formula containing the circular reference and displays the result:

	A5	▼		f_x	=(A5^3+12)/17	
	A	B	C	D	E	F
1	**In-Cell Iteration**					
2						
3	Cell A5 contains the formula and displays the final result.					
4						
5	0.7286					
6						

This approach has another very serious limitation: It always uses zero as the initial guess. That means it can find no more than one of the equation's roots. You cannot use this method to find multiple roots. Fortunately, Excel provides an alternative method: the *Solver*.

Note: If you are using Excel and following along in the text, you probably want to turn Iteration back off. Select Tools/Options... and then Calculation in the Options dialog box, and clear the Iteration check box.

6.3 INTRODUCTION TO EXCEL'S SOLVER

Excel provides an alternative to in-cell iteration that it calls the *Solver*. The Solver is not *installed* by default as part of the typical setup. Even if it was installed on your PC, it is not *activated* by default. So, you might need to activate the Solver (you can do this while running Excel), or you might have to go back to the Excel installation CD to install the Solver. If you install the Solver, you will then need to start Excel and activate the Solver.

If Solver... appears on the Tools menu, it has been installed and activated, and you are ready to use it. If it does not appear on the Tools menu, it has either not been installed or not been activated. First try activating the Solver, because this can be done without leaving Excel.

6.3.1   To Activate the Solver

Select Add-Ins... from the Tools menu, and choose Solver Add-In from the selection list on the Add-Ins window. Click OK to activate the Solver. If Solver Add-In does not appear in the Add-Ins selection list, it was not installed.

6.3.2   To Install the Solver

Run Setup from the Excel (or Office) installation CD, and choose Add or Remove Components (several steps into the installation). Use Change Options on Excel, and select the Solver. Complete the setup to finish the installation.

6.3.3 To Use the Solver

Once the Solver has been installed and activated, it can be used. The Solver requires that the equation be written in a variation of standard form 1; it allows you to leave a constant on the right side, not just zero. The sample equation in form 1 is $x^3 - 17x + 12 = 0$ and is ready to be used with the Solver.

The Solver requires an initial guess value. To begin using the Solver, first enter your initial guess value in a cell, and then enter the equation (actually, the left-hand side of the equation) into a second cell:

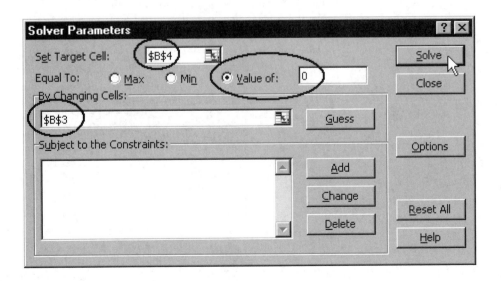

	B4	▼		f_x	=B3^3-17*B3+12	
	A		B		C	D
1	**Excel's Solver**					
2						
3	Guess:		3.6			
4	Formula:		-2.544			
5						

In the preceding spreadsheet, the initial guess (3.6) was entered into cell B3, and the formula

```
= B3^3-17*B3+12
```

was entered into cell B4. Excel evaluates the formula in cell B4 with the initial guess, but it does not iterate until you use the Solver.

Start the Solver by selecting Tools/Solver.... The Solver Parameters dialog will be displayed:

The cell references and value that must be changed have been circled on the figure. The *target cell* is the cell containing the formula: cell B4. When you use the button at the right of the Set Target Cell: field to return to the spreadsheet to select the target cell, Excel automatically adds dollar signs to the cell address to indicate an absolute cell address. This doesn't hurt anything, but absolute addressing is not required for simple problems.

We want the Solver to keep guessing until the target formula produces the result zero (because our equation was set equal to zero). We tell the Solver this by setting the "Equal To:" line to "Value of:" 0.

We tell the Solver where the guess value is located with the "By Changing Cells:" field. The value in this cell will be changed by the Solver during the solution process.

Note: The Solver can find roots, as we are using it here, and it can also be used to find maximum and minimum values, with and without constraints.

Once the required information has been set in the Solver Parameters dialog box, click on the Solve button to iterate for a solution. Once the iteration process is complete, Excel displays some options on how to handle the results:

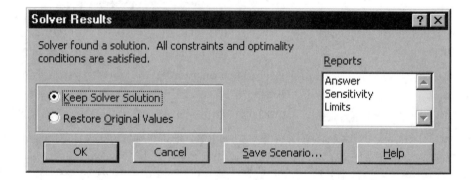

The last guess value is now in cell B3 (shown in the next spreadsheet), and you have the option either of leaving it there or of having Excel restore the original value. Excel will also prepare short reports (in a new spreadsheet) describing the answer, sensitivity, and limits (for problems with constraints). Press the OK button to return to the spreadsheet and see the solution:

	A	B	C	D
1	**Excel's Solver**			
2				
3	Guess:	3.710214		
4	Formula:	-1.6E-07		
5				

The Solver found the root at 3.71, and the formula now evaluates to a value almost, but not quite, zero (-1.6×10^{-7}). To search for other roots, change the initial guess and run the Solver again.

APPLICATIONS: ENGINEERING ECONOMICS

Internal Rate of Return on a Project

The Solver can be used to find the *internal rate of return* on a series of incomes and expenses (listed as negative incomes). Internal rates of return are a fairly common way of looking at the value of a proposed project:

- A negative internal rate of return on a proposed project indicates that the project will lose money and is not a good investment for a company.
- A small positive rate of return indicates the project will not make very much money, so the company might want to put its money somewhere else.
- A large positive internal rate of return is a good investment.

Common internal rate of return functions, like Excel's IRR() function, assume that incomes and expenses occur at regular intervals. When the incomes and expenses do not occur at regular intervals, you need to build the timing of the incomes and expenses into the calculation of the internal rate of return. The Solver and Excel's present value function, PV(), can be used to find the internal rate of return on non-uniformly-spaced incomes and expenses, such as of the following series of cash flows over a 10-year period:

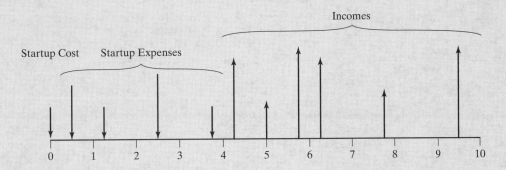

First, the times, incomes, and expenses are entered into a spreadsheet, along with an initial guess of the internal rate of return:

	A	B	C	D	E	F
1	**Internal Rate of Return - Non-Uniform Intervals**					
2		**IRR:**	2.0%	per year (guess value)		
3		**Net PV:**				
4						
5	**Time**	**Income**				
6	(years)					
7	0.00	-$375,000				
8	0.50	-$625,000				
9	1.25	-$375,000				
10	2.50	-$750,000				
11	3.75	-$375,000				
12	4.25	$1,000,000				
13	5.00	$500,000				
14	5.75	$1,125,000				
15	6.25	$1,000,000				
16	7.75	$625,000				
17	9.50	$1,125,000				
18						

The present value of each income or expense (indicated in the following spreadsheet as a negative income) is calculated by using Excel's PV() function. The arguments on the PV() function in cell C7 are the periodic interest rate (cell C2), the length of time between the present and the time of the income (cell A7), the periodic payment (0 in this example), and the amount of the future income (cell B7).

Note: The PV() function changes the sign of the result because it assumes that you will invest (an expense) at time zero to get a return (an income) in the future. In this case, we are using the PV() function just to move a future income to the present, so a minus sign is included on the future income in the formula to undo the sign change that the PV() function will introduce.

	C7		▼	fx =PV($C$2,A7,0,-B7)		
	A	B	C	D	E	F
1	**Internal Rate of Return - Non-Uniform Intervals**					
2		IRR:	2.0%	per year (guess value)		
3		Net PV:				
4						
5	Time	Income	PV			
6	(years)					
7	0.00	-$375,000	-$375,000			
8	0.50	-$625,000	-$618,842			
9	1.25	-$375,000	-$365,831			
10	2.50	-$750,000	-$713,774			
11	3.75	-$375,000	-$348,161			
12	4.25	$1,000,000	$919,283			
13	5.00	$500,000	$452,865			
14	5.75	$1,125,000	$1,003,926			
15	6.25	$1,000,000	$883,586			
16	7.75	$625,000	$536,079			
17	9.50	$1,125,000	$932,075			
18						

The net present value of the series of incomes and expenses is simply the sum of the PV values in cells C7 through C17:

	C3		▼	fx =SUM(C7:C17)		
	A	B	C	D	E	F
1	**Internal Rate of Return - Non-Uniform Intervals**					
2		IRR:	2.0%	per year (guess value)		
3		Net PV:	$2,306,205			
4						
5	Time	Income	PV			
6	(years)					
7	0.00	-$375,000	-$375,000			
8	0.50	-$625,000	-$618,842			
9	1.25	-$375,000	-$365,831			
10	2.50	-$750,000	-$713,774			
11	3.75	-$375,000	-$348,161			
12	4.25	$1,000,000	$919,283			
13	5.00	$500,000	$452,865			
14	5.75	$1,125,000	$1,003,926			
15	6.25	$1,000,000	$883,586			
16	7.75	$625,000	$536,079			
17	9.50	$1,125,000	$932,075			
18						

The internal rate of return is the interest rate that causes the present value of the incomes to just offset the present value of the expenses. That is, it is the interest rate that makes the net present value zero. We can ask the Solver to adjust the interest rate (cell C2) until the net present value (cell C3) is zero:

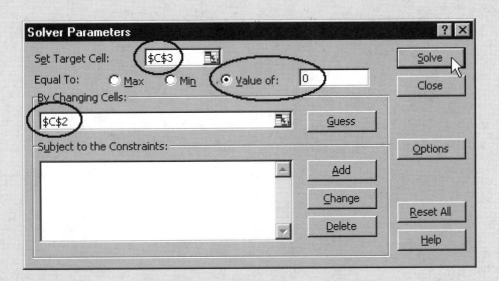

	A	B	C	D	E	F
1	**Internal Rate of Return - Non-Uniform Intervals**					
2		**IRR:**	17.4%	per year (guess value)		
3		**Net PV:**	$0			
4						
5	**Time**	**Income**	**PV**			
6	(years)					
7	0.00	-$375,000	-$375,000			
8	0.50	-$625,000	-$576,731			
9	1.25	-$375,000	-$306,735			
10	2.50	-$750,000	-$501,795			
11	3.75	-$375,000	-$205,225			
12	4.25	$1,000,000	$505,000			
13	5.00	$500,000	$223,821			
14	5.75	$1,125,000	$446,399			
15	6.25	$1,000,000	$366,154			
16	7.75	$625,000	$179,814			
17	9.50	$1,125,000	$244,299			
18						

This cash flow has an internal rate of return of 17.4%, so it is not a bad investment.

6.4 OPTIMIZATION USING THE SOLVER

The Solver can also be used for *optimization* problems. As a simple example, consider the equation

$$y = 10 + 8x - x^2. \tag{6.10}$$

This equation is graphed in the following figure and has the maximum 26 at $x = 4$:

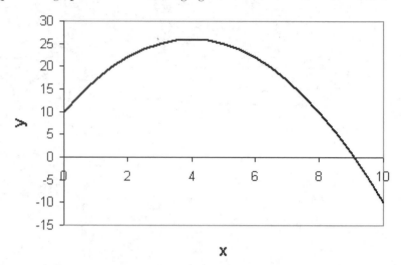

To find the maximum by using the Solver, the right side of the equation is entered in one cell and a guess for x in another:

	B4		f_x =10+8*B3-B3^2	
	A	B	C	D
1	**Optimization Using the Solver**			
2				
3	Guess:	2		
4	Equation:	22		
5				

Then the Solver is asked to find the maximum value by varying the value of x in cell B3:

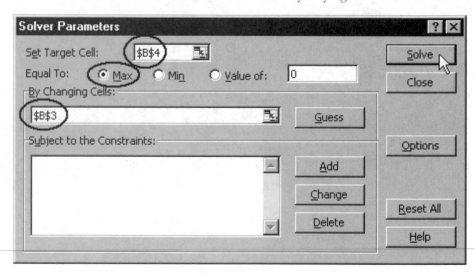

	B3	▼	f_x	3.99999997656181

	A	B	C	D
1	**Optimization Using the Solver**			
2				
3	Guess:	4		
4	Equation:	26		
5				

The Solver's result was not quite 4, (3.99999997656181). The Solver stopped when the guessed and computed values were within a predefined precision. The desired precision can be changed by using the Options button in the Solver Parameters dialog.

Adding Constraints   If you want the maximum value of y with $x \leq 3$, a *constraint* must be added to the system. This is done by using the constraints box on the Solver Parameters dialog box:

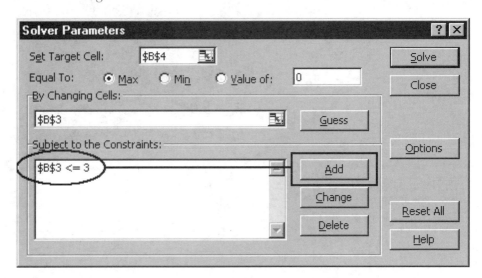

To include this constraint, click on the Add button at the right of the constraints box. This brings up the Add Constraint dialog box:

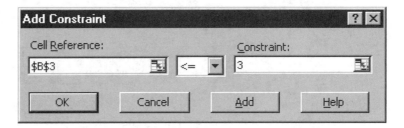

With the constraint, the maximum y-value is 25 at the x-value 3:

	A	B	C	D
1	**Optimization Using the Solver**			
2				
3	Guess:	3		
4	Equation:	25		
5				

Solving for Multiple Values   The Solver will change values in multiple cells to try to find your requested result. For example, the function $f(x, y) = \sin(x) \cdot \cos(y)$ has a maximum at $x = 1.5708$ (or $\pi/2$) and $y = 0$ in the following region ($-1 \le x \le 2$, $-1 \le y \le 2$):

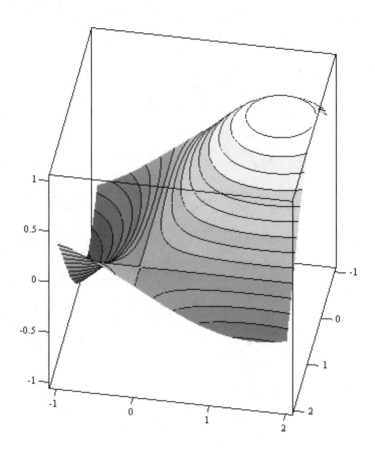

To use the Solver to find the values of x and y that maximize $f(x, y)$, enter guesses for x and y and the equation to be solved:

```
B10:=SIN(B8)*COS(C8)
```

	B10	▼	f_x =SIN(B8)*COS(C8)		
	A	B	C	D	E
6					
7		x	y		
8	Guesses:	0	0		
9					
10	f(x, y):	0			
11					

Then ask the Solver to find the values of x and y (by Changing Cells B8:C8) that maximize the value in cell B10 (which contains the function). The constraints limit the search region to $-1 \le x \le 2$, $-1 \le y \le 2$; otherwise, there are an infinite number of solutions (the Solver would find only one):

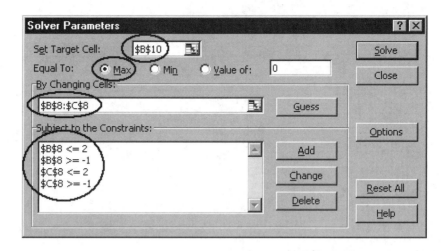

The solution that the Solver finds is the true maximum at $x = 1.5708$ (or $\pi/2$) and $y = 0$:

	A	B	C	D	E
6					
7		**x**	**y**		
8	**Guesses:**	1.570796	-3.2E-08		
9					
10	**f(x, y):**	1			
11					

This function also has a local maximum at $x = -1, y = 2$. We can try guess values near this local maximum to see whether the Solver can be fooled into finding the local maximum rather than the true maximum over the constraint region:

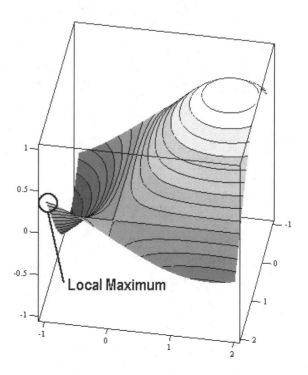

Guesses of $x = -0.9$ and $y = 1.8$ caused the Solver to find the following solution:

	A	B	C	D	E
6					
7		x	y		
8	Guesses:	-1	2		
9					
10	f(x, y):	0.350175			
11					

We fooled the Solver! Remember to test your solutions; visualization (such as the three-dimensional graph used here) is a good idea as well.

6.5 NONLINEAR REGRESSION

The idea behind *regression analysis* is to find coefficient values that minimize the total, unsigned (squared) error between a regression model (a function) and a data set. Linear-regression models can be solved directly for the coefficients. When the regression model is nonlinear in the coefficients, an iterative minimization approach can be used to find the coefficients. Excel's Solver can be used to solve the minimization problem.

The topic of *nonlinear regression* will be presented with an example: fitting experimental vapor pressure data to Antoine's equation. Data on the vapor pressure of water at various temperatures are easy to find. The data graphed here are from a website prepared by an honors chemistry class at the University of Idaho and have been used with the permission of the instructor, Dr. Tom Bitterwolf. The units on vapor pressure are millimeters of mercury, and temperature is in degrees Celsius:

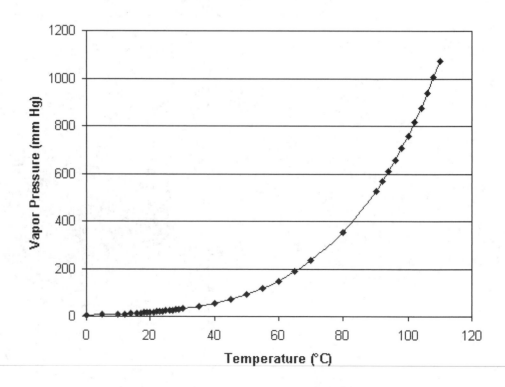

Note: Two data points, one at 150°C ($p = 3570.4$ mm Hg) and one at 200°C ($p* = 11659.2$ mm Hg), were left off this graph, to allow the curvature at the lower temperatures to be seen more clearly. These data points were included in the regression analysis that follows.*

There are several forms of Antoine's equation for vapor pressure in use. One common form is

$$\log(p*) = A - \frac{B}{C + T},$$ (6.11)

where A, B, and C are the coefficients used to fit the equation to vapor pressure ($p*$) data over various temperature (T) ranges. This version of Antoine's equation uses a base-10 logarithm. Antoine's equation is a dimensional equation; you must use specific units with Antoine's equation. The units come from the data used in the regression analysis used to find the coefficients.

The first step in solving for the Antoine equation coefficients is getting the data and initial guesses for the coefficients into a spreadsheet. Here, arbitrary values of 1 for each coefficient were used. If you have estimates for the size of each coefficient, you can use them. Better guesses will speed up the iteration, but the arbitrary guesses provide a good test of the Solver's abilities:

	A	B	C	D	E	F
1	**Vapor Pressure of Water**					
2		**Guesses**				
3	**A:**	1				
4	**B:**	1				
5	**C:**	1				
6						
7	**T (°C)**	**p* (mm Hg)**				
8	0	4.58				
9	5	6.54				
10	10	9.21				
11	12	10.52				
12	14	11.99				
13	16	13.63				
14	17	14.53				
15	18	15.48				
16	19	16.48				

Note: This data set contains 42 data points; only a few will be shown in the figures.

Regression analysis is designed to minimize the *sum of the squared error* (SSE) between the data values, y, and the *predicted values* calculated from the regression equation, y_p:

$$\text{SSE} = \sum_{i=1}^{N}(y_i - y_{p_i})^2.$$ (6.12)

In this example, we want to minimize the SSE between the logarithm of the experimental vapor pressure values and the value of $\log(p^*)$ predicted by the Antoine equation.

The next step is to take the logarithm of the experimental vapor pressures to generate the y-values for the regression:

C8		f_x =LOG(B8)			
A	B	C	D	E	F
1 Vapor Pressure of Water					
2	Guesses				
3 A:	1				
4 B:	1				
5 C:	1				
6					
7 T (°C)	p* (mm Hg)	y			
8 0	4.58	0.661			
9 5	6.54	0.816			
10 10	9.21	0.964			
11 12	10.52	1.022			
12 14	11.99	1.079			
13 16	13.63	1.134			
14 17	14.53	1.162			
15 18	15.48	1.190			
16 19	16.48	1.217			

Then calculate $\log(p^\circ)$, using the guesses in the Antoine equation, to create a column of y_P-values (predicted values):

D8		f_x =$B$3-($B$4/($B$5+A8))			
A	B	C	D	E	F
1 Vapor Pressure of Water					
2	Guesses				
3 A:	1				
4 B:	1				
5 C:	1				
6					
7 T (°C)	p* (mm Hg)	y	y_P		
8 0	4.58	0.661	0.000		
9 5	6.54	0.816	0.833		
10 10	9.21	0.964	0.909		
11 12	10.52	1.022	0.923		
12 14	11.99	1.079	0.933		
13 16	13.63	1.134	0.941		
14 17	14.53	1.162	0.944		
15 18	15.48	1.190	0.947		
16 19	16.48	1.217	0.950		

Each *error* is defined as the difference between the corresponding set of y-values:

$$\text{error}_i = y_i - y_{p_i} \qquad (6.13)$$

	E8	▼	f_x =C8-D8			
	A	B	C	D	E	F
1	**Vapor Pressure of Water**					
2		**Guesses**				
3	**A:**	1				
4	**B:**	1				
5	**C:**	1				
6						
7	**T (°C)**	**p* (mm Hg)**	**y**	**y_P**	**error**	**error²**
8	0	4.58	0.661	0.000	0.661	0.437
9	5	6.54	0.816	0.833	-0.018	0.000
10	10	9.21	0.964	0.909	0.055	0.003
11	12	10.52	1.022	0.923	0.099	0.010
12	14	11.99	1.079	0.933	0.145	0.021
13	16	13.63	1.134	0.941	0.193	0.037
14	17	14.53	1.162	0.944	0.218	0.047
15	18	15.48	1.190	0.947	0.242	0.059
16	19	16.48	1.217	0.950	0.267	0.071

If we just summed the individual errors, many of the positive errors would cancel out negative errors. To prevent this, the sum of the squared errors, or SSE, is calculated. This has been done by using Excel's SUM() function in cell D3 in the following spreadsheet:

	D3	▼	f_x =SUM(F8:F49)			
	A	B	C	D	E	F
1	**Vapor Pressure of Water**					
2		**Guesses**		**SSE**		
3	**A:**	1		69.347		
4	**B:**	1				
5	**C:**	1				
6						
7	**T (°C)**	**p* (mm Hg)**	**y**	**y_P**	**error**	**error²**
8	0	4.58	0.661	0.000	0.661	0.437
9	5	6.54	0.816	0.833	-0.018	0.000
10	10	9.21	0.964	0.909	0.055	0.003
11	12	10.52	1.022	0.923	0.099	0.010
12	14	11.99	1.079	0.933	0.145	0.021
13	16	13.63	1.134	0.941	0.193	0.037
14	17	14.53	1.162	0.944	0.218	0.047
15	18	15.48	1.190	0.947	0.242	0.059
16	19	16.48	1.217	0.950	0.267	0.071

Finally, we ask the Solver to find the values of A, B, and C that minimize the SSE:

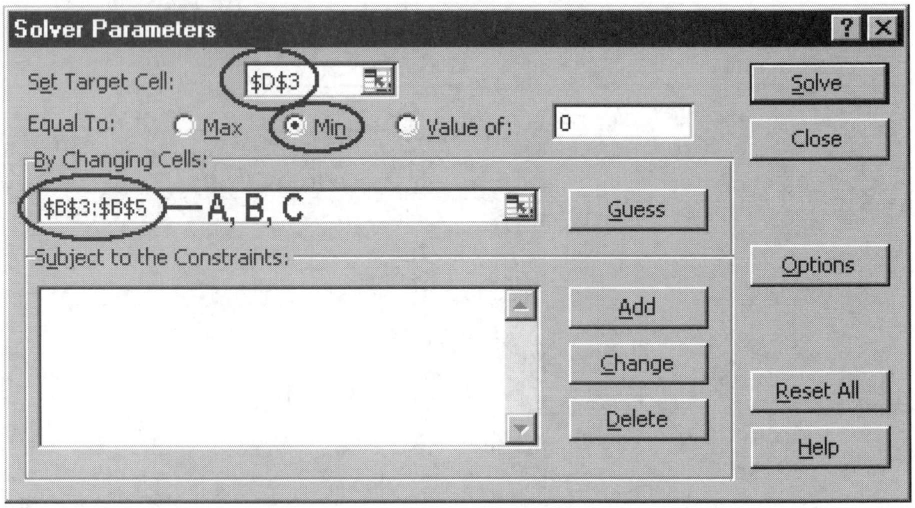

The Solver varies the values of A, B, and C (in cells B3 through B5) to find the minimum value of the SSE (in D3):

			D3		▼		f_x =SUM(F8:F49)		

	A	B	C	D	E	F
1	**Vapor Pressure of Water**					
2		**Guesses**		**SSE**		
3	**A:**	7.93428		0.000		
4	**B:**	1654.28				
5	**C:**	227.307				
6						
7	T (°C)	p* (mm Hg)	y	y_P	error	error^2
8	0	4.58	0.661	0.657	0.004	0.000
9	5	6.54	0.816	0.813	0.002	0.000
10	10	9.21	0.964	0.963	0.001	0.000
11	12	10.52	1.022	1.021	0.001	0.000
12	14	11.99	1.079	1.079	0.000	0.000
13	16	13.63	1.134	1.135	-0.001	0.000
14	17	14.53	1.162	1.163	-0.001	0.000
15	18	15.48	1.190	1.191	-0.001	0.000
16	19	16.48	1.217	1.218	-0.001	0.000

The minimum value of SSE was zero to the four significant figures displayed in cell D3, and the coefficient values were found to be

$$A = 7.93428,$$
$$B = 1654.28,$$
$$C = 227.307. \tag{6.14}$$

Reported values for Antoine coefficients for water are the following[3]:

---

[3]Lange's *Handbook of Chemistry*, 9th ed., Sandusky, OH: Handbook Publishers, Inc., 1956.

```
A = 7.96681   0.41% difference,
B = 1668.21   0.83% difference,
C = 228.00    0.30% difference.
```

However, the reported coefficients are for temperatures between 60 and 150°C; the data set contains values for temperatures between 0 and 200°C. One of the homework problems for this chapter asks you to reanalyze the portion of the data set between 60 and 150°C to allow a better comparison of the Solver's regression results and the published regression coefficients.

Summary of Regression Analysis Using the Solver The basic steps for using the Solver to perform a regression are the following:

1. Select a regression model (Antoine's equation in this example).
2. Enter the data set and initial guesses for the regression-model coefficients (A, B, and C).
3. Calculate the y-values for the regression model if needed. (They are frequently part of the data set.)
4. Calculate the y_P-values, using the regression model and the x-values from the data set.
5. Calculate the error values (error = $y - y_P$).
6. Calculate the squared errors.
7. Calculate the sum of the squared errors (SSE) by using the SUM() function.
8. Ask the Solver to minimize the SSE by changing the values of the coefficients in the regression model.
9. Test your result.

6.5.1   Testing the Regression Result

The last step, testing the result, is always a good idea after any regression analysis. The coefficients that give the "best" fit of the regression model to the data might still do a pretty poor job. The usual ways to test the result are to calculate a *coefficient of determination*, R^2 and then to plot the original data with the predicted values.

The coefficient of determination, usually called the R^2 value, is computed from the sum of the squared errors (SSE) and the *total sum of squares* (SSTo). SSTo is defined as

$$\text{SSTo} = \sum_{i=1}^{N}(y_i - \bar{y})^2, \tag{6.15}$$

where

y_i is the y-value for the ith data point in the data set, with i ranging from 1 to N,
$\bar{y}$ is the arithmetic average of the y-values in the data set and
N is the number of values in the data set.

The R^2 value is calculated as

$$R^2 = 1 - \frac{\text{SSE}}{\text{SSTo}}. \tag{6.16}$$

For a perfect fit, every predicted value is identical to the data set value, and SSE = 0. So, for a perfect fit, $R^2 = 1$.

In the spreadsheet, the average of the y-values is computed by the AVERAGE() function:

	E3		f_x =AVERAGE(C8:C49)				
	A	B	C	D	E	F	G
1	Vapor Pressure of Water						
2		Guesses		SSE	AVG	SSTo	R^2
3	A:	7.93428		0.000	1.959		
4	B:	1654.28					
5	C:	227.307					
6							
7	T ($^{\circ}$C)	p* (mm Hg)	y	y_P	error	error^2	
8	0	4.58	0.661	0.657	0.004	0.000	
9	5	6.54	0.816	0.813	0.002	0.000	
10	10	9.21	0.964	0.963	0.001	0.000	
11	12	10.52	1.022	1.021	0.001	0.000	

Then, the squared differences between the y-values and the average y-value are computed (column G), and the total sum of squares is computed (cell F3):

	F3		f_x =SUM(G8:G49)				
	A	B	C	D	E	F	G
1	Vapor Pressure of Water						
2		Guesses		SSE	AVG	SSTo	
3	A:	7.93428		0.000	1.959	29.159	
4	B:	1654.28					
5	C:	227.307					
6							
7	T ($^{\circ}$C)	p* (mm Hg)	y	y_P	error	error^2	$(y-y_{avg})^2$
8	0	4.58	0.661	0.657	0.004	0.000	1.6855
9	5	6.54	0.816	0.813	0.002	0.000	1.3078
10	10	9.21	0.964	0.963	0.001	0.000	0.9898
11	12	10.52	1.022	1.021	0.001	0.000	0.8782

The SSTo value in cell F3 was computed by using the SUM() function the values in column G:

F3:=SUM(G8:G49)

Finally, the R^2 value is computed from the SSE and SSTo values:

	G3	▼	f_x	=1-D3/F3		
	A	B	C	D	E	F	G
1	**Vapor Pressure of Water**						
2		**Guesses**		**SSE**	**AVG**	**SSTo**	R^2
3	**A:**	7.93428		0.000	1.959	29.159	0.999995
4	**B:**	1654.28					
5	**C:**	227.307					
6							
7	**T (°C)**	**p* (mm Hg)**	**y**	**y$_P$**	**error**	**error^2**	**(y-y$_{avg}$)^2**
8	0	4.58	0.661	0.657	0.004	0.000	1.6855
9	5	6.54	0.816	0.813	0.002	0.000	1.3078
10	10	9.21	0.964	0.963	0.001	0.000	0.9898
11	12	10.52	1.022	1.021	0.001	0.000	0.8782

The R^2 value of 0.999995 is very close to one, suggesting a good fit.

Using Plots to Validate Your Regression Results   Test the fit by using graphs; there are several that may be used.

1.   Plot the original data and the predicted values on the same graph.
The common practice is to graph data values by using markers and predicted values as a curve. This allows you to see whether the regression results go through the data points:

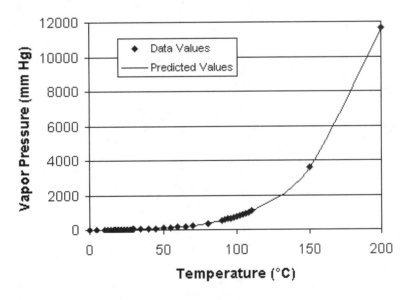

When the original vapor pressures and the vapor pressures calculated by using the Antoine equation with the regressed coefficients are plotted together, the values appear to agree. It looks as if the regression worked.

2.   Plot the original data against the predicted values.
In this plot, vapor pressure (data) is plotted against vapor pressure (predicted) (temperature has not been plotted):

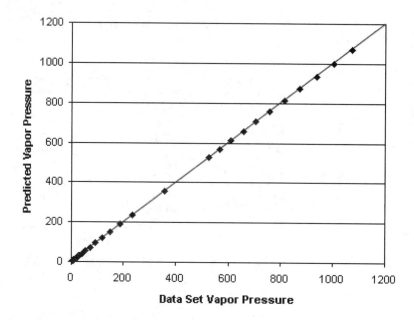

When plotted like this, the data points should follow a diagonal line across the graph. (The diagonal line was created by plotting y against y. The plot of y_p against y is shown by the data markers.) A poor regression fit will show up as scatter around the diagonal line, or patterns. If there were patterns in the scatter (e.g., more scatter at one end of the line, curvature), you might want to try a different regression model to see whether you could get a better fit.

3. Deviation Plot

In the field of regression analysis, it is common to refer to the difference between the data value (y) and the predicted value (y_P) as error, but, in other fields, this difference is referred to as a *deviation*. (The term error is often reserved for the difference between the data value and a known, or true value.) When these differences are plotted, the graph typically is called a *deviation plot*:

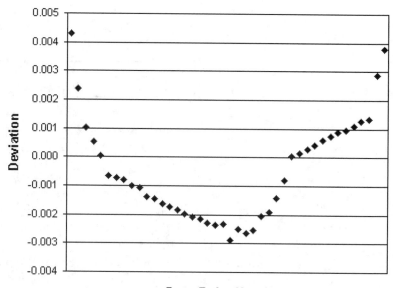

In this line plot of the deviations $[\log(p*) - \log(p*_P)]$, there is a clear pattern in the deviation values. The magnitude of the deviations is small, less than one percent of the $\log(p*)$ values, but if a high-precision fit is needed, you might want to look for a better regression model.

APPLICATIONS: THERMODYNAMICS

Bubble Point Temperature I

When a liquid mixture is heated at constant pressure, the temperature at which the first vapor bubble forms (incipient boiling) is called the *bubble point temperature*. At the bubble point, the liquid and vapor are in equilibrium, and the total pressure on the liquid is equal to the sum of the partial pressures of the components in the mixture, so

$$P = \sum_{i=1}^{N_{comp}} p_i \qquad (6.17)$$

where

P is the total pressure on the liquid,
p_i is the partial pressure of the ith component in the mixture, and
N_{comp} is the number of components in the mixture.

If the solution can be considered ideal, the partial pressures can be related to the vapor pressures of the components by

$$P = \sum_{i=1}^{N_{comp}} x_i p_i^* \qquad (6.18)$$

where

x_i is the mole fraction of component i in the mixture and
p_i^* is the vapor pressure of the ith component in the mixture.

There are good correlating equations, such as Antoine's equation, for vapor pressure as a function of temperature:

$$\log(p_i^*) = A_i - \frac{B_i}{T_{bp} + C_i} \qquad (6.19)$$

Solving for the vapor pressure and substituting into the preceding summation yields the fairly involved equation

$$P = \sum_{i=1}^{N_{comp}} x_i \cdot 10^{\left[A_i - \frac{B_i}{T_{bp} + C_i}\right]} \qquad (6.20)$$

where

A_i, B_i, C_i are the Antoine equation coefficients for component i and
T_{bp} is the bubble point temperature of the mixture

This equation can be used to find the bubble point temperature if the composition (x-values) and total pressure (P) are known. This is a fairly unpleasant summation to solve by hand, but it is a good candidate for Excel's Solver. We will ask the Solver to keep trying T_{bp}-values until the summation is equal to the total pressure.

Consider the following mixture of similar components (i.e., ideal solution is a reasonable assumption):[4]

COMPONENT	MOLE FRACTION	ANTOINE COEFFICIENTS		
		A	B	C
1-Propanol	0.3	7.74416	1437.686	198.463
1-Butanol	0.2	7.36366	1305.198	173.427
1-Pentanol	0.5	7.18246	1287.625	161.330

Use Excel's Solver to estimate the bubble point temperature for this mixture at a total pressure equal to 2 atmospheres (1520 mm Hg).

First, the known values are entered into a spreadsheet:

	A	B	C	D	E	F	G	H
1	**Bubble Point Temperature I**							
2								
3		**Pressure:**						
4		**T_bp:**						
5								
6								
7			**Mole Fraction**	**A**	**B**	**C**		
8		**Propanol:**	0.3	7.74416	1437.686	198.463		
9		**Butanol:**	0.2	7.36366	1305.198	173.427		
10		**Pentanol:**	0.5	7.18246	1287.625	161.330		
11								

Next, a guess for the bubble point temperature is entered; from that, the vapor pressures can be calculated:

G8		f_x	=10^(D8-E8/($C$4+F8))					
	A	B	C	D	E	F	G	H
1	**Bubble Point Temperature I**							
2								
3		**Pressure:**						
4		**T_bp:**	100	°C (guess)				
5								
6							**p***	
7			**Mole Fraction**	**A**	**B**	**C**	**(mm Hg)**	
8		**Propanol:**	0.3	7.74416	1437.686	198.463	845.7	
9		**Butanol:**	0.2	7.36366	1305.198	173.427	389.2	
10		**Pentanol:**	0.5	7.18246	1287.625	161.330	180.0	
11								

The total pressure can then be calculated by using the vapor pressures and the mole fractions and the equation

$$P = \sum_{i=1}^{N_{comp}} x_i p_i^*$$ (6.21)

---

[4]Data are from *Elementary Principles of Chemical Processes*, 3d ed., by R. M. Felder and R. W. Rousseau, New York: Wiley, 2000.

Thus, we get

	A	B	C	D	E	F	G	H
	C3	▼	*fx* =(C8*G8)+(C9*G9)+(C10*G10)					
1	**Bubble Point Temperature I**							
2								
3		**Pressure:**	421.5	mm Hg				
4		**T_bp:**	100	°C (guess)				
5								
6								**p***
7			**Mole Fraction**	**A**	**B**	**C**	(mm Hg)	
8		**Propanol:**	0.3	7.74416	1437.686	198.463	845.7	
9		**Butanol:**	0.2	7.36366	1305.198	173.427	389.2	
10		**Pentanol:**	0.5	7.18246	1287.625	161.330	180.0	
11								

The calculated total pressure is nowhere near the 1520 mm Hg we want, so we will ask Excel's Solver to try other temperature values until the desired total pressure is found:

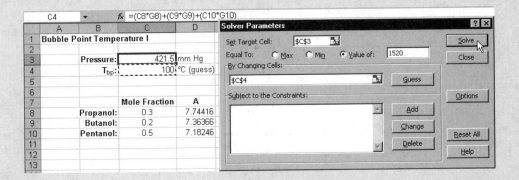

The final result is shown next. This mixture is expected to produce a vapor pressure equal to 1520 mm Hg (2 atm) at the temperature 137°C:

	A	B	C	D	E	F	G	H
1	**Bubble Point Temperature I**							
2								
3		**Pressure:**	1520.0	mm Hg				
4		**T_bp:**	137.0	°C				
5								
6								**p***
7			**Mole Fraction**	**A**	**B**	**C**	(mm Hg)	
8		**Propanol:**	0.3	7.74416	1437.686	198.463	2877.1	
9		**Butanol:**	0.2	7.36366	1305.198	173.427	1444.2	
10		**Pentanol:**	0.5	7.18246	1287.625	161.330	736.0	
11								

6.6   LINEAR PROGRAMMING

An engineering problem encountered fairly commonly is finding the maximum or minimum of some function, subject to one or more constraints. As a simple example that can be graphed to help visualize the problem, consider the minimization problem

Minimize: $3x_1 - 5x_2$

subject to the following constraints that place the solution within a two-dimensional region:

$$x_1 \geq 4$$
$$x_1 \leq 8$$
$$x_2 \geq 2$$
$$x_2 \leq 5$$

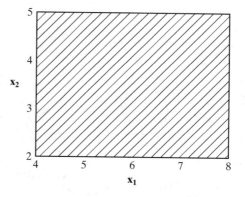

The following additional constraints further limit the solution domain to a band across the region:

$$x_2 \leq \tfrac{1}{2}x_1 + 1$$
$$x_2 \geq \tfrac{1}{4}x_1 + 1$$

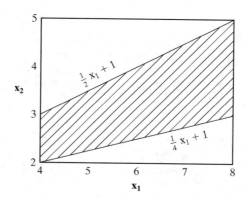

A linear function to be minimized subject to a series of linear constraints is a *linear programming* problem. A common nomenclature for linear-programming problems is this:

Minimize:	$[c]^T[x]$	(objective function)
Subject to:	$[A][x] \leq [b]$	(general constraints)
	$[x] \geq 0$	(nonnegativity constraint)

In this form, the function to be optimized (called the *objective function*) must be written as a minimization function. Additionally, each constraint must be written as a less-than-or-equal-to statement.

The example used here, written in this form, would look like this:

Minimize: $3x_1 - 5x_2$ (objective function—already a minimization function)

$$-x_1 \leq -4$$
$$x_1 \leq 8$$

Subject to: $-x_2 \leq -2$ (constraints written as less-than-or-equal-to statements)

$$x_2 \leq 5$$
$$-\tfrac{1}{2}x_1 + x_2 \leq 1$$
$$\tfrac{1}{4}x_1 - x_2 \leq -1$$

Nonnegativity constraints are not required in this problem, because positive values are enforced by the other constraints.

The vectors required to write the objective function as $[c]^T[x]$ are

$$[c] = \begin{bmatrix} 3 \\ 5 \end{bmatrix} \qquad [x] = \begin{bmatrix} x_1 \\ x_2 \end{bmatrix}.$$

The coefficient matrix for the constraints, $[A]$, and the constraint right-hand-side vector, $[b]$ are

$$[A] = \begin{bmatrix} -1 & 0 \\ 1 & 0 \\ 0 & -1 \\ 0 & 1 \\ \tfrac{1}{2} & -1 \\ \tfrac{1}{4} & -1 \end{bmatrix} \qquad [b] = \begin{bmatrix} -4 \\ 8 \\ -2 \\ 5 \\ -1 \\ -1 \end{bmatrix}.$$

Often, it is convenient to write the linear-programming problem in matrix form, but that preparation is not necessary in order to solve the problem in Excel. Furthermore, Excel's Solver, like most modern linear-programming packages, does not require that all of the constraints be written as less-than-or-equal-to statements. So, for this example, we will drop the matrices and simplify the constraints. The problem can then be written as follows:

Minimize: $3x_1 - 5x_2$ (objective function)

$$x_1 \geq 4$$
$$x_1 \leq 8$$

Subject to: $x_2 \geq 2$ (constraints)

$$x_2 \leq 5$$
$$\tfrac{1}{2}x_1 - x_2 \geq -1$$
$$\tfrac{1}{4}x_1 - x_2 \leq -1$$

This can be entered into an Excel spreadsheet as

	D6	▼		f_x =3*D3-5*D4				
	A	B	C	D	E	F	G	H	I

	A	B	C	D	E	F	G	H	I
1	Linear Programming								
2							initial feasible solution		
3		Decision Variables:	x_1:	6				Equations Used	
4			x_2:	3.5					
5									
6		Objective Function:		0.5				=3*D3-5*D4	
7									
8		Constraints:		6	>=	4		=D3	
9				6	<=	8		=D3	
10				3.5	>=	2		=D4	
11				3.5	<=	5		=D4	
12				-0.5	>=	-1		=(1/2)*D3-D4	
13				-2	<=	-1		=(1/4)*D3-D4	
14									

Initial feasible values for each decision variable (i.e., each x-value) must be provided. Providing starting values is standard practice any time an iterative solution process is used. The cells containing the values of each decision variable are used in cell D6 in the definition of the objective function ($=3°D3-5°D4$).

Cells D8:D13 also use the values of the decision variables to evaluate each constraint. The formulas used in cells D8:D13 are listed in cells H8:H13. The operators listed in column E and the values listed in column F are not used in the solution, but are included to make the constraints easier to read. The actual constraints are entered into the Solver as part of the solution process.

Once the problem has been set up with an initial feasible solution, as shown previously, start the Solver, using Tools/Solver ... and begin entering parameters in the Solver Parameters dialog:

- Set the target cell to the cell containing the evaluated objective function, cell D6.
- Select Min to minimize the objective function.
- Indicate the cells containing the decision variable values in the By Changing Cells: field (cells D3:D4)

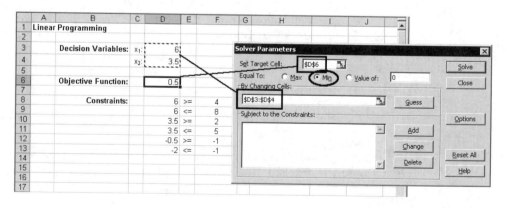

Then, begin to add constraints by clicking the Add button.

The first constraint is entered by indicating the cell reference that contains the formula that uses the decision variables (cells D3:D4) to evaluate the constraint. The first constraint cell reference is cell D8, which contains the formula $= D3$—that is, it contains the value of x_1. The constraint is $x_1 \geq 4$. The greater-than-or-equal-to operator is selected from the drop-down list of operators on the Add Constraint dialog, and the constraint value, 4, is entered into the Constraint field. (Or, you can reference cell F8 in the Constraint field.) These steps are captured in the following spreadsheet:

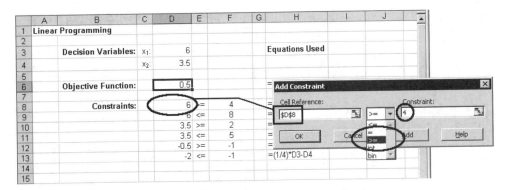

To conclude the first constraint and add another, click the Add button on the Add Constraint dialog.

In the same manner, the second constraint is entered by indicating the cell reference that contains the formula used to evaluate that constraint, cell D9. The less-than-or-equal-to operator is selected from the list, and the constraint value, 8, is entered in the Constraint field:

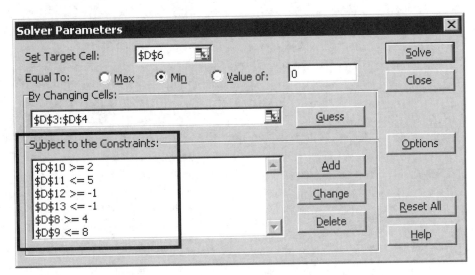

Each of the six constraints is added in the same way. After entering all of the constraints, click the OK button to return to the Solver Parameters dialog. The six constraints are shown in the dialog:

Next, click the Options button to set additional Solver options. In particular, select a linear model:

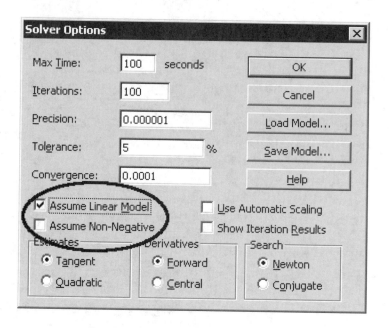

You can also enforce nonnegativity constraints by checking the Assume Non-Negative box on the Solver Options dialog. (This is not required for this example because the other constraints force nonnegative results.) Click OK to return to the Solver Parameters dialog:

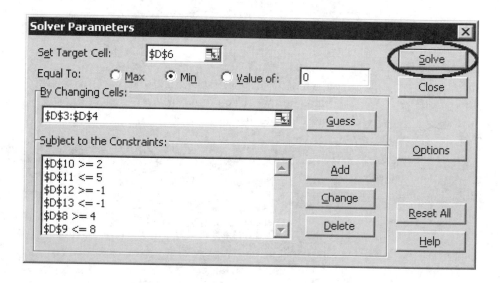

Click the Solve button to solve the linear-programming problem. The Solver then varies the decision values in cells D3 and D4 to minimize the objective function in cell D6. The results are displayed in the spreadsheet, and a Solver results window is displayed. Because iterative solutions can fail, the Solver Results window gives you an opportunity to restore the original values if needed:

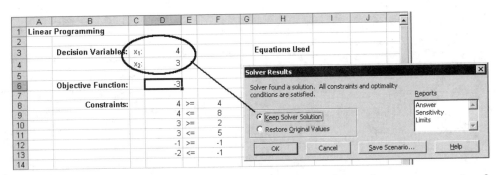

Here, the optimal solution is at the top-left corner of the solution domain, at $x_1 = 4$ and $x_2 = 3$:

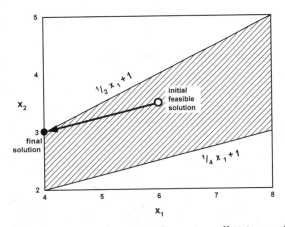

The minimum value of the objective function, shown in cell D6, was found to be -3. Also, in cells D8 through D13, each constraint has been evaluated with the optimal values of x_1 and x_2, and each constraint has been satisfied. Now that it appears that an optimal solution has been found, click OK to keep the Solver solution.

Many linear-programming problems are easy to set up and solve by using Excel's Solver; however, there are a couple of limitations. The Solver that comes with Excel

- is limited to 200 decision variables (we used 2 in this example), and
- cannot handle linear-programming problems where some or all of the decision variables are restricted to integer values.

Enhanced Solvers that can handle much larger problems and integer problems can be purchased from the company that created the Solver, if needed.

APPLICATIONS: LINEAR PROGRAMMING

As a slightly more complex linear-programming example, consider the problem of trying to find the least expensive way to meet five dietary minimum requirements. We may state the problem as follows:

Minimize: cost of food
Subject to: At least

- 1.5 mg thiamin (each day)
- 1.7 mg riboflavin
- 20 mg niacin
- 400 μg folic acid
- 60 mg vitamin C

The available foods and typical nutrient values (per serving) are listed in the following table:

FOOD #	NAME	THIAMIN	RIBOFLAVIN	NIACIN	FOLIC ACID	VITAMIN C	COST
		MG	MG	MG	μG	MG	$
1	Rice	0	0	4	11	0	0.25
2	Cabbage	0	0	0	23	124	0.17
3	Spinach	0	0.1	0	123	28	0.35
4	Peanuts	0.4	0	20	20	0	0.95
5	Or. Juice	0.2	0	0	109	64	0.75
6	Tuna	0	0.1	10	0	0	1.45
7	Turkey	0	0	10	0	0	0.85
8	Pork	0.8	0	6	0	0	1.85
9	Bread	0	0.8	0	0	0	0.15
10	Lima Beans	0.4	0	0	78	9	0.35

This problem has enough decision variables that matrix methods are helpful. The decision variables are stored in vector $[x]$, where x_1 through x_{10} represent the number of servings of each food. The $[c]$ vector is the vector of cost per serving for each food in dollars. The constraint right-hand-side vector, $[r]$, represents the daily minimum requirement of each nutrient. Then,

$$[c] = \begin{bmatrix} 0.25 \\ 0.17 \\ 0.35 \\ 0.95 \\ 0.75 \\ 1.45 \\ 0.85 \\ 1.85 \\ 0.15 \\ 0.35 \end{bmatrix} \qquad [r] = \begin{bmatrix} 1.5 \\ 1.7 \\ 20 \\ 400 \\ 60 \end{bmatrix}$$

Note: The units on the values in $[r]$ are mg per day (except for folic acid, which is in μg per day).

The $[A]$ matrix represents the constraint-coefficient matrix. Each of these matrices is shown in the following spreadsheet (initial guesses of 1 for each x have been included in the decision variable cells, O7:O16):

In matrix form, the problem statement is

Minimize: $[c]^T[x]$ (objective function—minimize total cost)
Subject to: $[A][x] \geq [b]$ (constraints—must meet or exceed daily minimum for each nutrient)

In the spreadsheet, the objective function is in cell O5:

```
O5: =MMULT(C5:L5,O7:O16)
```

This is an array function, so the function was concluded by pressing [Ctrl-Shift-Enter], and cell O5 contains the formula =MMULT(C5:L5, O7:O16), where the braces indicate an array.

The constraints are evaluated in cells C13 through C17. This is accomplished by multiplying matrices [A] and [x]. To do this, the five cells containing constraint evaluations (C13:C17) were selected, then the matrix multiplication function was entered as

```
=MMULT(C7:L11,O7:O16)
```
         this is =MMULT([A], [X])

followed by [Ctrl-Shift-Enter] to finish the array function and store {=MMULT (C7:L11,O7:O16)} in each cell in the [A] [x] vector (cells C13:C17).

At this point, the spreadsheet is using the initial guesses for each x to evaluate the objective function in cell O5 and the constraints in cells C13:C17. The initial guesses indicate that the total cost is $7.12 each day and that the riboflavin and folic acid requirements are not being met—that is, the 1s in the [x] vector do not represent an initial feasible solution. The Solver might be able to find a solution anyway.

The next step is to use the Solver to attempt to find optimal values for each x that minimize the total cost while satisfying each constraint. Start the Solver, using Tools/Solver … to open the Solver Parameters dialog.

The Set Target Cell field points to the cell containing the objective function, cell O5. Minimization is requested, and the decision variables in cells O7:O16 are indicated in the By Changing Cells field.

Next, the constraints must be entered. All five constraints are greater-than-or-equal-to constraints, so all five constraints can be entered at once. To do this, click Add to open the Add Constraint dialog.

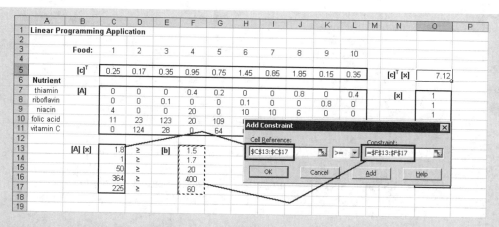

All five cells that contain constraint evaluations, cells C13:C17, are indicated in the Cell Reference field, and all five constraint values (in cells F13:F17) are indicated in the Constraint field. The greater-than-or-equal-to operator is selected from the drop-down operator list. The OK button is used to return to the Solver Parameters dialog:

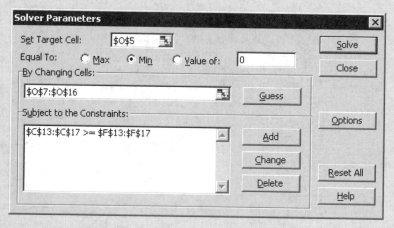

The Options button is used to open the Solver Options dialog so that a linear model can be requested, one with nonnegative decision variables (the nonnegativity constraint prevents the Solver from attempting to minimize cost by using negative servings):

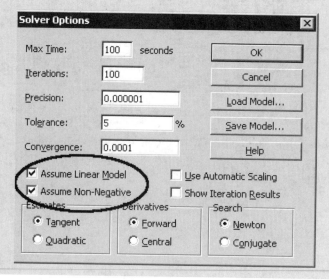

The OK button is used to return to the Solver Parameters dialog. Clicking the Solve button causes the Solver to seek a new set of decision variables that minimizes the objective function while satisfying each constraint—that is, the Solver looks for the number of servings of each food that meets or exceeds the minimum daily requirement for each nutrient at minimum cost:

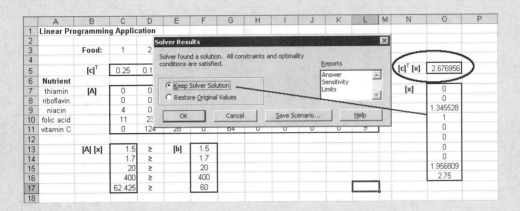

The Solver's optimum solution costs $2.68 each day and includes the following foods:

FOOD #	NAME	SERVINGS
1	Rice	0
2	Cabbage	0
3	Spinach	1.35
4	Peanuts	1
5	Or. Juice	0
6	Tuna	0
7	Turkey	0
8	Pork	0
9	Bread	1.96
10	Lima Beans	2.75

Notice that the Solver did not return integer numbers of servings—the version of the Solver that comes with Excel does not solve integer programming models—but portions of servings are certainly possible, so this noninteger solution is valid for this problem.

Only four of the foods are actually required to meet all of the minimum daily requirements. These four foods provide the following nutrients:

NUTRIENT	AMT. FROM FOOD	DAILY REQUIREMENT
	mg	mg
Thiamin	1.5	1.5
Riboflavin	1.7	1.7
Niacin	20	20
Folic Acid	0.400	0.400
Vitamin C	62.4	60

All of the constraints have been met minimally except for Vitamin C, which is slightly in excess of the daily minimum.

This is the minimum cost solution that provides the required amounts of each nutrient; it might be acceptable, unless you really don't like one of the foods. If that is the case, you might want to consider optimizing on a different objective function–one related to taste rather than cost. That is the subject of one of the homework problems at the end of this chapter.

KEY TERMS

Bubble point temperature	Guessed value	Predicted value
Circular reference	In-cell iteration	Regression analysis
Coefficient of determination (R^2)	Internal rate of return	Roots
Computed value	Iteration	Standard form (of equation
Constraint	Iterative solution	to be solved)
Deviation	Linear programming	Sum of the squared error (SSE)
Deviation plot	Nonlinear regression	Solver
Direct substitution	Objective function	Target cell
Diverging	Optimization	Tolerance
Error	Precision setting	Total sum of squares (SSTo)

SUMMARY

Equation Standard Forms

Form 1 Rearrange the equation to get all of the terms on one side, equal to zero:

$$x^3 - 17x + 12 = 0. \tag{6.22}$$

Form 2 Get one instance of the desired variable on the left side of the equation.

$$x = \frac{x^3 + 12}{17}. \tag{6.23}$$

This form is often called "guess and check" form, with the calculated variable x_C by itself and the guessed variable x_G included in the right side of the equation:

$$x_G = \frac{x_G^3 + 12}{17}. \tag{6.24}$$

Iterative Methods

DIRECT SUBSTITUTION

Put the equation into "guess and check" form (form 2) with a calculated value of the desired quantity on the left and a function of the desired quantity (the guess value) on the right:

$$x_C = f(x_G).$$

With the direct-substitution method, the calculated value at the end of one iteration becomes the guess value for the next iteration.

Benefits This is easy to do in a spreadsheet.

Drawbacks It tends to diverge and cannot find all roots.

IN-CELL ITERATION

Use "guess and check" form (form 2) for the equation. The $f(x_G)$ goes in the cell, and the x_C is the displayed result of the calculation.

In-cell iteration is not enabled by default. Enable iteration by selecting Tools/ Options/ and Iterations from the calculation panel.

Benefits   It is straightforward.

Drawbacks   It gives no indication of whether it stopped iterating because it found a root or because it reached the maximum number of iterations. (Check your results!) And, it always uses the initial guess zero and so cannot find multiple roots.

SOLVER

Needs a constant on one side of the equation. (Basically, form 1 is used, but it can have any constant on the right, not just zero.)

This capability might not be installed or activated by default. Installation requires the Excel Installation CD. To activate it, select Tools/Add-Ins and check the Solver Add-In box in the list of available add-ins. (If Solver Add-In is not listed, it needs to be installed.)

Benefits   The Solver is easy to use, powerful, and fast, can handle constraints, and can maximize and minimize.

Optimization by Using the Solver   Select Max or Min on the "Equal to": line of the Solver Parameters dialog box.

Include constraints by clicking the Add button on the Subject to Constraints frame. Constraints must be constructed by using the Add Constraint dialog box.

To optimize on multiple values, include multiple cell references in the By Changing Cells field.

NonlinearRegression   Nonlinear regression is an optimization problem that seeks to minimize the sum of the squared error, SSE, between dependent values predicted by a regression model (y_P) and those from the data set (y):

$$\text{SSE} = \left[\sum_{i=1}^{N} (y_i - y_{p_i})^2 \right]. \tag{6.25}$$

Here,

y_p is the value predicted by the regression model and
y is the data set value.

The model coefficients go in the By Changing Cells field, and the cell containing the sum of the squared errors (SSE) is the target.

Linear Programming   In linear-programming problems, a linear objective function is optimized subject to one or more linear constraints. Because all of the equations involved are linear, the problem can be written concisely in matrix form.

Minimize: $[c]^T[x]$ (objective function)
Subject to: $[A][x] \leq [b]$ (constraints)

Additionally, the decision variables are often constrained to be nonnegative values.

Linear-programming problems are readily solved by using Excel's Solver with the following parameters:

FIELD	CONTENTS
Target Cell	Objective Function, $[c]^T[x]$
Equal To	Min
By Changing Cells	Decision Variables, $[x]$ *(Initial values for each decision variable must be specified in the spreadsheet.)*
Subject to the Constraints	Constraints, $[A][x] \leq [b]$ *(Nonnegativity can be imposed using the Options button and selecting Assume Non-Negative on the Solver Options dialog.)*

Problems

Iterative Solutions I

1. The three solutions to the following equation, $x = 1$, $x = 3$, and $x = 7$, are evident by inspection:

$$(x - 1)(x - 3)(x - 7) = 0. \tag{6.26}$$

The same equation can be written as a polynomial:

$$x^3 - 11x^2 + 31x - 21 = 0. \tag{6.27}$$

a. Use Excel's Solver to find the three solutions to this polynomial.

b. Use the Solver to find the solutions to the following polynomial:

$$x^3 - 10x^2 + \underset{30x}{35x} - 27 = 0. \tag{6.28}$$

Iterative Solutions II

2. A graph of a J_1 Bessel function between $x = 0$ and 10 shows that there are three roots (solutions) to the equation:

$$J_1(x) = 0. \tag{6.29}$$

The solutions are at $x = 0$ and near $x = 4$ and $x = 7$. Use the Solver with Excel's BESSELJ() function to find more precise values for the roots near 4 and 7.

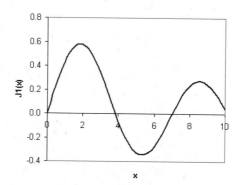

Nonideal Gas Equation

3. Determine the molar volume $\hat{V}$ of ammonia at 300°C and 1,200 kPa by using

a. the ideal gas equation

$$P\hat{V} = RT; \tag{6.30}$$

b. the Soave-Redlich-Kwong (SRK) equation[5] and Excel's Solver:

$$P = \frac{RT}{\hat{V} - b} - \frac{\alpha a}{\hat{V}(\hat{V} + b)} \tag{6.31}$$

---

[5]From *Elementary Principles of Chemical Processes*, 3d ed., R. M. Feldar and R. W. Rousseau, New York: Wiley, 2000.

Here,
α, a, and b are coefficients for the SRK equation, defined as

$$a = 0.42747 \frac{(RT_C)^2}{P_C},$$

$$b = 0.08664 \frac{RT_C}{P_C},$$ (6.32)

$$\alpha = \left[1 + m\left(1 - \sqrt{T_r}\right)\right]^2,$$

$$m = 0.48508 + 1.55171\,\omega - 0.1561\,\omega^2,$$

$$T_r = \frac{T}{T_c},$$

P is the absolute pressure (1200 kPa),
P_C is the critical pressure of the gas (11,280 kPa for ammonia),
T is the absolute temperature (300 C + 273 = 573 K),
T_C is the critical temperature of the gas (405.5 K for ammonia),   (6.33)
ω is the Pitzer acentric factor for the gas (0.250 for ammonia),
$\hat{V}$ is the molar volume (liters per mole)
R is the ideal gas constant [8.314 (liter Pa)/ (mole K)].

Internal Rate of Return

4.   Use Excel's NPV() function with an arbitrary (guess) interest rate to calculate the net present value of the following cash flow. Then, use Excel's Solver to find the interest rate that makes the net present value zero.

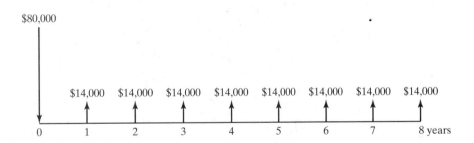

Bubble Point Temperature II

5.   Compute the bubble point temperature for a mixture containing 10 mole % 1-propanol, 70 mole % 1-butanol, and 20 mole % 1-pentanol at a pressure equal to 1000 mm Hg.

Bubble Point Temperature III

6.   Use Excel's Solver to estimate the bubble point temperature of a mixture of four hydrocarbons at 3 atmospheres pressure (2280 mm Hg). Composition and other required data for the mixture are tabulated as follows:[6]

---

[6]Data are from *Elementary Principles of Chemical Processes*, 3d ed., by R. M. Felder and R. W. Rousseau, New York: Wiley, 2000.

COMPONENT	MOLE FRACTION	ANTOINE COEFFICIENTS		
		A	B	C
n-Octane	0.1	6.98174	1351.756	209.100
i-Octane	0.5	6.88814	1319.529	211.625
n-Nonane	0.3	6.93764	1430.459	201.808
n-Decane	0.1	6.95707	1503.568	194.738

Nonlinear Regression

7. Rework the nonlinear regression example used in this chapter (fitting the Antoine equation to vapor pressure data) using only data in the temperature range $60 \leq T \leq 150°C$. (The complete data set is listed next, only a portion is needed for this problem.)

    a. Use the Solver to find the best-fit values for the coefficients in Antoine's equation (A, B, and C).

    b. Compare your results with the following published values:[7]

T (°C)	P* (MM HG)	T (°C)	P* (MM HG)	COEFFICIENT	PUBLISHED VALUE
0	4.58	40	55.3	A	7.96681
5	6.54	45	71.9	B	1668.21
10	9.21	50	92.5	C	228.000
12	10.52	55	118		
14	11.99	60	149.4		
16	13.63	65	187.5		
17	14.53	70	233.7		
18	15.48	80	355.1		
19	16.48	90	525.8		
20	17.54	92	567		
21	18.65	94	610.9		
22	19.83	96	657.6		
23	21.07	98	707.3		
24	22.38	100	760		
25	23.76	102	815.9		
26	25.21	104	875.1		
27	26.74	106	937.9		
28	28.35	108	1004.4		
29	30.04	110	1074.6		
30	31.82	150	3570.4		
35	42.2	200	11659.2		

Nonlinear Regression

8. A decaying sine wave is a common waveform. It occurs in oscillating circuits, controlled process responses, and wave dynamics. A decaying sine wave may have the form

$$y_p = e^{-Ax} \sin(Bx)$$

---

[7]Lange's Handbook of Chemistry, 9th ed., Sandusky, OH: Handbook Publishers, Inc., 1956.

where A and B are model coefficients, typically obtained by regression analysis. Now

- Perform a nonlinear regression on the data shown next to compute the values of coefficients A and B.
- Plot the regression equation with the original data values to verify visually that the regression model fits the data.
- Calculate the R^2 value for the regression result.

The plotted data values are listed in the table that follows, but they are also available at the text website,

http://www.coe.montana.edu/che/excel

The values on the website can be imported into your Excel spreadsheet:

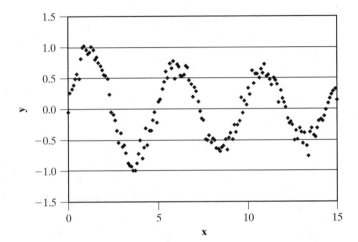

Data Values for Problem 8

X	Y	X	Y	X	Y	X	Y	X	Y
0.0	−0.05	3.0	−0.60	6.0	0.73	9.0	−0.38	12.0	0.10
0.1	0.26	3.1	−0.58	6.1	0.69	9.1	−0.49	12.1	0.02
0.2	0.32	3.2	−0.72	6.2	0.55	9.2	−0.26	12.2	−0.15
0.3	0.39	3.3	−0.88	6.3	0.53	9.3	−0.38	12.3	−0.24
0.4	0.50	3.4	−0.92	6.4	0.56	9.4	−0.26	12.4	−0.21
0.5	0.56	3.5	−0.93	6.5	0.70	9.5	−0.19	12.5	−0.25
0.6	0.49	3.6	−1.00	6.6	0.68	9.6	0.18	12.6	−0.50
0.7	0.81	3.7	−0.99	6.7	0.47	9.7	−0.07	12.7	−0.35
0.8	1.00	3.8	−0.88	6.8	0.40	9.8	0.13	12.8	−0.49
0.9	1.02	3.9	−0.72	6.9	0.20	9.9	0.07	12.9	−0.31
1.0	0.96	4.0	−0.51	7.0	0.35	10.0	0.33	13.0	−0.38
1.1	0.90	4.1	−0.80	7.1	0.28	10.1	0.23	13.1	−0.55
1.2	0.92	4.2	−0.62	7.2	0.11	10.2	0.63	13.2	−0.39
1.3	1.00	4.3	−0.31	7.3	−0.04	10.3	0.41	13.3	−0.59
1.4	0.95	4.4	−0.58	7.4	−0.15	10.4	0.57	13.4	−0.76
1.5	0.80	4.5	−0.35	7.5	−0.19	10.5	0.57	13.5	−0.37
1.6	0.84	4.6	−0.35	7.6	−0.50	10.6	0.50	13.6	−0.31
1.7	0.74	4.7	−0.18	7.7	−0.51	10.7	0.65	13.7	−0.43
1.8	0.70	4.8	−0.06	7.8	−0.43	10.8	0.59	13.8	−0.45
1.9	0.63	4.9	−0.22	7.9	−0.54	10.9	0.73	13.9	−0.31
2.0	0.56	5.0	0.12	8.0	−0.48	11.0	0.53	14.0	−0.19

Data Values for Problem 8 (Continued)

2.1	0.54	5.1	0.15	8.1	−0.50	11.1	0.56	14.1	−0.17
2.2	0.49	5.2	0.32	8.2	−0.63	11.2	0.48	14.2	−0.20
2.3	0.23	5.3	0.43	8.3	−0.64	11.3	0.39	14.3	−0.12
2.4	−0.05	5.4	0.60	8.4	−0.68	11.4	0.50	14.4	−0.02
2.5	−0.09	5.5	0.51	8.5	−0.63	11.5	0.46	14.5	0.09
2.6	−0.19	5.6	0.73	8.6	−0.59	11.6	0.40	14.6	0.16
2.7	−0.34	5.7	0.66	8.7	−0.49	11.7	0.11	14.7	0.23
2.8	−0.54	5.8	0.78	8.8	−0.66	11.8	0.29	14.8	0.28
2.9	−0.39	5.9	0.49	8.9	−0.46	11.9	0.17	14.9	0.32

Linear Programming

9. Solve the following linear-programming problem:

Maximize: $12 x_1 + 21 x_2 + 7 x_3 + 15 x_4$

Subject to:

$$4 x_1 + 3 x_2 + 2 x_3 + 5 x_4 \leq 400$$
$$2 x_1 + 1.5 x_2 + 3.3 x_3 + 2 x_4 \leq 150$$
$$x_1 \geq 45$$
$$x_2 \leq 100$$
$$x_1, x_2, x_3, x_4 \geq 0 \quad \text{(nonnegative)}$$

Rework the linear-programming application example presented in this chapter with a new objective function. This time, the objective will be to minimize a "distaste factor" rather than cost. The distaste factor can have a value between 1 and 10, where

1—I don't mind this food at all; serve it up!
4—I can eat some of it, but tire of it quickly.
8—I would prefer not to eat this food.
10—I'll gag if I have to eat this food.

The distaste factors are obviously specific to each individual, but a set of distaste factors has been assigned to each food in the following table:

FOOD #		1	2	3	4	5	6	7	8	9	10
	NAME	RICE	CABBAGE	SPINACH	PEANUTS	OR. JUICE	TUNA	TURKEY	PORK	BREAD	LIMA BEANS
Nutrient											
Thiamin	mg	0	0	0	0.4	0.2	0	0	0.8	0	0.4
Riboflavin	mg	0	0	0.1	0	0	0.1	0	0	0.8	0
Niacin	mg	4	0	0	20	0	10	10	6	0	0
Folic Acid	μg	11	23	123	20	109	0	0	0	0	78
Vitamin C	mg	0	124	28	0	64	0	0	0	0	9
Distaste Factor		1	4	3	1	1	2	1	3	1	7
Cost	$	0.25	0.17	0.35	0.95	0.75	1.45	0.85	1.85	0.15	0.35

(a.) Solve the linear programming problem to minimize distaste factor while still satisfying each of the nutritional constraints listed in the application example. How many servings of each type of food are needed, and what is the overall distaste level?

(b.) Create your own set of distaste factors for these foods and repeat part (a.).

7

Using Macros in Excel

7.1 INTRODUCTION

Macros are usually short programs or sets of recorded keystrokes (stored as a program) that can be reused as needed. There are basically two reasons for writing a macro:

- Excel doesn't automatically do what you need.
  The macros you write can be used to extend Excel's capabilities. You can use Excel's functions and any you write in your macros.
- You need to perform the same task (usually a multistep task) over and over again. Complex operations can be combined in a macro and invoked with a simple keystroke.

There are two ways to create a macro in Excel: You can record a set of keystrokes or write a Visual Basic for Applications (VBA)® subprogram. Programming macros gives you a lot of control over the result, but takes more effort than recording.

7.1.1 Formulas, Functions, Macros, VBA Programs: Which One and When?

Excel provides a lot of ways to get work done:

- Formulas
- Functions
- Macros
- VBA Programs

There are situations in which each of these approaches is preferred over the others, but there is a lot of overlap—situations where more than one approach could be used. In some cases, any of these approaches could be used. The following discussion is intended to provide some very general guidelines for selecting the most appropriate approach for some common types of problems.

OBJECTIVES

After reading this chapter, you will know

- What an Excel macro is and how one can be created and used
- How macros have been used to create computer viruses, and the precautions that Excel users should take to protect their work from macro viruses
- How to create recorded macros
- How to write simple macros directly by using Excel's built-in Visual Basic programming environment

Formulas work directly with the data on the spreadsheet, using basic math operations. They tend to be short, simple calculations. If they get very complex they become hard to enter, edit, and debug. They can be copied from one cell to another cell or cell range.

Conclusion: Formulas are preferred when the needed operation is a simple math operation using nearby cells as operands or when the same operation will be performed on many cells.

Functions accept *arguments* (or *parameters*), can perform complex sequences of math or other operations on the arguments, and return a result. The arguments provide a lot of flexibility: They can contain a value, a cell reference, or another function. Because functions are used in formulas, they can also be copied from one cell to another or many others.

Built-In Functions are included with Excel to handle many common tasks. They are available to any workbook and spreadsheet. They can be used in combination (one function can call another). There is little on the down side, except that there might be no built-in function that meets your need.

User-Written Functions can be written whenever needed. They can be tailored to meet very specific needs. But the writing takes time, and the functions typically are available only in the workbook in which they are stored.

Conclusion: Built-in functions should be used if they are available. User-written functions typically are used for very specific, complex calculations.

Macros can use information from nearby cells (e.g., add three to the value in the cell to the left), but they do not accept arguments. They can output results to the active cell or a related cell (e.g., place the result in the cell three positions to the right), but do not actually return a value. Macros are often used to store sequences of nonmathematical operations. Many of the operations that used to be coded into macros have been replaced by the pop-up menus available by clicking the right mouse button and using the Format Painter. Macros can be tied to a keystroke, so they are easy to call when needed, but typically they are available only in the workbook in which they are stored. They cannot be copied from one cell to another.

Conclusion: Macros are most commonly used for repeated sequences of commands, often nonmathematical commands.

Of the options available within Excel, *VBA Programs*, offer the most flexibility and power. You can create graphical user interfaces (GUIs) to get information from the user. You can use standard programming methods to make decisions, control operations, and store values. The down side is that programming requires much more effort than the other approaches listed here.

Conclusion: Usually, VBA programs are written for special applications that require either speed to handle lots of calculations, complex structures to decide how to handle various situations, or a more elaborate user interface (i.e., a graphical user interface, or GUI) than is possible with the simpler methods.

7.2 MACROS AND VIRUSES

Because Excel macros are stored as VBA Subs and VBA is a programming language, it provides individuals who are so inclined the opportunity to create computer viruses as macros. One of the requirements for creating a virus is some way to *automatically* activate computer code. Once upon a time, the ability to automatically run a macro when you started Excel was a good thing: It allowed you to customize your spreadsheets. But the virus writers made use of this feature to activate their viruses and do their damage. All Excel users who might obtain a workbook from an unknown source, such as the Internet, e-mail, or a disk, should take security against macro viruses very seriously.

Excel now provides four levels of security against macro viruses:

- Very High: Only macros installed in trusted locations will be allowed to run. All other signed and unsigned macros are disabled.
- High: Only digitally signed macros from trusted sources are allowed to run; all others are disabled.
- Medium: If a macro is encountered that is from a source that is not included in your trusted source list, Excel displays a warning and asks whether it should disable the macro. When in doubt, you should say "yes" and have the macro disabled.
- Low: All macros from any source are allowed to run without warning.

If your computer is connected to the outside world (even if only via disk), you should have the macro virus protection set to "medium" or higher.

To change your macro security level, use Tools/options ... to open the options dialog then select Macro Security ... from the Security panel.

7.3 RECORDED MACROS

Recording of keystrokes is the easiest way to create a macro, so we will consider *recorded macros* first.

When you ask Excel to record a macro, it actually writes a program in VBA, using Basic statements that are equivalent to the commands you enter via the keyboard or mouse. Because the macro is stored as a program, after recording, you can still edit the program. Recording is a good way to create simple macros, and it is also an effective way to begin writing a VBA *programmed macro*.

We'll need an example to demonstrate the process of recording a macro. As an easy example, consider a macro that converts a temperature in degrees Fahrenheit to degrees Celsius. The conversion equation is

$$T_C = \frac{T_F - 32}{1.8}. \tag{7.1}$$

Some benchmarks will also be helpful in making sure the macro is working correctly. Here are four well-known temperatures in the two temperature systems:

212°F =	100°C	(boiling point of water)
98.6°F =	37°C	(human body temperature)
32°F =	0°C	(freezing point of water)
−40°F =	−40°C	(equivalency temperature)

7.3.1 Recording a Macro

We begin by entering a Fahrenheit temperature into a cell (A4) and selecting the adjacent cell (B4):

	A	B	C
1	Temp.	Temp.	
2	(°F)	(°C)	
3			
4	212		
5			

Then we begin recording the macro:

1. Tell Excel you want to record a macro by selecting Tools/Macro/Record New Macro:

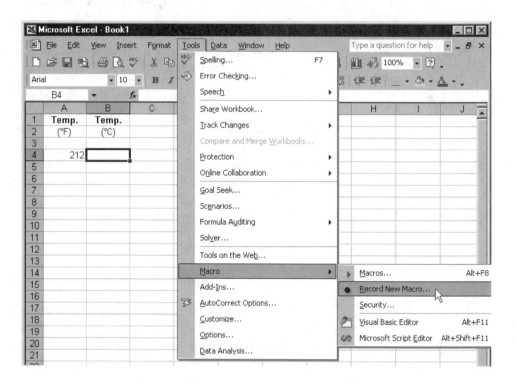

The Record Macro dialog box will be displayed.

2. Establish the macro name, *shortcut key* (optional), storage location, and description (optional). Here the macro has been named "F_to_C." The shortcut key was set as [Ctrl-f]. (Use a lowercase "f"; shortcut keys are case sensitive.) The macro will be stored in "This Workbook", which means it is available to any sheet in the current workbook, but unavailable to any other workbook. Other options include "New Workbook" and "Personal Macro Workbook."

   • If the macro was stored in a new workbook, the macro would be available only if the new workbook is open.
   • The Personal Macro Workbook is designed to be a place to store macros that you want to be available to use from any workbook.

If your macro will be used frequently from more than one workbook, you probably want to save the macro in your Personal Macro Workbook. However, if the macro will be used only with the data in a particular workbook, you probably want to store the macro in the workbook with the data:

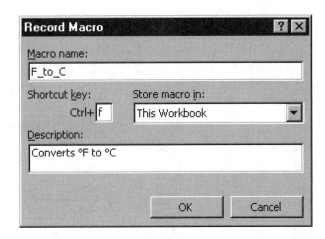

3. Select absolute or relative cell referencing.

If you want the temperature conversion macro to use the value "in the cell to the left of the currently selected cell," then you need to record the macro using *relative referencing*. If you want the temperature conversion macro to use the value "in cell A4," then you want to use *absolute referencing*. Relative referencing is more common. To indicate that relative cell referencing should be used, be sure that the Relative Reference button on the macro recording control panel is selected. If it is not selected, the macro will be recorded with absolute cell addresses.

	A	B	C	D	E	F
1	**Temp.**	**Temp.**				
2	(°F)	(°C)				
3						
4	212					
5						
6						
7						
8				Stop		
9						
10						
11				Relative Reference		
12						

4. Enter the conversion equation in cell B4:

× ✓ *fx* =(A4-32)/1.8

	A	B	C	D	E
1	**Temp.**	**Temp.**			
2	(°F)	(°C)			
3					
4	212	=(A4-32)/1.8			
5					
6					
7					
8				Stop	
9					
10					

This formula has been entered just as is any formula in Excel. The only difference is that the macro recorder is running as the formula is entered. Once the conversion formula has been completed, press the [enter] key. (The macro recorder is still running.)

5. Stop the macro recorder by pressing the Stop Recording button on the macro recorder control panel:

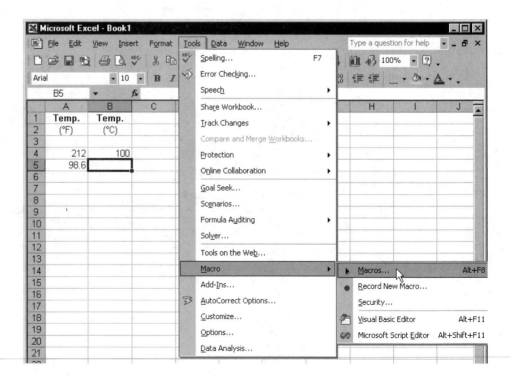

At this point, the macro has been recorded. The next step is to see whether it works. We'll use our temperature benchmarks to test the macro.

7.3.2 Testing the Recorded Macro

1. First, the benchmark temperature 98.6°F is entered in cell A5. Then cell B5 is selected. The list of currently available macros is displayed if you select Tools/Macro/Macros . . . :

2. The F_to_C macro is run by selecting the macro name and pressing the Run button:

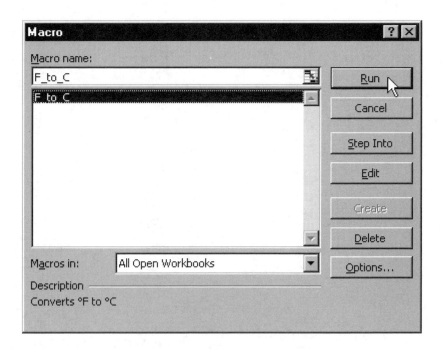

3. The value in cell A5 is converted from Fahrenheit to Celsius, and the result is placed in cell B5. (*Note: Cell B5 contains the conversion formula* = (A5– 32)/1.8).

The other temperature benchmarks were converted by using the assigned shortcut key [Ctrl-f] rather than selecting Tools/Macro:

	A	B	C	D
	Temp.	Temp.		
	(°F)	(°C)		
1				
2				
3				
4	212	100		
5	98.6	37		
6	32	0		
7	-40	-40		
8				

B7   ▼   *fx* =(A7-32)/1.8

7.3.3 Using Absolute Cell References When Recording Macros

If the macro had been recorded with absolute cell references (with the Relative Reference button deselected), then the macro would have been recorded using absolute addresses such as "A4" rather than "the cell to the left." The first conversion would have been unchanged:

	A	B	C
1	Temp.	Temp.	
2	(°F)	(°C)	
3			
4	212	100	
5			
6			

But the next time the macro was used (in cell B5), it would appear to fail:

	A	B	C
1	Temp.	Temp.	
2	(°F)	(°C)	
3			
4	212	100	
5	98.6		
6			

Actually, the macro (recorded with absolute referencing selected) did its job: It converted the 212°F in cell A4 to 100°C and stored the result in cell B4. Because the macro was recorded with absolute addresses, then, no matter which cell is currently selected when the macro is run, the macro will always convert the value in cell A4 and put the result in cell B4.

7.3.4 Including an Absolute Cell Reference in a Macro Recorded Using Relative Referencing

If you want to include an absolute address in a macro recorded with relative referencing, use the dollar signs in the absolute address. For example, if the constant (32) were moved out of the conversion formula and into cell C2, the temperature conversion macro should be recorded with relative referencing selected and absolute address $C$2 used in the recorded formula. Press [F4] immediately after entering C2 to add the dollar signs as the macro is being recorded:

	A	B	C	D
	▼ ✕ ✓ fx	=(A4-$C$2)/1.8		
1	Temp.	Temp.	Constant	
2	(°F)	(°C)	32	
3				
4	212	=(A4-$C$2)/1.8		
5				
6				
7				
8			Stop ▼ ✕	
9				
10				

7.3.5   Editing a Recorded Macro

Recorded macros are stored as VBA programs. The macro recorded with relative addressing on was stored like this:

```
Sub F_to_C()
'
' F_to_C Macro
' Converts °F to °C
'
' Keyboard Shortcut: Ctrl+f
'
    ActiveCell.Select
    ActiveCell.FormulaR1C1 = "=(RC[-1]-32)/1.8"
    ActiveCell.Offset(1, 0).Range("A1").Select
End Sub
```

The VBA subprogram was given the assigned macro name, F_to_C. The single quote at the left of a line indicates a comment line in the program. Comment lines are ignored when the macro is run, but are included to provide information about the macro.

The three operational lines do the following:

COMMAND	DESCRIPTION
`ActiveCell.Select`	The active cell is the cell from which the macro was run. This command selects the active cell. This is the programming equivalent to clicking on the cell before entering a formula.

`ActiveCell.FormulaR1C1 =`
`"=(RC[-1]-32)/1.8"`
`1) ActiveCell.FormulaR1C1 =`
`2) "=(RC-1232)/1.8"`
	This statement has two pieces:
	(1) Assign the text string on the right side of the equation to the active cell as a formula, using R1C1-style addressing.
	(2) The formula applied to the active cell should include the cell in the current row (R) and one column to the left (`C[-1]`) in the temperature conversion equation.

`ActiveCell.Offset(1, 0)` `.Range("A1").Select`	From the current active cell, move one row down and zero columns right (`Offset(1,0)`) and make that the selected cell. (This is Excel's way of telling the cursor to move down one row after entering the formula.)

The macro recorded with absolute cell referencing looks quite different:

```
Sub F_to_C_abs()
'
' F_to_C_abs Macro
' Converts °F to °C — USING ABSOLUTE CELL REFERENCES
'
' Keyboard Shortcut: Ctrl+a
'
    Range("B4").Select
    ActiveCell.FormulaR1C1 = "=(RC[-1]-32)/1.8"
    Range("B5").Select
End Sub
```

The statement: `Range("B4").Select` always causes cell B4 to be made the currently selected cell before performing the conversion. That is why, no matter which cell you select before running the macro, the value in the cell to the left of cell B4 (from the RC[−1], which refers to cell A4) will always be used, and the result will always be placed in the currently selected cell, B4. After the formula is stored in cell B4, the Range ("B5").Select statement always moves the cursor to cell B5.

The macros can be edited as VBA programs by using the *VBA editor*. To open the editor, simply select Tools/Macro/Macros... to set the list of currently available macros:

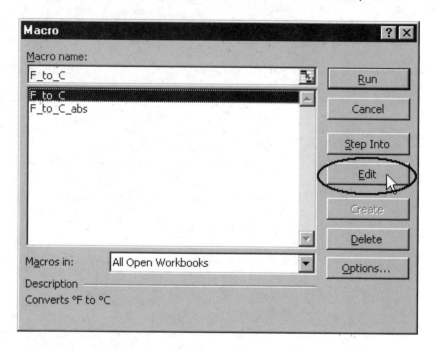

Press the Edit button to open the VBA editor. The program code for the currently selected macro will be displayed in the editor. You can modify the existing program (e.g., correct errors) or create entirely new macros. The macro for converting temperatures in degrees Celsius to degrees Fahrenheit was created by copying Sub F_to_C() and modifying it as follows:

```
Sub C_to_F()                                  'changed from F_to_C()
'
' C_to_F Macro                                'changed from F_to_C Macro
' Converts °C to °F                           'changed from °F to °C
'
' Keyboard Shortcut: Ctrl+c                    'changed from Ctrl+f
'
  ActiveCell.Select
  ActiveCell.FormulaR1C1 = "=RC[−1]*1.8+32"' changed from
"=(RC[−1]−32)/1.8"
    ActiveCell.Offset(1, 0).Range("A1").Select
End Sub
```

Creating a new *Sub* (subprogram) in VBA makes the macro available to the worksheet. The macro name will be the same as the name of the VBA Sub. However, the shortcut key and description are not assigned to the macro when the macro is created by using VBA. To assign the shortcut key and description, select Tools/Macro/Macros... from the Excel worksheet, select the new macro from the list of available macros, and press the Options button:

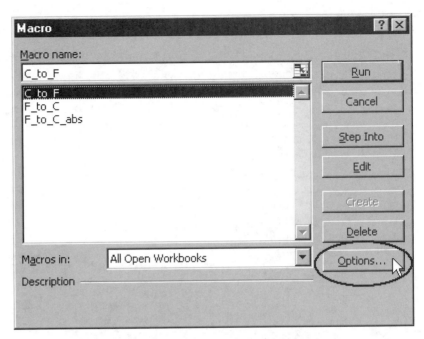

On the Macro Options dialog box, you can set the shortcut key and description:

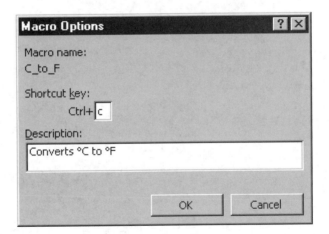

Finally, we test the new macro in the worksheet by entering 100 in cell A4, selecting cell B4, and running the C_to_F macro by using the [Ctrl-c] shortcut. It worked!

	A	B	C
1	**Temp.**	**Temp.**	
2	(°C)	(°F)	
3			
4	100	212	
5			
6			

EXAMPLE 7.1

ADDING A TRENDLINE TO A GRAPH

Macros are commonly used for automating tasks. Adding a trendline to a graph is easy enough, but if you are always adding a linear trendline to each of your graphs and always want the equation of the trendline and the R^2 value to be displayed, you can record a macro that will do the task in a couple of keystrokes.

First, start with some data and an XY Scatter plot. Then click on the data series to select it:

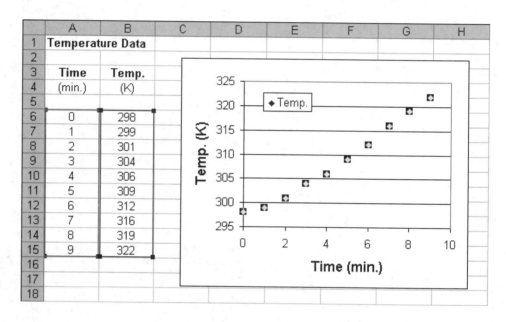

Step 1. Start Recording the Macro

Select Tools/Macro/Record New Macro . . . to begin the recording:

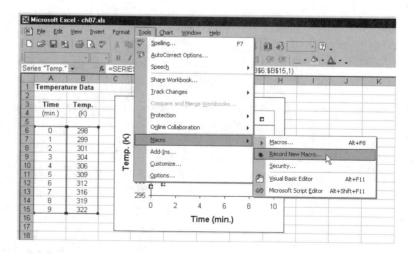

Then give the macro a name, assign it to a Shortcut key (if desired), and add a description (optional):

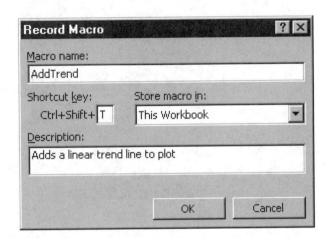

Step 2. Add the Trendline to the Graph

First, right click on any marker; then, select Add Trendline from the pop-up menu:

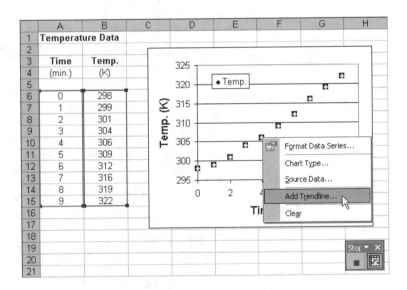

Then choose a linear trendline from the Type panel:

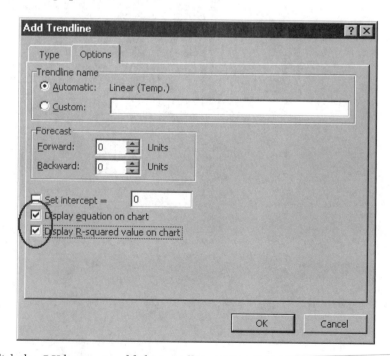

Using the Options panel, ask that the equation of the trendline and the R^2 value be displayed on the graph:

Click the OK button to add the trendline.

Step 3. Stop the Macro Recorder

Now, selecting any graph and pressing [Ctrl-Shift-T] is all it takes to add a linear trend-line and display the equation of the line and the R^2 value on the graph.

EXAMPLE 7.2

A MACRO FOR LINEAR INTERPOLATION

A linear-interpolation macro might be useful for interpolating in data tables. The formula that the macro will create will depend on how the data to be interpolated are arranged. We will assume that the data are arranged in columns, with the known x values on the left and y values, including a space for the unknown value, on the right. The x values will include a low, middle, and high value. The low and high y values will be known, and we will solve for the middle y-value by linear interpolation:

X VALUES (KNOWNS)	Y VALUES (LOW AND HIGH VALUES ARE KNOWN)
x_{LOW}	y_{LOW}
x_{MID}	$[y_{\text{MID}}]$
x_{HIGH}	y_{HIGH}

A quick way to write a linear-interpolation equation uses the ratios of differences of x-values and y-values:

$$\frac{x_{\text{MID}} - x_{\text{LOW}}}{x_{\text{HIGH}} - x_{\text{LOW}}} = \frac{[y_{\text{MID}}] - y_{\text{LOW}}}{y_{\text{HIGH}} - y_{\text{LOW}}}. \tag{7.2}$$

Solving for the unknown, y_{MID}, gives

$$y_{\text{MID}} = y_{\text{LOW}} + (y_{\text{HIGH}} - y_{\text{LOW}}) \times \left[\frac{x_{\text{MID}} - x_{\text{LOW}}}{x_{\text{HIGH}} - x_{\text{LOW}}} \right]. \tag{7.3}$$

We'll develop the macro with a really obvious interpolation to make sure the macro works correctly:

	A	B	C	D
1				
2		x values	y values	
3				
4		0	1	
5		50		
6		100	3	
7				

The macro should interpolate and get the value 2 for the unknown *y*-value in cell C5:

1. To write the macro, first click on cell C5 to select it (as shown before). Then begin recording the macro.

2. Give the macro a name, and assign it to a shortcut key:

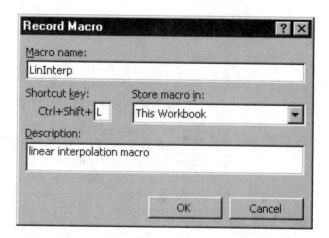

3. Be sure that the relative reference button is selected:

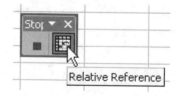

4. Enter the interpolation formula by clicking on the cells containing the following values:

```
= 1+(3-1)*((50-0)/(100-0)).
```

Excel will take care of the cell addresses, but the formula is actually =C4+(C6−C4)*((B5−B4)/(B6−B4)):

fx =C4+(C6-C4)*((B5-B4)/(B6-B4))

	A	B	C	D	E	F
1						
2		x values	y values			
3						
4		0	1			
5		50	=C4+(C6-C4)*((B5-B4)/(B6-B4))			
6		100	3			
7					Stop	
8						
9						
10						

5.   Press [enter] to complete the formula, then stop the macro recorder:

	A	B	C	D	E
1					
2		x values	y values		
3					
4		0	1		
5		50	2		
6		100	3		
7				Stop	
8					
9					
10					Stop Recording
11					

The interpolation formula got the value 2 correctly, but the real test is how the interpolation will work on other tables. Here are a couple of examples of the interpolation macro in use.

Sine Data between 0 and 90°: Interpolate for Sin(45°)   Original Data:

	A	B	C	D	E	F
1						
2		x values	y values			
3		(degrees)	(sin(x))			
4						
5		0	0.000			
6		10	0.174			
7		20	0.342			
8		30	0.500			
9		40	0.643			
10		50	0.766			
11		60	0.866			
12		70	0.940			
13		80	0.985			
14		90	1.000			
15						

Insert a row to add the *x*-value 45°:

	A	B	C	D	E	F
1						
2		**x values**	**y values**			
3		(degrees)	(sin(x))			
4						
5		0	0.000			
6		10	0.174			
7		20	0.342			
8		30	0.500			
9		40	0.643			
10		45				
11		50	0.766			
12		60	0.866			
13		70	0.940			
14		80	0.985			
15		90	1.000			
16						

Use the LinInterp macro to interpolate for sin(45°):

C10	▼		f_x	=C9+(C11-C9)*((B10-B9)/(B11-B9))		
	A	B	C	D	E	F
1						
2		**x values**	**y values**			
3		(degrees)	(sin(x))			
4						
5		0	0.000			
6		10	0.174			
7		20	0.342			
8		30	0.500			
9		40	0.643			
10		45	0.704			
11		50	0.766			
12		60	0.866			
13		70	0.940			
14		80	0.985			
15		90	1.000			
16						

The actual value is 0.707, so linear interpolation on this nonlinear data resulted in a 0.4% error.

Superheated Steam Enthalpy Data: Interpolate for Specific Enthalpy at 510 K   The original data are

	A	B	C	D	E	F
1	Steam Enthalpies at 5 bars					
2						
3		Temperature	Specific Enthalpy			
4		(K)	(kJ/kg)			
5						
6		300	3065			
7		350	3168			
8		400	3272			
9		450	3379			
10		500	3484			
11		550	3592			
12		600	3702			
13						

With the row inserted for 510 K, we get

	A	B	C	D	E	F
1	Steam Enthalpies at 5 bars					
2						
3		Temperature	Specific Enthalpy			
4		(K)	(kJ/kg)			
5						
6		300	3065			
7		350	3168			
8		400	3272			
9		450	3379			
10		500	3484			
11		510				
12		550	3592			
13		600	3702			
14						

Performing the linear interpolation, we have

C11 f_x =C10+(C12-C10)*((B11-B10)/(B12-B10))

	A	B	C	D	E	F
1	Steam Enthalpies at 5 bars					
2						
3		Temperature	Specific Enthalpy			
4		(K)	(kJ/kg)			
5						
6		300	3065			
7		350	3168			
8		400	3272			
9		450	3379			
10		500	3484			
11		510	3506			
12		550	3592			
13		600	3702			
14						

APPLICATIONS: STATICS

Resolving Forces into Components: Part 1

A standard procedure in determining the net force on an object is breaking down, or *resolving*, the applied forces into their components in each coordinate direction. In this example, a 400-N force is pulling on a stationary hook at an angle of 20° from the *x*-axis. A second force of 200 N acts at 135° from the *x*-axis.

  Record a macro in Excel that can be used to resolve forces into their components. Then compute the magnitude and direction of the resultant force. The following is a sketch of the problem:

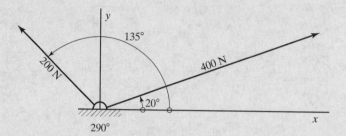

1.   We begin by entering the first force and its angles from the *x*-axis:

	A	B	C	D	E
1	Force	Ang. From X	X-comp.	Y-comp.	
2	(N)	(degrees)	(N)	(N)	
3					
4	400	20			
5					

2.   Then make cell C4 active (i.e., click on it) and start the macro recorder by selecting Tools/Macro/Record New Macro....

3.   Give the new macro a name and shortcut key (if desired). When the capital "R" was entered as the shortcut key, Excel automatically added the "Shift" to the shortcut key designation:

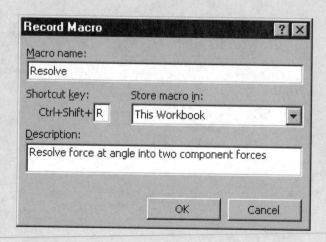

4. Remember to use relative cell references:

	A	B	C	D	E	F
1	**Force**	**Ang. From X**	**X-comp.**	**Y-comp.**		
2	(N)	(degrees)	(N)	(N)		
3						
4	400	20				
5						
6					Stop ▼ ✕	
7						
8						
9					Relative Reference	
10						

5. Enter the following formula in cell C4 to compute the *x*-component of the 400-N force:

   `=A4*COS(RADIANS(B4))`

   The `RADIANS()` function is built into Excel; it converts angles in degrees into angles in radians. It is included here because the `COS()` function requires angles in radians.

6. While still recording the macro, click on cell D4, and enter the formula for computing the *y*-component of the 400-N force.

   `=A4*SIN(RADIANS(B4))`

7. Stop the macro recorder.

The new macro wasn't useful at all for finding the components of the 400-N force, because the required formulas were typed into cells C4 and D4 as the macro was recorded. The value of the macro starts to become apparent during calculation of the components of the 200-N force.

First, the force and angles are entered into the spreadsheet, and cell C5 is selected:

	A	B	C	D	E
1	**Force**	**Ang. From X**	**X-comp.**	**Y-comp.**	
2	(N)	(degrees)	(N)	(N)	
3					
4	400	20	375.9	136.8	
5	200	135			
6					

Then the macro is run, either from the macro list (Tools/Macro/Macros ...) or by using the assigned shortcut ([Ctrl-Shift-R] in this example). The *x*- and *y*-components are calculated by the macro:

	A	B	C	D	E
1	**Force**	**Ang. From X**	**X-comp.**	**Y-comp.**	
2	(N)	(degrees)	(N)	(N)	
3					
4	400	20	375.9	136.8	
5	200	135	-141.4	141.4	
6					

To find the magnitude and direction of the resultant force, first sum the x-components and y-components of the forces:

C10	▼	f_x	=DEGREES(ATAN(D7/C7))	
	A	B	C	D	E
1	**Force**	**Ang. From X**	**X-comp.**	**Y-comp.**	
2	(N)	(degrees)	(N)	(N)	
3					
4	400	20	375.9	136.8	
5	200	135	-141.4	141.4	
6					
7		**Sums:**	234.5	278.2	
8					
9		**Resultant Force:**	363.8	N	
10		**Resultant Angle from X:**	49.9	°	
11					

Then compute the magnitude of the resultant force as

$$F_{\text{res}} = \sqrt{\left(\sum F_x\right)^2 + \left(\sum F_y\right)^2},$$

(7.4)

or, as entered in cell C9, =SQRT (C7^2 + D7^2).

Finally, compute the angle of the resultant from the x-axis as

$$\theta_{\text{res}} = \text{ATAN}\left(\frac{\sum F_y}{\sum F_x}\right),$$

(7.5)

which is entered in cell C10 as: =DEGREES(ATAN(D7/C7)). Here, Excel's DEGREES() function was used to convert the radians returned by the ATAN() function into degrees.

Some of the homework problems at the end of this chapter will provide opportunities to use the Resolve macro.

7.4 PROGRAMMED MACROS (VBA)

Excel stores macros as VBA Subs (subprograms), so an alternative to recording macros is simply to write the programs yourself. If you create a VBA Sub without any arguments from within an Excel workbook, it will appear in the list of available macros in that workbook.

7.4.1 Starting Visual Basic for Applications (VBA)

Excel comes with Visual Basic for Applications (VBA) built in. To start the VBA editor, select Tools/Macro/Visual Basic Editor:

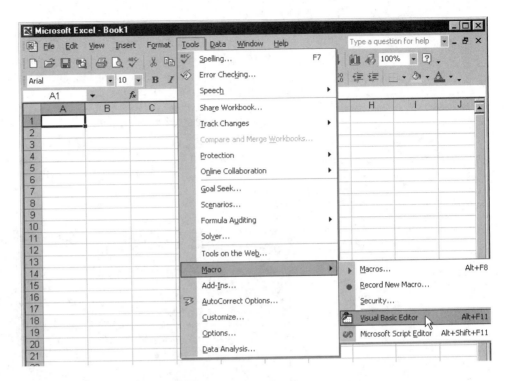

The editor opens in a new window:

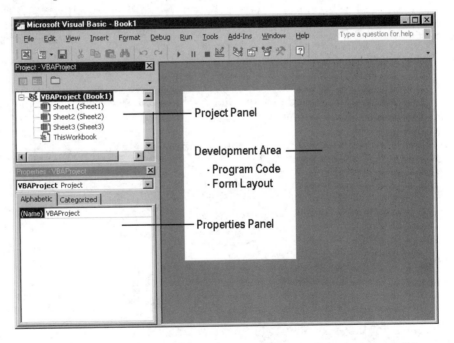

The VBA editor is a multipanel window, and you can control the layout of the various panels. The three standard panels are shown here. The main area (empty in this figure) is the *development area*. When you are adding buttons and text fields to a form, the form is displayed in the development area. This panel is also used for writing program code. If you have already recorded one or more macros, the VBA code for the macros will be displayed in the development area.

The *project panel* lists all of the items in the project. By default, a workbook contains three sheets (spreadsheets). By selecting a project item, you can connect program code with that item (only). For example, you could have Sheet1 and Sheet2 respond differently to mouse clicks. If you want to program a macro, you should write the code in a *module*. A module is simply a project item that stores program code, such as Subs and functions.

7.4.2 Modules, Subs, and Functions

If you have already recorded a macro and stored it in "This Workbook," then Excel has already added a module and named it Module1. If there is no Module1 listed in the project list, then you should insert a module before writing the macro.

Inserting a Module   To insert a module, select Insert/Module from the VBA menu:

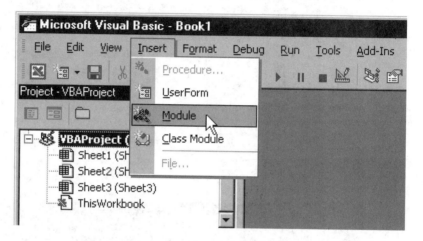

A module named Module1 will then be added to the project list. (You can click twice on the name in the project list to rename the module.) Once the module has been added to the project, the development area shows the code stored in the module (which is empty for now):

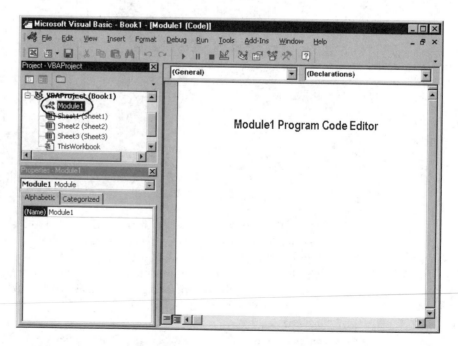

Once you have Module1 in your project, you have a place to write the macro. A macro is just a VBA subprogram without arguments. VBA will help you with the subprogram syntax if you insert the Sub into the module using the Insert menu.

Inserting the Sub Procedure into the Module   To have VBA create the framework for the new macro, select Insert/Procedure . . . from the VBA menu:

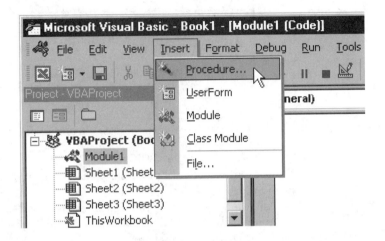

The Add Procedure dialog box is displayed:

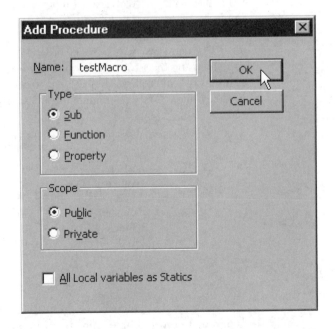

Every procedure (e.g., a Sub or a function) must have a unique name. Here, the name testMacro has been entered into the dialog box. For a macro, the Sub type is used. If the macro is going to be called from outside of the module (e.g., from one of the sheets), then the scope must be set to *Public*. Finally, you must decide whether you need static variables. By default, VBA resets all variables to zero each time a Sub or function is used. If you want the variables used in a Sub or function to retain their values (not be reinitialized), then check the "All Local variables as Statics" box. For example, if you wanted to count the number of times a Sub was used, the variable inside the Sub that holds the count would have to be declared *static* or the counter would be reset each time the Sub was used.

When you click the OK button, VBA creates the skeleton of the new Sub in the development area. The program code you write goes between the `Public Sub test-Macro()` line and the `End Sub` line:

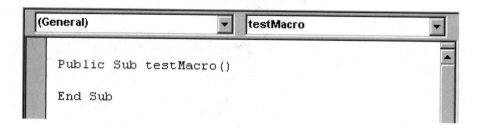

```
(General)                        testMacro

        Public Sub testMacro()

        End Sub
```

Now that we have the basic structure of the macro, what can we do with it?

7.4.3 A Simple Border Macro

For starters, let's just put a heavy, black line around the active cell. To do so, we'll use the *BorderAround method* on the *ActiveCell* object. The ActiveCell is the currently selected cell when the macro is run. The BorderAround method needs to know what style of line to use, how thick to draw the line (i.e., the line weight), and what color to make the line.

There are several predefined constants in VBA used to specify line style: xlContinuous, xlDash, xlDashDot, xlDashDotDot, xlDot, xlDouble, xlSlantDashDot, and xlLineStyle None. For example, to create a simple solid line, use `LineStyle := xlContinuous`. Similarly, VBA uses predefined constants to specify the line weight: xlHairline, xlThin, xlMedium, and xlThick.

There are two ways to specify the line color: Specify the *ColorIndex*; or set the color value by using the `RGB()` function. The ColorIndex is a code ranging from 1 to 56 that identifies specific colors. The best way to select colors is to use the VBA help files and look up ColorIndex. A few commonly used ColorIndex values are listed in the following table:

	COLORINDEX	RGB VALUES
Black	1	0, 0, 0
White	2	255, 255, 255
Red	3	255, 0, 0
Green	4	0, 255, 0
Blue	5	0, 0, 255
Yellow	6	255, 255, 0
Magenta	7	255, 0, 255
Cyan	8	0, 255, 255

Alternatively, you can use the `RGB(red,green,blue)` function to set the levels of red, green, and blue individually in the range from 0 (least intense) to 255 (brightest). The RGB values required to generate some common colors are listed in the preceding table.

Including the statement

```
ActiveCell.BorderAround LineStyle:=xlContinuous,
    Weight:=xlThick, ColorIndex:=1
```

in the testMacro Sub, will cause a continuous, heavy-weight, black line to be drawn around the cell that is currently active when the macro is run:

```
(General)                                          testMacro

Public Sub testMacro()

    ActiveCell.BorderAround LineStyle:=xlContinuous, Weight:=xlThick, ColorIndex:=1

End Sub
```

7.4.4   Running the Macro

Excel assumes that a Sub with no parameters (i.e., one with no variables inside the parentheses after the Sub name) is a macro; it includes the Sub name in the list of Macros that is displayed when you select Tools/Macro/Macros . . . :

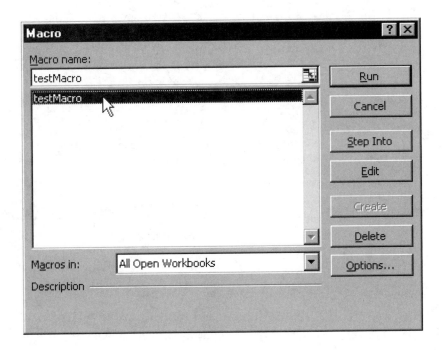

When you double click on the testMacro name in the list (or select the testMacro name and click the Run button), the testMacro Sub runs, and a black border is drawn around the currently active cell (cell B3 in the following figure):

	A	B	C	D	E
1					
2					
3					
4					
5					

If you plan to use the macro a lot, you probably want to assign a shortcut key. You can do this from the Macro list dialog by selecting the testMacro name in the list and clicking the Options . . . button:

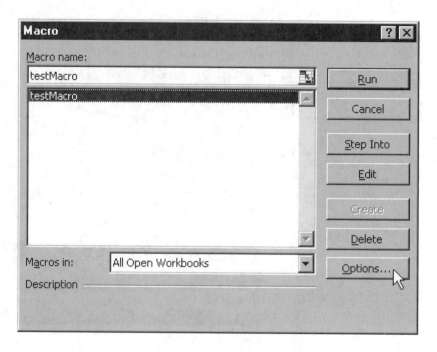

On the Macro Options page, select a shortcut key; you can also add a brief description for this macro, as shown here:

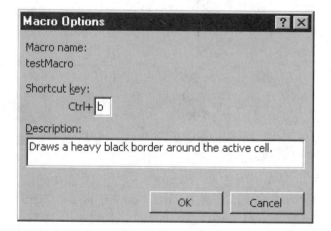

7.4.5 Saving Your Macro Project

Under the procedure described previously, the macro is housed in Module1, which is part of the Excel workbook. To save the macro, simply save the Excel workbook. The macro developed here might be saved in a workbook file named testMacro.xls.

7.4.6 Using a Programmed Macro from Another Workbook

Because macros are stored with workbooks, a macro you write (or record) in one workbook will not be immediately available for use in another workbook. However, Excel makes it easy to use macros from other workbooks by showing all of the macros in all open workbooks in the macro list that is displayed by using Tools/Macro/Macros.... To use a macro from one workbook in another, just make sure that both workbooks are open.

For example, if a new workbook is opened (Book 2) and the testMacro.xls workbook is still open, the testMacro macro is available to the new workbook (Book 2):

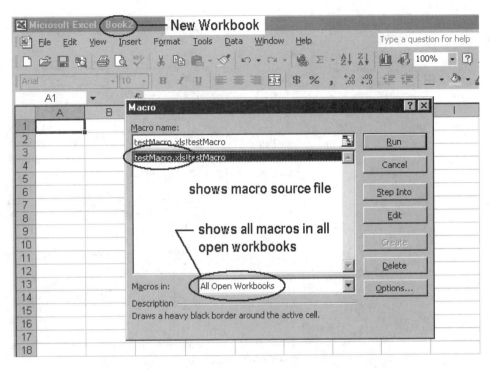

The list of available macros shows where the macro is stored if it is not from the displayed workbook. In the preceding figure, the source of the testMacro macro is shown as file testMacro.xls.

Note: It is possible for the shortcut keys assigned to macros in multiple workbooks to conflict when the files are open together. The shortcut keys for the macros in the displayed workbook are used if there is a conflict. The macro names will never conflict, because the macro name is shown with the workbook file name, as file name!macro name. In the previous figure, the macro named testMacro in file testMacro.xls was listed as testMacro.xls!testMacro.

7.4.7   Common Macro Commands

In a single chapter, we cannot begin to describe the range of commands available through VBA, but there are a few very common tasks that can be accomplished through VBA commands. These include the following:

- changing the values in the active cell;
- changing the properties of the active cell;
- moving the active cell;
- using values from cells near the active cell to calculate a value;
- using values from fixed cell locations to calculate a value;
- selecting a range of cells;
- changing the values in the selected range of cells;
- changing the properties of the selected range of cells.

Each of these basic operations will be discussed briefly.

Changing the Values in the Active Cell   The ActiveCell object tells VBA where the cell currently selected is located and provides access to the contents of the cell and the properties associated with the cell. If you click on a cell in a workbook, the selection box moves to that cell, and the cell becomes the active cell. In the following figure, cell B3 is the active cell:

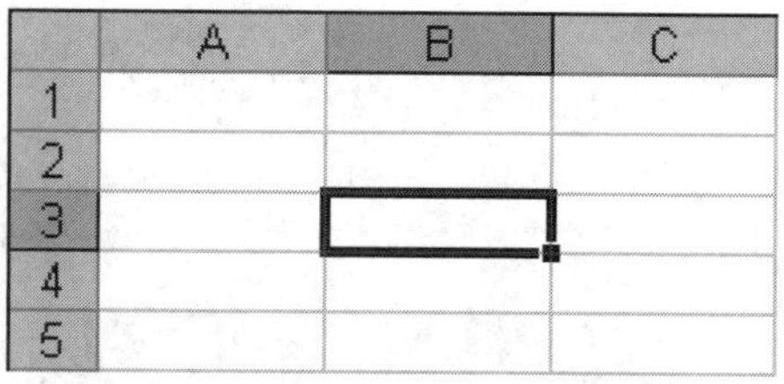

When multiple cells are selected, the active cell is indicated by a background colored differently from the rest of the selected cells. In the following figure, cell B3 is still the active cell; cells B3:B5 comprise the selected range:

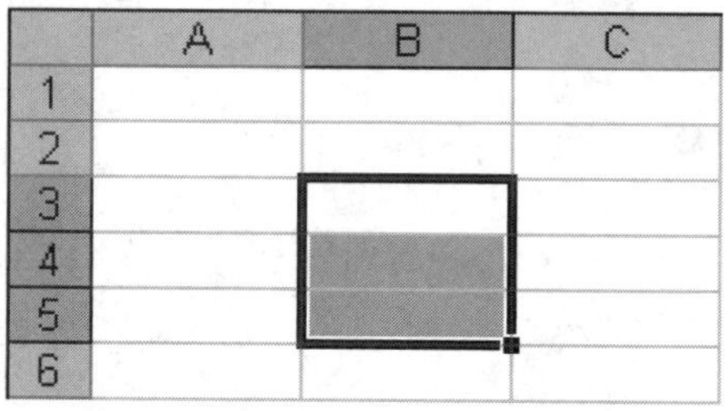

If your macro uses the ActiveCell object, only the active cell will be changed by the macro, not the selected range.

To access or change the contents of the active cell, use ActiveCell.value. The following macro, called setValue, checks the current value of the selected cell and then changes it. If the current value is a number greater than or equal to 5, then the macro sets the value to 100. If the current value is less than 5, then the macro puts a text string, "less than five," in the cell. It's not obvious why you would want to do this, but the macro does illustrate how your Sub can *use* the value of the active cell (in the If statement), as well as change the value, either to a numeric or to a text value:

```
Public Sub setValue()
    If ActiveCell.Value >= 5 Then
        ActiveCell.Value = 100
    Else
        ActiveCell.Value = "Less than five"
    End If
End Sub
```

EXAMPLE 7.3

MODIFIED LINEAR INTERPOLATION WITH RETURNED VALUE

Sometimes, instead of a formula, you might just want the interpolated value returned. Here's a quick modification of the LinInterp macro that causes it to leave the interpolated value in the active cell.

The following is the original LinInterp macro:

```
Sub LinInterp()
'
'LinInterp Macro
'linear interpolation
'
'Keyboard Shortcut: Ctrl+Shift+L
'
    ActiveCell.FormulaR1C1 = _
```

```
        "=R[-1]C+(R[1]C-R[-1]C)*((RC[-1]-R[-1]C[-1])/(R[1]C[-1]-
    R[-1]C[-1]))"
    ' <insert new line here>
        ActiveCell.Offset(11, 6).Range("A1").Select
    End Sub
```

The formula entered into the active cell is fairly complicated, but we're not going to modify that. Instead, we will simply allow Excel to put the formula into the cell (as it already does) and grab the value from the cell and assign it back to the cell—effectively overwriting the formula with the calculated value.

To accomplish this, simply replace

```
    '<insert new line here>
```

with

```
        ActiveCell.Value=ActiveCell.Value
```

The `ActiveCell.Value` on the right reads the current value in the cell (after the formula was entered). The `ActiveCell.Value=` on the left sets the cell's value to whatever is on the right side of the equal sign (in this case the cell's own value, thus replacing the formula as the cell's contents).

After this modification, the final result in the Sine Data example looks like this:

	C10	▼	f_x	0.704416026402759	
	A	B	C	D	E
1					
2		x values	y values		
3		(degrees)	(sin(x))		
4					
5		0	0.000		
6		10	0.174		
7		20	0.342		
8		30	0.500		
9		40	0.643		
10		45	0.704		
11		50	0.766		
12		60	0.866		
13		70	0.940		
14		80	0.985		
15		90	1.000		
16					

The spreadsheet looks about the same, but cell C10 now contains a value rather than the interpolation formula.

Changing the Properties of the Active Cell   The macro developed earlier, called testMacro, was an example of changing the properties of the active cell. In that macro, the border of the cell was changed by using the BorderAround method. Common cell properties that can be changed include those in the following table:

ActiveCell.Borders	.Color or .ColorIndex	Color: = RGB(red, blue, green), where red, blue, and green are values between 0 and 255
		ColorIndex: = val, where val is an integer between 0 and 55

	.LineStyle	xlContinuous, xlDash, xlDashDot, xlDashDotDot, xlDot, xlDouble, xlSlantDashDot, xlLineStyleNone
	.Weight	xlHairline, xlThin, xlMedium, xlThick
ActiveCell.font	.Bold	True or False
	.Color *or* .ColorIndex	(Same as described above for ActiveCell.Borders)
	.Italic	True or False
	.Name	A font name in quotes, such as "Courier"
	.Size	Numeric values (8–12 are common)
	.Subscript	True or False
	.Superscript	True or False
ActiveCell.Interior	.Color or .ColorIndex	(Same as described above for ActiveCell.Borders)
	.Pattern	xlPatternAutomatic, xlPatternChecker, xlPatternCrissCross, xlPatternDown, xlPatternGray16, xlPatternGray25, xlPatternGray50, xlPatternGray75, xlPatternGray8, xlPatternGrid, xlPatternHorizontal, xlPatternLightDown, xlPatternLightHorizontal, xlPatternLightUp, xlPatternLightVertical, xlPatternNone, xlPatternSemiGray75, xlPatternSolid, xlPatternUp, xlPatternVertical
	.PatternColor or .PatternColorIndex	(Same as described above for ActiveCell.Borders)

The following Sub macro changes the active cell's font color to blue [RGB(0,0,255)], its font name to "Courier," its Font size to 24, and its interior color to yellow (ColorIndex = 6) with a green (PatternColorIndex = 4) diagonal stripe (Pattern = xlPatternLightUp):

```
Public Sub setProperty()
    ActiveCell.Font.Color = RGB(0, 0, 255)
    ActiveCell.Font.Name = "Courier"
    ActiveCell.Font.Size = 24
    ActiveCell.Interior.ColorIndex = 6
    ActiveCell.Interior.Pattern = xlPatternLightUp
    ActiveCell.Interior.PatternColorIndex = 4
End Sub
```

Making a Different Cell the Active Cell   The Offset property of the ActiveCell object allows you to select cells a specified number of rows and columns away from the active cell. The Activate property causes the selected cell to become the active cell.

The following Sub macro finds a cell 2 rows below and one row to the right of the currently active cell and makes the new cell the active cell:

```
Public Sub moveActive()
    Dim rowOffset As Integer
    Dim colOffset As Integer
```

```
        rowOffset = 2
        colOffset = 1
        ActiveCell.Offset(rowOffset, colOffset).Activate
    End Sub
```

The following Sub is functionally equivalent:

```
Public Sub moveActive()
    ActiveCell.Offset(2, 1).Activate
End Sub
```

In the spreadsheet shown next, the cell labeled "old" is selected before the running of the moveActive macro. After we run the macro, the cell labeled "new" is active.
Before running the macro, we have

	A	B	C	D
1				
2				
3		old		
4				
5			new	
6				

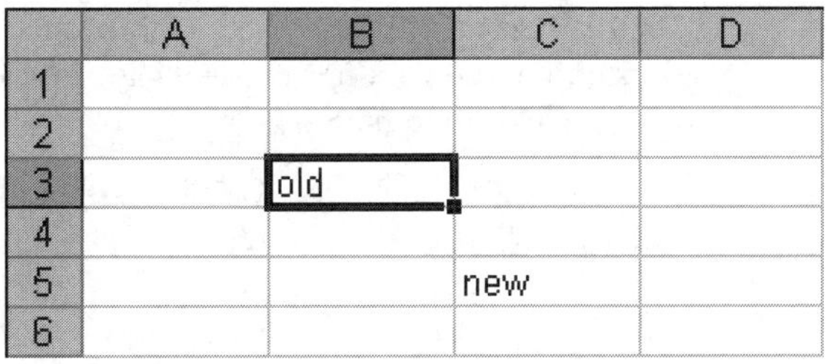

After running the macro, we have

	A	B	C	D
1				
2				
3		old		
4				
5			new	
6				

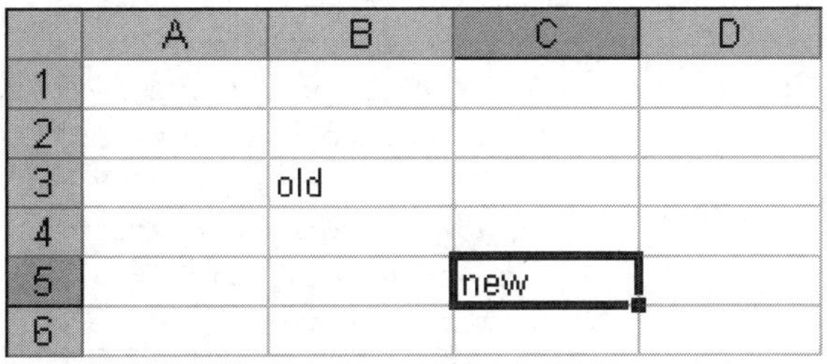

The most common use of this technique is to move the selection box to the next cell after completing a task—to prepare for the next task. The following macro takes the length in inches in the active cell, converts it to centimeters, marks the new units in the adjacent cell, and moves the active cell down to the next row to prepare for the next value to be entered or converted:

```
Public Sub convertUnits()
' convert inches in active cell to cm, then store new value
back in active cell
    ActiveCell.Value = ActiveCell.Value * 2.54
' mark the units as "cm" in adjacent column
    ActiveCell.Offset(0, 1).Value = "cm"
' move selection box down to prepare for next conversion
    ActiveCell.Offset(1, 0).Activate
    End Sub
```

In this spreadsheet, the first three values have already been converted:

	A	B
1	**Length**	
2		
3	2.54	cm
4	5.08	cm
5	7.62	cm
6	4	in
7	5	in
8	6	in
9	7	in
10		

Using Values from Cells near the Active Cell to Calculate a Value The Offset property is also used to grab values from nearby cells to perform calculations. For example, the volumes of ideal gas at various temperatures can be calculated by running the calcVolume macro:

```
Public Sub calcVolume()
    Dim P As Single
    Dim V As Single
    Dim N As Single
    Dim R As Single
    Dim T As Single
    R = 0.08206     ' liter atm / gmol K
    P = 1#          ' atm
    N = 1#          ' gmol
' get temperature from cell to left of active cell
    T = ActiveCell.Offset(0, -1).Value
' calculate volume using ideal gas law
    V = N * R * T / P
' assign volume to active cell
    ActiveCell.Value = V
' move active cell down one row to prepare for next
  calculation
    ActiveCell.Offset(1, 0).Activate
End Sub
```

In the following spreadsheet, the calcVolume macro has been run in cells B6 through B8 (admittedly, it is easier to calculate these volumes without using a macro):

	A	B	C	D
1	Pressure:	1.00	atm	
2	Moles:	1.00	gram mole	
3				
4	**Temp (K)**	**Volume (liters)**		
5				
6	200	16.41		
7	250	20.52		
8	300	24.62		
9	350			
10				

Using Values from Fixed Cell Locations   The pressure and number of moles were specified in the spreadsheet and then set in the macro. Specifying the values in two places is a good way to create errors. Someone might change the pressure on the spreadsheet and think that they were recalculating the volumes with the new pressure, but the pressure value in the macro would still be 1 atm. To fix this, we need to get the values for P and N from cells B1 and B2, respectively. This is done by using the Cells property:

```
P = Cells(1, 2).Value   ' pressure in cell B1 (row 1, column 2)
N = Cells(2, 2).Value   ' moles in cell B2 (row 2, column 2)
```

Note that `Cells(1,2).Value` refers to the value in the cell in row 1, column 2—or cell B2 using the standard cell-referencing nomenclature.

Selecting a Range of Cells   The setRangeProp Sub macro uses the Range property to set the background of cells B2 through C5 to a light-gray pattern and uses the Select property to select the entire range:

```
Public Sub setRangeProp()
    Range("B2:C5").Interior.Pattern = xlPatternGray16
    Range("B2:C5").Select
End Sub
```

The result is

Changing the Values in a Selected Range of Cells   If you have selected a range of cells before running a macro, you can use the RangeSelection.Address property of the active window (ActiveWindow property) to make changes to all of the selected cells. Here, the setRangeValues macro sets the value of each cell in the selected range to 1:

```
Public Sub setRangeValues()
    Range(ActiveWindow.RangeSelection.Address).Value = 1
End Sub
```

Before running the setRangeValues macro, we have

After running the macro, we have

Changing the Properties of a Selected Range of Cells If a range of cells is selected before the running of the macro, use `Range(ActiveWindow.Range Selection.Address)` to access the range of cells, and then append the desired property change. The following example changes the background pattern of the selected range to diagonal stripes:

```
Public Sub setSelRangeProp()
    Range(ActiveWindow.RangeSelection.Address).Interior.Pattern
= xlPatternUp
End Sub
```

Before running the setSelRangeProp macro, we have

After running the macro and clicking outside of the shaded region, we have

APPLICATIONS: STATICS

Calculating Resultant Forces and Angles: Part 2

We will use VBA to write a macro to take the x- and y-components of force and determine the resultant force and the direction of the resultant force vector.

The resultant force is computed as

$$F_{res} = \sqrt{\left(\sum F_x\right)^2 + \left(\sum F_y\right)^2}, \tag{7.6}$$

and the angle of the resultant from the x-axis is

$$\theta_{res} = \text{Atan}\left(\frac{\sum F_y}{\sum F_x}\right). \tag{7.7}$$

For this macro, we will assume that the *x*- and *y*-components of force will be available in adjacent cells, and we will place the calculated results in cells to the right of the force components. In the following spreadsheet, the force components are in cells D5 and E5. The active cell is cell F5, which will contain the computed resultant force. The computed angle will be put in cell G5. This gives us

	A	B	C	D	E	F	G	H
1	Force	Ang. from X	Ang. from Y	X-comp	Y-comp	Resultant	Ang. from X	
2	400	20	290	375.9	136.8	**(N)**	**(°)**	
3	200	135	45	-141.4	141.4			
4								
5			Sums:	234.5	278.2			
6								

The VBA Sub required to compute the resultant force and angle might look as follows:

```
Public Sub Resultant()
    Dim Fx As Single
    Dim Fy As Single
    Dim Fres As Single
    Dim Theta As Single
    Dim Pi As Single
    Pi = 3.1416
'   the cell containing Fx is two columns to the left of the active cell
    Fx = ActiveCell.Offset(0, -2).Value
'   the cell containing Fy is one column to the left of the active cell
    Fy = ActiveCell.Offset(0, -1).Value
'   calculate the resultant force, store result in active cell
    Fres = Sqr(Fx ^ 2 + Fy ^ 2)          ' Note: Sqr() is a VBA function, not an
    Excel function
    ActiveCell.Value = Fres
'   calculate the angle in radians, convert to degrees, then store in cell
'   one column to the right of the active cell
    Theta = (Atn(Fy / Fx)) * (180 / Pi) ' Note: Atn is a VBA function
    ActiveCell.Offset(0, 1).Value = Theta
End Sub
```

Notice that this macro uses VBA's `Sqr()` and `Atn()` functions rather than Excel's `SQRT()` and `ATAN()` functions. Also, VBA does not have a `DEGREES()` function, so the angle in radians was converted to degrees by using `(180/Pi)`.

The resulting spreadsheet looks as follows:

F5   fx 363.842102050781

	A	B	C	D	E	F	G	H
1	Force	Ang. from X	Ang. from Y	X-comp	Y-comp	Resultant	Ang. from X	
2	400	20	290	375.9	136.8	**(N)**	**(°)**	
3	200	135	45	-141.4	141.4			
4								
5			Sums:	234.5	278.2	363.8	49.9	
6								

Note that the results in cells F5 and G5 are values, not formulas. VBA did the calculation, not Excel.

KEY TERMS

Absolute referencing
ActiveCell object
Argument
BorderAround method
Built-in function
ColorIndex
Development area
Formula

Function
Macro
Module
Parameter
Programmed macro
Project panel
Public (scope)
Recorded macro

Relative cell referencing
Shortcut key
Static (variables)
Sub (subprogram)
User-written function
VBA editor
VBA program

SUMMARY

Recorded Macros

1. Tell Excel you want to record a macro by selecting Tools/Macro/Record New Macro.
2. Establish the macro name, shortcut key (optional), storage location, and description (optional).
3. Select absolute or relative cell referencing.
4. Record the macro.
5. Stop the recorder.

You can edit a recorded macro, using the Visual Basic editor from Excel. Use Tools/Macro/Visual Basic Editor.

Programmed Macros

Macros are Visual Basic Subs (subprograms) that receive no arguments. They are stored in a module that is saved as part of a Visual Basic project with the Excel workbook.

Inserting a Module into the Visual Basic Project to hold programmed macros:
From the Visual Basic Editor, select Insert/Module.
Inserting a Sub into the Module:
Click on the module in the Visual Basic Editor to make it active. Then select Insert/Procedure.

Common Macro Commands

In a single chapter, we cannot begin to describe the range of commands available through VBA, but there are a few very common tasks that can be accomplished through VBA commands. These include the following:

- To change the values in the active cell, use `ActiveCell.Value`.
- To change the properties, use the following:

ActiveCell.Borders	.Color or .ColorIndex	Color: = RGB(red, blue, green), where red, blue, and green are values between 0 and 255 ColorIndex: = val, where val is an integer between 0 and 55
	.LineStyle	xlContinuous, xlDash, xlDashDot, xlDashDotDot, xlDot, xlDouble, xlSlantDashDot, xlLineStyleNone
	.Weight	xlHairline, xlThin, xlMedium, xlThick
ActiveCell.Font	.Bold	True or False
	.Color *or* .ColorIndex	(Same as described above for ActiveCell.Borders)
	.Italic	True or False

	.Name	A font name in quotes, such as "Courier"
	.Size	Numeric values (8–12 are common)
	.Subscript	True or False
	.Superscript	True or False
ActiveCell.Interior	.Color or .ColorIndex	(Same as described above for ActiveCell.Borders)
	.Pattern	xlPatternAutomatic, xlPatternChecker, xlPatternCrissCross, xlPatternDown, xlPatternGray16, xlPatternGray25, xlPatternGray50, xlPatternGray75, xlPatternGray8, xlPatternGrid, xlPatternHorizontal, xlPatternLightDown, xlPatternLightHorizontal, xlPatternLightUp, xlPatternLightVertical, xlPatternNone, xlPatternSemiGray75, xlPatternSolid, xlPatternUp, xlPatternVertical
	.PatternColor or .PatternColorIndex	(Same as described above for ActiveCell.Borders)

- To move the active cell, use `ActiveCell.Offset(rowOffset, colOffset).Activate`.
- To use values from cells near the active cell to calculate a value, use `ActiveCell.Offset(rowOffset, colOffset).Value`.
- To use values from fixed cell locations, use `Cells(row, col).Value`.
- To select a range of cells (e.g., B2:C5), use `Range("B2:C5").Select`.
- To change the values in the selected range of cells, use

`Range(ActiveWindow.RangeSelection.Address).Value = newValue`.

Problems

Note: The first few problems are attempts to create some potentially useful macros for students. These need to be stored in an accessible location if they are to be useful. One solution is to store commonly used macros in a worksheet file designated for that purpose, such as Macros.xls. All macros in currently open worksheets are available to all open worksheets, so you could simply open Macros.xls to have all of your macros available for use.

Recorded Macro: Name and Date

1. Record a macro that will insert your name and the current date in the active cell. The current date is available via the TODAY() function.

Recorded Macro: Highlight Your Results

2. Record a macro that will add a colored background to the selected range and add a border.

Recorded Macro: Insert the Gravitational Constant

3. Record a macro that will insert the gravitational constant g_c in the active cell. The commonly used value of g_c is

$$g_c = 32.174 \frac{ft\, lb_m}{lb_f\, s^2}. \tag{7.8}$$

Recorded Macro: Insert the Ideal-Gas Constant

4. Record a macro that will insert the ideal-gas constant R in the active cell. The value of R depends on the units. Select the value you use most frequently:

IDEAL-GAS CONSTANTS	
8.314	$\dfrac{m^3\,Pa}{gmol\,K}$
0.08314	$\dfrac{L\,bar}{gmol\,K}$
0.08206	$\dfrac{L\,atm}{gmol\,K}$
0.7302	$\dfrac{ft^3\,atm}{lbmol\,°R}$
10.73	$\dfrac{ft^3\,psia}{lbmol\,°R}$
8.314	$\dfrac{J}{gmol\,K}$
1.987	$\dfrac{BTU}{lbmol\,°R}$

Computing Ideal-Gas Volume

5. Create a macro (either by recording or programming) that will compute the volume of an ideal gas when given the absolute temperature, the absolute pressure, the number of moles, and an ideal-gas constant. Use your macro to complete the following table:

Ideal Gas Volume				
TEMPERATURE	PRESSURE	MOLES	GAS CONSTANT	VOLUME
273	1	1	0.08206	
500	1	1	0.08206	
273	10	1	0.08206	
273	1	10	0.08206	
500	1	10	0.08206	
500	1	1	0.7302	

What are the units of volume in each row?

Computing Gas Heat Capacity I

6. There are standard equations for calculating the heat capacity (specific heat) of a gas at a specified temperature. One common form of a heat-capacity equation is a simple third-order polynomial in T:[1]

$$C_p = a + bT + cT^2 + dT^2. \tag{7.9}$$

If the coefficients a, b, c, and d are known for a particular gas, you can calculate the heat capacity of the gas at any T (within an allowable range). The coefficients for a few common gases are listed in the following table:

---

[1]From *Elementary Principles of Chemical Processes*, 3d ed., R. M. Felder and R. W. Rousseau, New York: Wiley, 2000.

Heat Capacity Coefficients

GAS	a	b	c	d	UNITS ON T	VALID RANGE
Air	28.94×10^{-3}	0.4147×10^{-5}	0.3191×10^{-8}	-1.965×10^{-12}	°C	0–1500°C
CO_2	36.11×10^{-3}	4.233×10^{-5}	-2.887×10^{-8}	7.464×10^{-12}	°C	0–1500°C
CH_4	34.31×10^{-3}	5.469×10^{-5}	0.3661×10^{-8}	-11.00×10^{-12}	°C	0–1200°C
H_2O	33.46×10^{-3}	0.6880×10^{-5}	0.7607×10^{-8}	-3.593×10^{-12}	°C	0–1500°C

The units on the heat capacity values computed from this equation are kJ/gmol°C.

Create a macro (either by recording or programming) that will compute the heat capacity of a gas, using coefficients and temperature values in nearby cells, such as in the following table:

TEMPERATURE	a	b	c	d	c_P
300	33.46×10^{-3}	0.6880×10^{-5}	0.7604×10^{-8}	-3.593×10^{-12}	

Use your macro to find the heat capacity of water vapor at 300°C (shown above), 500°C, and 750°C. Does the heat capacity of steam change significantly with temperature?

Resolving Forces into Components I

7. Create a macro to compute the horizontal and vertical components for the following forces:

a. A 400-N force acting at 40° from horizontal.

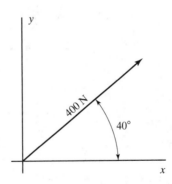

b. A 400-N force acting at 75° from horizontal.

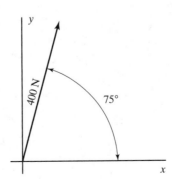

c. A 2000-N force acting at 210° from horizontal.

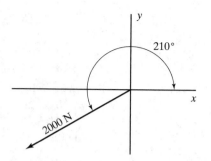

Calculating Resultant Force and Angle

8.   Create one or two macros to calculate the horizontal and vertical components and the resultant force and angle for the following forces:

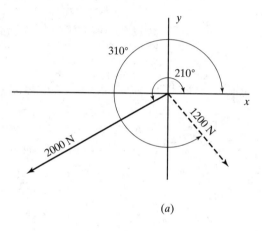

(a)

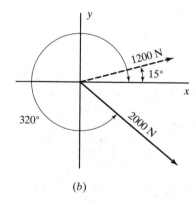

(b)

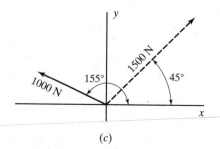

(c)

8

Programming in Excel with VBA

8.1 INTRODUCTION

Many years ago, Microsoft made the decision to use a version of Visual Basic, called Visual Basic for Applications, or VBA®, as the macro language in their spreadsheet, word processor, and so on. That decision gave macro programming a lot of power in these applications.

VBA has developed over time into a very powerful programming language. The "A" in VBA is important; VBA is a version of Visual Basic that is highly integrated with the Application—that is, it is a version of Visual Basic that has been designed to work well with Excel. The presence of VBA imbedded in Excel puts a programming environment on the engineer's desktop. This chapter will introduce VBA and how it can be used with Excel to accomplish programming tasks.

8.2 VISUAL BASIC FOR APPLICATIONS (VBA) OVERVIEW

Visual Basic for Applications is a complete programming language residing within Microsoft applications such as Excel. Running within Excel, VBA can take input from a worksheet, calculate results, and send the results back to a worksheet. Or, you can create forms to request input from the user and display the results.

Starting the VBA Editor

You write VBA code within the *Visual Basic Editor*. To start the editor, use Tools/Macro/Visual Basic Editor.

OBJECTIVES

After reading this chapter, you will know

- How to obtain access to and use the VBA programming environment
- The fundamental elements of programming
- How to read and create programming flowcharts
- How to create and edit functions and subprograms
- How to create and use forms for data entry and for presenting results
- How to use an Excel macro to open a form from the Excel workbook

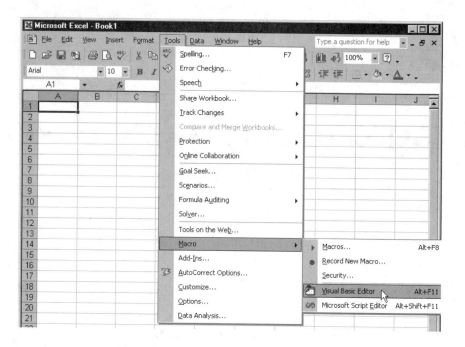

The editor opens in a new window.

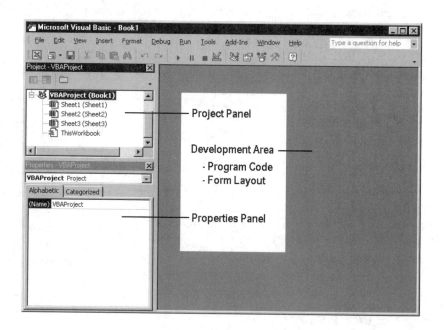

The editor is a multipaneled environment. The three standard panels are shown here. The main area (empty in this figure) is the *development area*. When you are adding buttons and text fields to a form, the form is displayed in the development area. This panel is also used for writing program code. If you have already recorded one or more macros, the VBA code for the macros will be displayed in the development area.

The *project panel* lists all of the objects in the project. By default, a VBA project contains three sheets (spreadsheets) and a workbook. By selecting a project item, you can connect program code with that item (only). For example, you could have Sheet1 and Sheet2 respond differently to mouse clicks. Code stored with the workbook object is available to each sheet in the workbook.

The *properties panel* displays all of the properties associated with a selected object. Changes made in the properties panel (e.g., font changes) change the properties of the object. Also, changes made to an object in the development area (e.g., resizing, moving) are reflected in the object's property values.

8.3   PROJECTS, FORMS, AND MODULES

A *VB project* is a collection of programming pieces that are used together to create a complete program. A typical project includes

- one or more *forms* to collect and present information to the user; and
- one or more *modules* to hold variable definitions and program code.

An Excel VBA project also includes

- the *workbook* from which the VBA program is created—this is listed as This Workbook in the project list; and
- the individual *sheets* that comprise the workbook.

By including a module, the workbook, and each individual sheet in the project, you have a great deal of control over access to your programs:

- Program elements stored in a module generally are accessible from any project source—that is, a function stored in a module (for example) can be accessed from a form stored in the project or from any sheet of the workbook.
- Program elements stored in the workbook are available to any of the sheets, but generally are not available to the rest of the project.
- Program elements stored with a specific sheet generally are available only to that sheet. This allows the programmer to write multiple versions of a function (same function name, stored with different sheets) to create specialized responses for each sheet. For example, you might create two different Create-Header( ) functions. One might create a simple title, date, and author header on spreadsheets designed to be used within the company, whereas another version would create a more elaborate header, including a logo and company contact information, on spreadsheets intended for customers and clients.

You can add additional objects to the project. Program code that is not specifically tied to an object (such as a sheet, form, or form object) normally is stored in a *module*. A module is simply an object that stores program code, such as variable definitions, Subs (subprograms) and functions. You can also insert *form objects* to create graphical user interfaces for your programs.

8.3.1   Inserting a Module

To insert a module, select Inset/Module from the VBA menu. A module named Module1 will be added to the project list. (You can right click on the name in the project list to rename the module.)

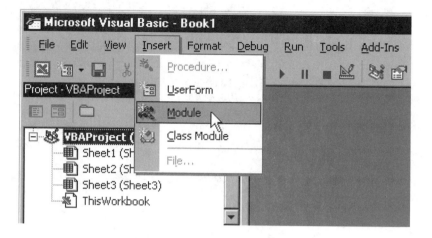

Once the module has been added to the project, the development area shows the code stored in the module (empty, for now).

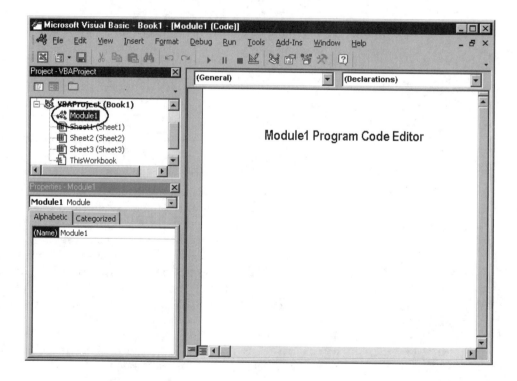

Once you have Module1 in your project, you have a place to write a function. A function is a piece of program code that accepts values through an argument list, does something with the values, and then returns a result. The result can be a single value, a range of values, or an array. (Excel treats arrays differently from simple ranges of cells; arrays are collections of cells that must be kept together.)

8.3.2   Creating a Function

Later in the chapter, we will present the fundamental elements of all programming languages; first, however, we will create a very simple function just to illustrate the process used to create VBA functions.

If you store the code for a function in a VBA module, your function will be available to any of the sheets in the project—that is, you can use your function in any cell on any of the sheets in the workbook. The process for creating a new function is the following:

1. Use Insert/Procedure . . . to

    a. declare the function,
    b. create the first and last lines of the function, and
    c. store the function code in a module.

2. Enter the required lines of program code to accomplish the desired task.

3. Return to the spreadsheet and use the function in the same way that Excel's built-in functions are used.

4. Return to the VBA editor as needed to debug the program code in the function.

As an example, we will write a calcVolume() function that is intended to calculate the volume occupied by an ideal gas. The pressure, temperature, and number of moles of gas are values that must be passed into the function.

Use Insert/Procedure . . . to Create the Function

To insert the starting and ending lines of a function into a module, first select the module in the project panel, then click in the development area for the module (to let the editor know where to insert the function). Then insert the function, using Insert/Procedure . . . from the VB main menu.

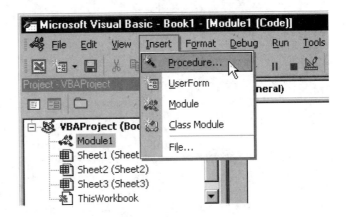

The Add Procedure dialog will be displayed.

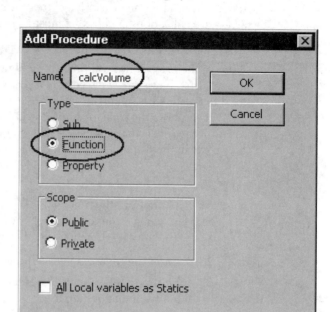

Enter a name for the new function, and set the Type to "Function". Accept the default scope, *"Public,"* so that the function will be available outside the module (to your spreadsheets, for example), and leave the "All Local variables as Statics" box unchecked—the variables in this function do not need to retain their value between calls.

When you click the OK button, the editor will insert the first and last lines of the function into the module.

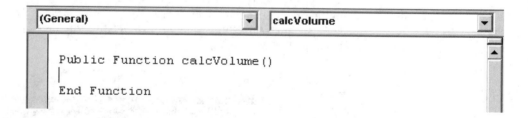

Add Required Program Code

Right now, the function is not receiving any values; we want it to receive the pressure, P, temperature, T, and number of moles of gas, N. These variable names must be added inside the parentheses after the function name (i.e., inside the argument list).

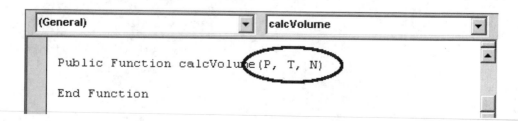

The ideal gas constant is needed for the calculation, so a variable, R, is assigned the value 0.08206. This is the gas constant with units of liter atm/mole Kelvin.

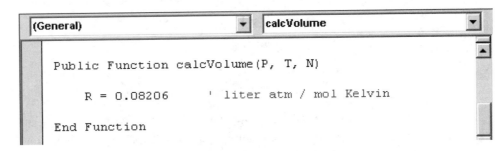

Finally, we calculate the volume and assign the result to the variable calcVolume. In VB, the value of the variable with the same name as the function name is the value that is returned when the End Function statement is reached.

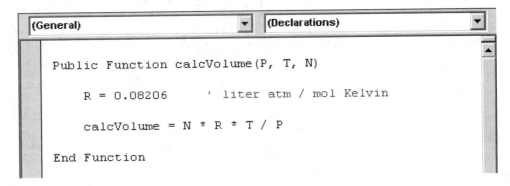

Return to the Spreadsheet and Use the Function

The calcVolume() function is now complete; it can be called from any spreadsheet in the workbook.

	C7	▼	f_x	=calcVolume(C3,C4,C5)	
	A	B	C	D	E
1	**Calculating the Volume of an Ideal Gas**				
2					
3		**Pressure:**	1	atm	
4		**Temperature:**	273.15	K	
5		**Number of Moles:**	1	gmole	
6					
7		**Volume:**	22.415	liters	
8					

Notice that the calcVolume() function requires specific units on the *P*, *T*, and N values and returns a volume in liters. Those units are shown in column *D*.

Note: It is possible that the calcVolume() function, as written here, will not work for you. The variable R was used without being explicitly declared. Depending on how the options are set in your VB environment, this practice might not be allowed. Under more precise programming techniques, the calcVolume() function would be written as follows:

```
(General)                                          calcVolume

Public Function calcVolume(P As Single, T As Single, N As Single) As Single

    Dim R As Single

    R = 0.08206      ' liter atm / mol Kelvin

    calcVolume = N * R * T / P

End Function
```

The more complete version declares variable R through the Dim R statement and indicates that each of the variables (*P, T, N, R,* and calcVolume) is of the "Single" (i.e., single-precision) data type. More information on data types will be presented later in this chapter.

8.4 FLOWCHARTS

A *flowchart* is a visual depiction of a program's operation. It is designed to show, step by step, what a program does. Typically, it is created before the writing of the program; it is used by the programmer to assist in the development of the program and by others to help them understand how the program works.

There are standard symbols used in computer flowcharts. These include the following:

Symbol	Name	Usage
	Terminator	Indicates the start or end of a program.
	Operation	Indicates a computation step.
	Data	Indicates an input or output step.
	Decision	Indicates a decision point in a program.
	Connector	Indicates that the flowchart continues in another location.

These symbols are connected by arrows to indicate how the steps are connected and the order in which the steps occur.

Flowcharting Example

As an example of a simple flowchart, consider a thermostat function that is designed to return a code value that is used to turn a heater on or off. Specifically,

- If the temperature is below 23°C, return the value 1 to indicate that the heater should be activated.
- If the temperature is above 25°C, return the value -1 to indicate that the heater should be shut off.

- If the temperature is between 23°C and 25°C, return the value 0 to indicate that there should be no change in the heater's status.

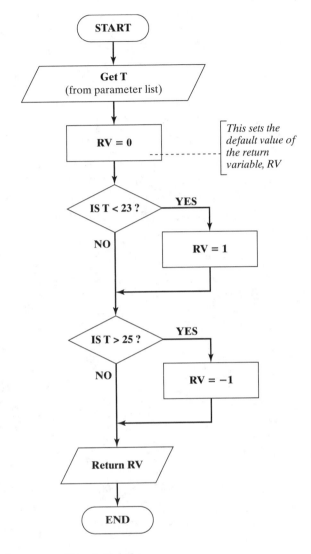

The complete function will look like this:

```
Public Function thermostat(T As Single) as Integer
    Dim RV As Integer
    RV = 0                  ' assign a default value to RV
    If T < 23 Then RV = 1
    If T > 25 Then RV = -1
    thermostat = RV    ' assign the return value to the return variable
End Function
```

This function's flowchart is shown in the preceding figure. It indicates that the function starts and the value of T is passed in from the parameter list. Then, the first line of the function is an operation step, namely, assign variable RV the value 0:

```
RV = 0                  ' assign a default value to RV
```

Note: RV stands for return value, *the value that will be returned when the function terminates.*

This step ensures that the return variable, RV, has a value no matter what T value is received. This step is called *assigning a default value* to RV.

The next line is a decision step, indicated by the diamond symbol on the flowchart:

```
If T < 23 Then RV = 1
```

The "If T < 23" portion of the second line is the condition that is checked, and the "Then RV = 1" is the operation that is performed if the condition is found to be true.

The next line of the program is another decision step:

```
If T > 25 Then RV = -1
```

The condition "If T > 25" is tested, and, if the condition is found to be true, the operation "RV = −1" is performed. Just before the program ends, the value of RV is returned through variable `thermostat` so that it is available to the worksheet:

```
thermostat = RV    ' assign the return value to the return variable
```

Note: The thermostat variable could have been used throughout the function in place of RV.

You might have noticed that the `thermostat` function tests T twice. Even if it has already been found that T < 23, it still checks to see whether T > 25. This is not a particularly efficient way to write this program. The last steps of a better version might look like this:

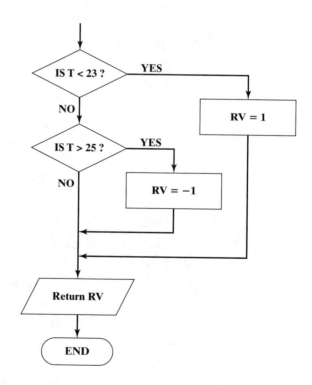

This illustrates two of the reasons to flowchart:

1.   to help identify inefficient programming; and
2.   to indicate how the program should be written.

PRACTICE!

Flowcharts may be used to depict virtually any multistep process and are frequently used to illustrate a *decision tree* (process leading to a particular decision). This **Practice!** exercise is about creating a flowchart to show a common decision process: discovering whether someone has a fever and should take some medicine and whether that medicine should be aspirin.

Disclaimer: There is no universally accepted criterion for deciding whether someone's body temperature is high enough to require medication. The values given here are sometimes used, but certainly are no replacement for good medical advice.

- For babies less than a year old, a temperature over 38.3°C (101°F) should receive medication.
- For children less than 12 years old who are not babies a temperature over 38.9°C (102°F) should receive medication.
- For people over 12 years old, a temperature over 38.3°C (101°F) should receive medication.

Oral or ear temperatures are assumed in all cases. The threshold is slightly higher for children because their body temperatures fluctuate more than baby or adult temperatures.

Aspirin is a good fever reducer, but you should not give it to children less than 12 years old (some say less than 19 years) because of the risk of a rare but serious illness known as Reye's syndrome. Acetaminophen and ibuprofen are alternatives for young people.

Create a flowchart that illustrates the input values and the decision steps necessary to learn

a. whether the person's fever is high enough to warrant medication, and

b. whether the medication should be aspirin.

How would your flowchart need to be modified to include a check for whether the person's temperature is above 41.1°C (106°F)—a level that requires immediate medical assistance?

8.5   FUNDAMENTAL ELEMENTS OF PROGRAMMING

There are a few elements that are common to all programming languages. These include the following:

• **Data**	Single-valued variables and array variables are used to hold data.
• **Input**	Getting information into the program is an essential first step in most cases.
• **Operations**	These may be as simple as addition and subtraction, but operations are essential elements of programming.
• **Output**	Once you have a result, you need to do something with it. This usually means assigning the result to a variable, saving it to a file, or displaying the result on the screen. Only the first option is available in Mathcad programs.
• **Conditional Execution**	The ability to have a program decide how to respond to a situation is a very important and powerful aspect of programming.
• **Loops**	Loop structures make repetitive calculations easy to perform in a program.
• **Functions**	The ability to create reusable code elements that (typically) perform a single task is considered an integral part of a modern programming language.

8.5.1 Data

What a program does is manipulate data, which is stored in variables. The data values can

- come from cells on the worksheet,
- come from parameter values passed into the function through the parameter list (argument list),
- be read from an external file,
- be computed by calculations within the program.

Variables can hold a single value or an array of values.

Using Spreadsheet Cell Values in Functions

When you are working in Excel, your data normally is stored in the cells of the spreadsheets. Being able to use this data in your VBA functions is essential—it is also pretty easy. There are two ways to get spreadsheet values into a VBA program:

1. Pass the data into a function through a parameter list.
2. Access a cell's contents directly from a VBA function.

Both methods are useful, but using a parameter list is more common and is presented first.

Passing Values through a Parameter List

The items in the parentheses after the function name are the function's *arguments* or *parameters*. It is good programming practice to pass all of the information required by a function into the function through the parameter list. This allows the function to be self-contained and ready to be used anywhere.

When the calcVolume( ) function was used from the spreadsheet in Section 8.3, the function received the pressure, P, temperature, T, and number of moles, N, from spreadsheet cells C3, C4, and C5.

C7	▼	f_x =calcVolume(C3,C4,C5)			
	A	B	C	D	E
1	Calculating the Volume of an Ideal Gas				
2					
3		Pressure:	1	atm	
4		Temperature:	273.15	K	
5	Number of Moles:		1	gmole	
6					
7		Volume:	22.415	liters	
8					

By default in VBA, argument values are passed to a function by reference, not by value. *Passing by reference* means that the memory address of the variable is sent to the function, and the function can look up the value when it needs it. It can also change the value at that memory address. So, when values are passed by reference, the function can change the value. You can explicitly request that variables be passed by reference by including ByRef before the variable name in the argument list, but if ByRef is omitted, the default will be used, so the variable will be passed by reference anyway.

There is one exception to this that is very important for Excel programmers:

Values stored in spreadsheet cell locations and sent to a function through the argument list can be changed within the function, but the value stored in the spreadsheet cell will not be changed. The return value from the function can (and usually does) change the value in a spreadsheet cell, but the cell values sent to the function through the argument list will not be changed by the function call. This means that values stored in cells are passed to functions by value.

Passing by value means the actual value is sent to the function, not the memory address. The function can change the value it receives, but the value stored in memory is not changed, because the function does not know where it is stored. Passing by value is not the default in VBA, except for values stored in cell addresses (as described earlier). You can request that a value be passed by value by including `ByVal` before the variable name in the argument list.

While you are writing a function, the variables you include in a parameter list and then use in the body of the function are there only to indicate how the various parameter values should be manipulated—that is, while you are defining the function, the variables in the parameter list are placeholders, or dummy variables. Also, if the variables used in the new function definition already have their own definitions in another function, it doesn't matter. You can use any variable names you want, but well-named variables will make your program easier to understand.

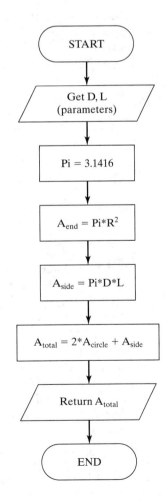

START

Get D, L
(parameters)

$Pi = 3.1416$

$A_{end} = Pi*R^2$

$A_{side} = Pi*D*L$

$A_{total} = 2*A_{circle} + A_{side}$

Return A_{total}

END

The following two program definitions are functionally equivalent, although the first one is preferred because the variable names have more meaning:

```
Public Function CylinderArea_1(D As Single, L As Single) As Single

    Dim AreaEnd As Single
    Dim AreaSide As Single
    Dim AreaTotal As Single
    Dim Pi As Single

    Pi = 3.1416
    AreaEnd = Pi * (D / 2) ^ 2
    AreaSide = Pi * D * L
    AreaTotal = 2 * AreaEnd + AreaSide

    CylinderArea_1 = AreaTotal     ' set return variable value

End Function


Public Function CylinderArea_2(v1 As Single, v2 As Single) As Single

    Dim v3 As Single
    Dim v4 As Single
    Dim v5 As Single
    Dim v6 As Single

    v6 = 3.1416
    v3 = v6 * (v1 / 2) ^ 2
    v4 = v6 * v1 * v2
    v5 = 2 * v3 + v4

    CylinderArea_2 = v5     ' set return variable value

End Function
```

Here is a quick summary of passing data into a function by means of a parameter list:

- By default, passing information into a function through a parameter causes the memory address of the variables in the parameter list, not their values, to be passed into the function. This allows the function to access the value when needed and to change the value stored in memory. To have a variable's value passed into a function instead if its address, use the ByVal keyword in front of the variable name in the parameter list.
- Values sent into VBA functions from spreadsheet cells are passed by value—a function can use cell values passed into a function, but cannot change the value in the cells referenced as parameters. (There are VBA program statements that will allow a function to change a cell's contents.)
- Using a parameter list is the preferred way to pass data from a spreadsheet into a function.
- Parameter lists help make functions self-contained so that they can more easily be reused in other workbooks.

Calculating a Resultant Force and Angle

Force-balance problems often require resolving multiple force vectors into horizontal and vertical force components. The solution may then be obtained by summing the force components in each direction and solving for the resultant force and angle. This multistep solution process can be written as an Excel function.

The resultant (vh, vv) function receives two cell ranges: vh, a vector of horizontal force components; and vv, a vector of corresponding vertical force components. The solution process, as shown by the flowchart, requires that the

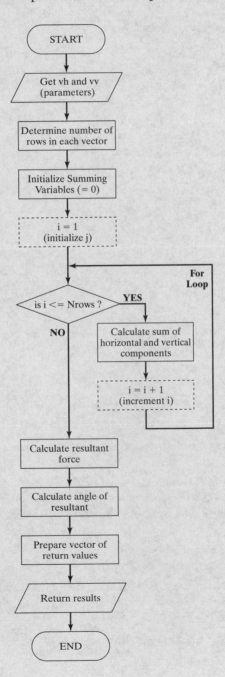

components in each direction be summed, the magnitude of the resultant force be computed by using the Pythagorean theorem, and the angle of the resultant force be calculated by using the VB atn() function. Here is the code:

```
Public Function resultant(vh As Range, vv As Range) As Variant

    Dim i As Integer
    Dim Nrows As Integer
    Dim SumVh As Single
    Dim sumVv As Single
    Dim RF As Single
    Dim Rtheta As Single
    Dim Result(2) As Single

    Nrows = vh.Cells.Count

    SumVh = 0
    sumVv = 0

    For i = 1 To Nrows
       SumVh = SumVh + vh(i)
       sumVv = sumVv + vv(i)
    Next i

    RF = Sqr(SumVh ^ 2 + sumVv ^ 2)
'   Rtheta = Atn(sumVv / SumVh)                     ' radians
    Rtheta = Atn(sumVv / SumVh) * 180 / 3.1416      ' degrees

    Result(0) = RF
    Result(1) = Rtheta

    resultant = Result

  End Function
```

Notes:

1. Using the number of cells in vh (rather than the number of rows) to establish the number of rows means that the vh and vv vectors can be either row or column vectors.

2. There are two lines of code for calculating Rtheta. The first leaves the angle in radians and has a single quote at the beginning of the line to turn the line into a comment (noncalculated line). The second Rtheta calculation is active and leaves the calculated angle in degrees.

3. This is an array function, because it returns two values. As with any array function, in order to use it, the size of the result must be selected before the entering of the formula containing the function, and the formula is entered by using [Ctrl-Shift-Enter].

4. Excel's default is to place the returned array values side by side (a row vector). You can use the Excel Transpose() function from VBA to create a column vector if desired. If this is done, the resultant variable is assigned values as follows:

```
    resultant = Application.WorksheetFunction.Transpose(Result)
```

5. Getting the resultant force and angle linked together as an array may not be the handiest way to use these results in other calculations. An alternative to the resultant() function is to create two functions, one to calculate the resultant force, and one to calculate the resultant angle.

6. The flowchart boxes with dashed-line borders represent parts of the For . . . Loop that are automatically handled by VBA.

When the resultant function is used in the spreadsheet, the magnitude and direction of the resultant vector are returned:

	B14	▼	f_x {=resultant(B6:B10,C6:C10)}		
	A	B	C	D	E	F
1	Calculating Resultant Force and Angle					
2						
3		vH	vV			
4		(N)	(N)			
5						
6		-50	50			
7		35	125			
8		75	270			
9		85	100			
10		25	120			
11						
12		Force	Angle			
13		(N)	(degrees)			
14		686.4	75.7			
15						

Accessing a Cell's Contents Directly

VBA also provides ways of accessing cells directly, either for reading values from cells or for placing formulas or values into cells. The following is a short list of available methods.

Obtaining a Value from a Specific Cell

```
CellVal = Workbooks("Book1").
Sheets("Sheet1").Range("B5")
.Value
```

There are two referencing methods available. One uses the "A1" style used in the spreadsheets, the other uses row and column values.

```
CellVal = Worksheets("Sheet1").
Cells(5, 2).Value
```

Range("B5") and Cells(5,2) refer to the same cell.

Assigning a Value to a Specific Cell

```
Workbooks("Book1").Sheets("Sheet1").
Range("B5").Value = CellVal
Worksheets("Sheet1").Cells(5, 2).
Value = CellVal
```

As used here, CellVal is a VBA variable that has already been assigned a value. These statements assign the value of CellVal to cell B5.

Making a Specific Sheet the Active Sheet

```
Worksheets("Sheet1").Activate
```

Selecting a Range of Cells

```
Worksheets("Sheet1").Activate
Range("A1:D4").Select
```

First be sure that the correct sheet is active. Then select the desired cell range.

Activating a Particular Cell

```
Worksheets("Sheet1").Activate
Range("A1:D4").Select
Range("B2").Activate
```

First be sure that the correct sheet is active. Then select the desired cell range. Then activate the desired cell.

Only one cell can be the active cell, even if multiple cells are selected.

Reading a Value from the Active Cell

```
CellVal = ActiveCell.Value
```

Assigning a Value to the Active Cell

```
ActiveCell.Value = CellVal
```

As used here, CellVal is a VBA variable that has already been assigned a value. These statements assign the values of CellVal to cell B5.

8.5.2 Input

There are a variety of *input sources* available on computers: keyboard, disk drives, tape drive, a mouse, microphone, and more. Because Excel functions are housed within a workbook, there are basically two available input sources: data available in the worksheet itself, and data in files. The use of worksheet data was presented in the previous section, so this section deals only with reading data files.

All of the standard Visual Basic file access statements are available in VBA. The process is as follows:

1. Open a data file for input, and assign a unit ID.
2. Read data from the file by using the unit ID.
3. Close the file.

In the following example, data are read from file C:\MyData.txt and stored in two array variables, A and B:

```
Public Function getData()
    Dim A, B
    Dim i as Integer
    Open "C:\MyData.txt" For Input As #1  ' Open file for input as
                                             Unit 1
    I = 1                                  ' initialize I
    Do While Not EOF(1)                    ' Loop until end of file
      Input #1, A(I), B(I)                 ' Read data
    Loop
    Close #1                               ' Close the file
End Function
```

8.5.3 Operations

VBA uses the typical mathematical *operators* as listed in the following table:

Standard Math Operators		
SYMBOL	**NAME**	**SHORTCUT KEY**
+	Addition	+
−	Subtraction	−
°	Multiplication	[Shift-8]
/	Division	/
^	Exponentiation	[Shift-6]

Operator Precedence Rules

VBA evaluates expressions from left to right (starting at the assignment operator, $=$), following standard *operator precedence rules*:

Operator Precedence		
PRECEDENCE	**OPERATOR**	**NAME**
First	$\wedge$	Exponentiation
Second	$\cdot$, /	Multiplication, Division
Third	$+$, $-$	Addition, Subtraction

For example, you might see the following equation in a function:

$$C = A \cdot B + E \cdot F.$$

You would need to know that VBA multiples before it adds (operator precedence) in order to understand that the equation would be evaluated as

$$C = (A \cdot B) + (E \cdot F).$$

It is a good idea to include the parentheses to make the order of evaluation clear.

Assignment Operator, $=$

Variables may be defined and assigned values inside a function by using the assignment operator, $=$

$$C = A \cdot B + E \cdot F.$$

8.5.4 Output

Output from a VBA function can be handled in several ways:

- Information can be returned to the cell(s) in which the function was called.
- The VBA function can place values or formulas directly into cells.
- The VBA function can send data to a file.

Using the Function's Return Value

VBA uses the function name as the variable name that contains the function's return value. For example, in the calcVolume function, "calcVolume" is both the function name and the name of the variable that holds the function's return value:

```
Public Function calcVolume(P As Single, T As Single, N As Single) As Single

  Dim R As Single

  R = 0.08206      ' liter atm / mol Kelvin
  calcVolume = N * R * T / P    ' assign the result to the return variable
End Function
```

You can return multiple values from a function by returning them as an array.

Writing Information Directly into a Cell

It is possible to write a value either to a specific cell or to the active cell. VBA also allows a formula to be written to either a specific cell or the active cell.

Assigning a Value to a Specific Cell

Workbooks("Book1").Sheets("Sheet1").
Range("B5").Value = CellVal

Worksheets("Sheet1").Cells(5, 2).
Value = CellVal

As used here, CellVal is a VBA variable that has already been assigned a value. These statements assign the values of CellVal to cell B5.

Assigning a Value to the Active Cell

Worksheets("Sheet1").Activate
ActiveCell.Value = CellVal

As used here, CellVal is a VBA variable that has already been assigned a value.

Writing a Formula to a Cell

Worksheets("Sheet1").Range("C5").
Formula = "=B5^2"

This places the formula "=B5^2" in cell C5 on Sheet 1.

Worksheets("Sheet1").ActiveCell.
Formula = "=B5^2"

This places the formula "=B5^2" in the currently active cell on Sheet 1.

Sending Data to a File

The process used to write data to a file is very similar to that used to read a file:

1. Open a data file for output, and assign a unit ID.

2. Write data to the file by using the unit ID.

3. Close the file.

In the following example, data are written to file C:\MyData.txt from two array variables, A and B. The values in arrays A and B, and the number of rows (Nrows) must be set within the program before calling function saveData().

```
Public Function saveData()
  Dim i as Integer
                                        ' A and B each contain Nrows of Data
  Open "C:\MyData.txt" For Output As #1 ' Open file for output as Unit 1
  For I = 1 to Nrows                    ' Loop through all of the data
    Output #1, A(I), B(I)               ' Write data
  Next I
  Close #1                              ' Close the file
End Function
```

EXAMPLE 8.1

Linear Regression of a Data Set

Here is an example of using an Excel VBA function to perform a series of operations on a data set and then return the results as a vector. This program will perform a linear regression (using Excel's built-in functions) on x and y values stored in two vectors and return the slope, intercept, and R^2 values. First, the data vectors must be entered into the spreadsheet:

	A	B	C	D	E	F
1	Linear Regression					
2						
3		x	y			
4		1	2			
5		2	5			
6		3	8			
7		4	13			
8		5	17			
9						

We could calculate the slope, intercept, and R^2 value by using three of Excel's built-in functions: Slope(), Intercept(), and Correl().

Note: The Correl() function returns R, which must be squared to obtain R^2.

Or, we can combine these three steps into a single function that calculates all three results at the same time . . .

```
Public Function regress(vX As Range, vY As Range) As Variant
   Dim Slope As Single
   Dim Intercept As Single
   Dim R2 As Single
   Dim Result(3)

   Slope = Application.WorksheetFunction.Slope(vY, vX)
   Intercept = Application.WorksheetFunction.Intercept(vY, vX)
   R2 = Application.WorksheetFunction.Correl(vY, vX) ^ 2

   Result(0) = Slope
   Result(1) = Intercept
   Result(2) = R2

   regress = Result

End Function
```

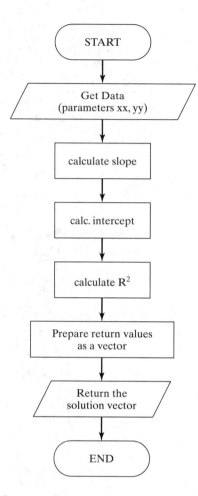

The three computed results are collected in the Result array, then returned as a vector through the regress variable:

	B11		f_x {=regress(B4:B8,C4:C8)}			
	A	B	C	D	E	F
1	**Linear Regression**					
2						
3		**x**	**y**			
4		1	2			
5		2	5			
6		3	8			
7		4	13			
8		5	17			
9						
10		**Slope**	**Intercept**	**R^2**		
11		3.8	-2.4	0.989		
12						

There is little need to build the Slope(), Intercept(), and Correl() functions into a function to handle a single data set, but it might be convenient to have a function like regress() available if you regularly need to perform a linear regression on many sets of data.

8.5.5　Conditional Execution

It is extremely important for a program to be able to perform certain calculations under specific conditions. For example, in order to determine the density of water at a specific temperature and pressure, you first have to find out whether water is a solid, liquid, or gas at those conditions. A program would use *conditional execution* statements to select the appropriate equation for density.

If Statement

The classic conditional execution statement is the If statement. An If statement is used to select from two options by means of the result of a calculated (logical) condition. In the following example, the temperature is checked to see whether freezing is a concern:

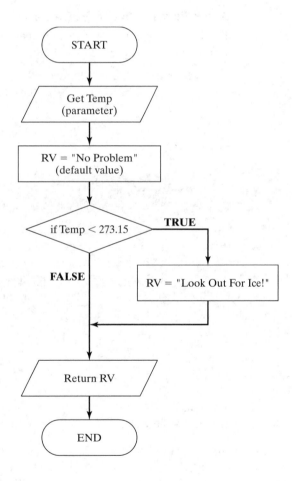

```
Public Function checkForIce(Temp As Single) As String

    Dim RV As String

    RV = "No Problem"

    If Temp < 273.25 Then RV = "Look Out For Ice!"

    checkForIce = RV

End Function
```

	D5	▼		*fx* =checkforice(B5)			
	A	B	C	D	E	F	G
1	Checking For Ice						
2							
3		Temp. (K)		Function Response		Formula	
4							
5		100		Look Out For Ice!		=checkforice(B5)	
6		150		Look Out For Ice!		=checkforice(B6)	
7		200		Look Out For Ice!		=checkforice(B7)	
8		250		Look Out For Ice!		=checkforice(B8)	
9		300		No Problem		=checkforice(B9)	
10		350		No Problem		=checkforice(B10)	
11							

Block Form If the If Statement

The If statement used in checkForIce() was a one-line version of the If statement. An alternative form is the If block. In an If block, the check for ice would look like this:

```
If Temp < 273.25 Then
        RV = "Look Out For Ice!"
End If
```

An If block allows multiple lines of code to be executed when the If condition evaluates to True.

APPLICATION: COMPUTING THE CORRECT KINETIC-ENERGY CORRECTION FACTOR, α, FOR A PARTICULAR FLOW

The mechanical-energy balance is an equation that is commonly used by engineers for working out the pump power required to move a fluid through a piping system. One of the terms in the equation accounts for the change in kinetic energy of the fluid and includes a *kinetic-energy correction factor*, α. The value of α is 2 for fully developed *laminar flow*, but approximately 1.05 for fully developed *turbulent flow*. In order to learn whether the flow is laminar or turbulent, we must calculate the *Reynolds number*, defined as

$$\text{Re} = \frac{DV_{avg}\,\rho}{\mu};$$

where

D is the inside diameter of the pipe,
V_{avg} is the average fluid velocity,
ρ is the density of the fluid, and
μ is the *absolute viscosity* of the fluid at the system temperature.

If the value of the Reynolds number is 2100 or less, then we will have laminar flow. If it is at least 6000, we will have turbulent flow. If the Reynolds number is between 2100 and 6000, the flow is in a transition region and the value of α cannot be forecast precisely. We can write a short VBA function to first calculate the Reynolds number and then use two if statements to set the value of α according to the value of the Reynolds number:

```
Public Function setAlpha(D As Single, Vavg As Single, Rho As Single, _
Mu As Single) As Variant
   Dim Re As Single
   setAlpha = "Cannot Determine"         ' set default response
```

```
    Re = (D * Vavg * Rho) / Mu

    If Re > 6000 Then setAlpha = 1.05

    If Re <= 2100 Then setAlpha = 2

End Function
```

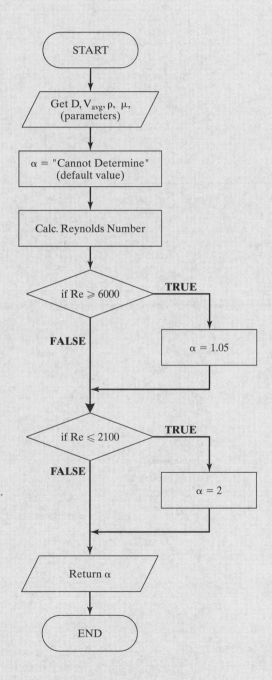

In the following example, the flow of a fluid having density equal to 950 kg/m^3 and viscosity equal to 0.012 poise, in a 2-inch pipe at an average velocity of 3 m/s, was found to be turbulent, so $\alpha = 1.05$ (the Reynolds number is 120,600):

	F11	▼		f_x	=setalpha(F5,F6,F7,F8)	
	A	B	C	D	E	F	G
1	**Selecting the Correct Kinetic Energy Correction Factor**						
2							
3			Original			Consistent	
4			Units			Units	
5		D:	2	inches		0.0508	m
6		Vavg:	3	m/s		3	m/s
7		Rho:	950	kg/m3		950	kg/m3
8		Mu:	0.012	poise		0.0012	Pa sec
9							
10					Re:	120650	
11					α:	1.05	
12							

Notice that the setAlpha return value was initially assigned the text string "Cannot Determine". This is the default case; if both of the If statements evaluate to false, the returned α value will be the warning text string.

Note: The return variable setAlpha was declared to be of type Variant so that the return "value" could be either the text phrase "Cannot Determine" or a numerical value. The variant data type is an "anything goes" type and will accept either text strings or numerical values.

Next, the Reynolds number is calculated. Then, an If statement is used to see whether the Reynolds number is less than or equal to 2100. If it is, then the flow is laminar and setAlpha is given the value 2. The next If statement checks to see whether the flow is turbulent (Re > 6000) and, if so, sets setAlpha = 1.05.

If the velocity is lowered to 0.1 m/s, then the Reynolds number falls below 6000, and the program indicates this by sending back the warning string shown:

	A	B	C	D	E	F	G	H
1	**Selecting the Correct Kinetic Energy Correction Factor**							
2								
3			Original			Consistent		
4			Units			Units		
5		D:	2 inches			0.0508	m	
6		Vavg:	0.1	m/s		0.1	m/s	
7		Rho:	950 kg/m3			950	kg/m3	
8		Mu:	0.012	poise		0.0012	Pa sec	
9								
10					Re:	4022		
11					α:	Cannot Determine		
12								

Else Statement

The Else statement is used in conjunction with an If statement when you want the program to do something else when the condition in the If statement evaluates to false. For example, we might rewrite the CheckForIce() function to test for temperatures below freezing, but provide an Else to set the text string to "No Problem" when freezing is not a concern:

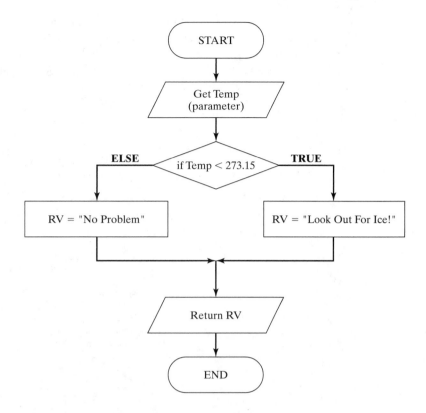

```
Public Function checkForIce(Temp As Single) As String

    If Temp < 273.25 Then
        checkForIce = "Look Out For Ice!"
    Else
        checkForIce = "No Problem"
    End If

End Function
```

This version is functionally equivalent to the earlier version, but the code might be more easily read by someone unfamiliar with programming. In the earlier version, the return value was set to "No Problem" and then overwritten by "Look Out For Ice!" if the temperature was below freezing. In this version of the program, there is no overwriting; the return value is set to one text string or the other by the result of the If condition.

On Error GoTo Statement for Error Trapping

The On Error GoTo statement is used for *error trapping* and provides an alternative program-flow path when certain conditions will cause errors in a program. Typical error situations are divisions when the denominator is zero and attempting to open a file that does not exist on the specified drive.

Error trapping involves turning on error handling just before the error-prone step, providing an alternative program-flow path in case an error is found, and turning off the error handling if the error is not detected. The statements used to accomplish these steps are the following:

On Error GoTo *MyErrorHandler*	*activates error handling*
On Error GoTo 0	*deactivates error handling*

The MyErrorHandler is actually an argument to the On Error GoTo statement; it is a line label indicating where the program flow should go if an error is detected. For example, the getData() function could be modified to provide error trapping in case the file cannot be opened. Also, return variable getData now returns a code value: 1 means that the file was read successfully, -1 that the file could not be read.

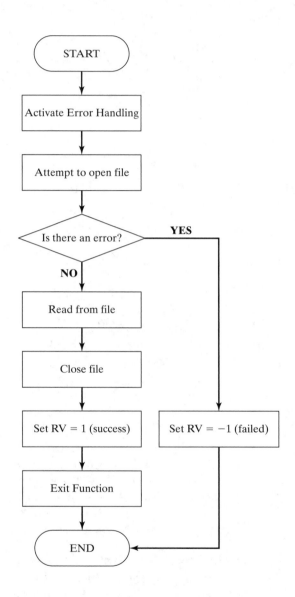

```
Public Function getData() As Integer

  Dim A, B
  Dim i As Integer

  On Error GoTo MyErrorHandler    ' turn on error handling
  Open "C:\MyData.txt" For Input As #1 ' Open file for input as Unit 1
  On Error GoTo 0                        ' turn off error handling
```

```
i = 1                              ' initialize i
Do While Not EOF(1)                ' Loop until end of file
    Input #1, A(i), B(i)           ' Read data
Loop
Close #1                           ' Close the file
getData = 1                        ' 1 = success, file was read
Exit Function                      ' if all goes well, exit here
MyErrorHandler:
    ' if you get here, the file could not be opened
getData = -1               ' -1 = failure, file could not be read
End Function
```

8.5.6 Loops

Loop structures are used to perform calculations over and over again. There are several instances in which these repetitive calculations are desirable—for example,

- You want to repeat a series of calculations for each value in a data set or matrix.
- You want to perform an iterative (guess and check) calculation until the difference between the guessed value and the calculated value is within a preset tolerance.
- You want to move through the rows of data in an array until you find a value that meets a particular criterion.

There are three loop structures supported in VBA:

- For...Next loops
- Do...Loop loops
- For Each...Next loops.

For...Next Loops For...Next loops use a counter to loop a specified number of times:

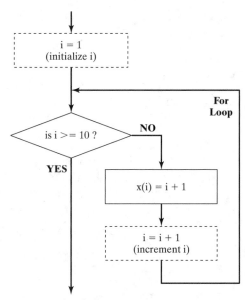

```
For i = 1 To 10
    x(i) = i +1
Next i
```

Loops can run backwards, or use nonunity steps if the step is specified. This loop decrements i by 2 each time through the loop and stops when i is less than or equal to zero:

```
For i = 10 To 0 Step -2
    x(i) = i +1
Next i
```

There is an Exit For statement that can be used to send the program flow out of a For..Next loop:

```
For i = 1 To 10
    x(i) = i +1
    If x(i) = 4 then Exit For
Next i
```

Do...Loop Loops   Do...Loop (or Do While) loops are used when you want to loop until some condition is met:

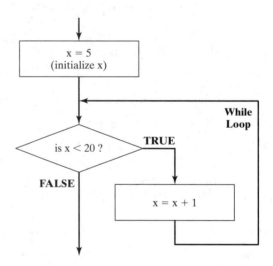

```
x = 5
Do While x < 20
    x = x +1
Loop
```

The Do While...Loop shown above evaluates the condition (is $x < 20$) before entering the loop. Because of this, the following loop would never execute, and x would never be incremented:

```
x = 50
Do While x < 20        ' since x > 20 before the loop, the loop
                         never executes at all
        x = x +1
Loop
```

Alternatively, you can use a Do...Loop While structure:

```
x = 50
Do
    x = x +1
Loop While x < 20          ' the loop always executes at least one
```

With the evaluation (While $x < 20$) placed after the Loop, the preceding loop always executes at least once.

There are also two Do...Loop loops that use an Until instead of a While. Again, the two versions vary the position of the condition evaluation:

```
x = 5
Do Until x < 20            ' since x < 20 before the loop, the loop
                             never executes at all
    x = x -1
Loop
```

This loop will never quit (watch out for this type of error):

```
x = 50
Do Until x < 20            ' since x > 20 the loop executes, but
                             since x is increased inside the loop,
                             the loop never terminates

    x = x +1
Loop
x = 50
Do
    x = x -1
Loop Until x < 20          ' since the condition follows the loop,
                             this loop always executes at least
                             once
```

There is an Exit Do statement that can be used to send the program flow out of a Do...Loop.

For Each . . . Next Loops   For Each...Next loops are specialized loop structures designed for use with *collections*. A collection is a set of related objects. A For Each...Next loop is designed to step through each object in the collection. In VBA, a common use of this loop structure is for stepping through each value in a cell range, because a Range is a type of collection in VBA. For example, the following loop counts the number of cells in the range A1:D10 that have values greater than 100:

```
Counter = 0
For Each cellObject In Worksheets("Sheet1").Range("A1:D10").Cells
        If cellObject.Value > 100 Then counter = counter + 1
Next
```

Note: cellObject is actually a variable name, and you can use any name you want. It is simply a way of identifying each element in the collection. In the For Each statement, cellObject is assigned a value corresponding to a specific element (i.e., cell) in the cell range. Then, in the If cellObject.value >100 statement, cellObject is used to identify the particular cell being tested.

8.5.7   Functions

Functions are an indispensable part of modern programming because they allow a program to be broken down into pieces, each of which ideally handles a single task (i.e., performs a single function). The programmer can then call upon the functions as needed to complete a more complex calculation.

Functions are an indispensable part of Excel as well. Excel provides built-in functions that can be used as needed to complete a lot of computational tasks. If Excel's built-in functions cannot perform a calculation, you can write your own function.

The Excel spreadsheet provides such a convenient place in which to call functions that, most of the time, you will never need to write a complete program. The most common type of programming in Excel is simply writing additional functions for use in the spreadsheets.

The Difference between Subs (Subprograms) and Functions

Subprograms or Subs are basically functions that don't return values (except through the argument list). Both have basically the same structure, with a start line and an end line that are created by the editor when the Sub or function is inserted. Both can have argument lists. But a function is designed to return a value, and a Sub isn't.

In Excel, a macro is simply a Sub that does not accept any arguments. This means you can write a Sub with no arguments in VBA, and Excel will recognize the no-argument Sub and automatically display your Sub in the list of available macros.

8.5.8 Declaring Data Types for Variables

VBA provides a lot of *data types* to describe the type of value being stored and tell VBA how much memory to allocate for each variable. If you do not explicitly declare variables to be of a particular type, VBA declares them to be Variant by default, which is an "anything goes" data type that can hold either numbers or text strings. *Variant*s are handy, but they use a lot of memory and can slow down your program.

Some commonly used data types include those in the following table:

DATA TYPE	DESCRIPTION	EXAMPLES	USE
Single	Single precision real numbers (4 bytes) $-3.402823E^{38}$ to $3.402823E^{38}$	12.6, 128.432, -836.5	General, low-precision math
Double	Double-precision real numbers (8 bytes) $-1.79769313486231E^{308}$ to $1.79769313486231E^{308}$	12.631245, -836.50001246	General, high-precision math
Integer	Small integer values (2 bytes) $-32,768$ to 32,767	1, 2, 2148, 16324	Counters, index variables
Long	Long integer values (4 bytes) $-2,147,483,648$ to 2,147,483,647	32768, 64000, 1280000	Used whenever an integer value could exceed 32,767
Boolean	Logical values (2 bytes) True or False	True, False	Status variables
String	Text strings	"yes", "the result is: ", "C:\My Documents"	Words and phrases, file names
Date	Date values (8 bytes) Floating point number representing days, hour, minutes, and seconds since 1/1/100	Values are stored and used as floating point numbers, but displayed as dates and times.	Dates and times

8.5.9   The Scope of Variables and Functions

The *scope* of a variable describes its availability to other objects. A variable declared within a function (or Sub) by using a Dim statement is available only within that function (or Sub). An example of this is the variable R in the calcVolume() function:

```
Public Function calcVolume(P As Single, T As Single, N As Single) As Single

  Dim R As Single

  R = 0.08206      ' liter atm / mol Kelvin

  calcVolume = N * R * T / P

End Function
```

Variable *R* was defined within the calcVolume() function by using a Dim statement and so can be used anywhere within the calcVolume() function, but no other function or object would know that variable *R* even exists.

The statement

```
Public Function calcVolume()
```

declares the scope of the calcVolume() function to be Public. This means that other functions or objects can call the calcVolume() function. (Technically, even objects in other workbooks can call the calcVolume() function, but they must specifically refer to the function within its defining workbook.)

As a general rule, variables that will be used only within a function should have *local scope* (defined with a Dim statement within the function and available only within the function). Variables that must be known to multiple functions (or *Subs*) must have *public scope*, and these are generally defined within the "Declarations" area of the "General" section of the project's module.

Declaring a Variable That Will Be Available to All Functions in a Module

When you insert a module into your VBA project, the module contains a Declarations section. All Public variables are declared in that section.

Example: Declare a variable named convFactor to be of type Double and Public in scope (i.e., available to any function or Sub in the module.)

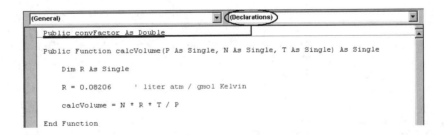

In this example, the Declarations section contains only a single line. It will expand as additional variables are declared. The editor shows all of the module's program code together, to make editing easier, but, if you click on the calcVolume() function code, the editor will show you that you have left the Declarations section and are now editing the CalcVolume section.

Declaring a Variable within a Function with Local Scope

To declare a local variable (a variable that cannot be used except within the function in which it is declared), you use a Dim statement inside the function. Declaration statements, such as Dim statements, are the first statements in a function.

Examples

```
Public Function calcValue(Dx as Single) as Single

        Dim ABC as Integer
        Public XYZ as Long       - this statement doesn't work here!
        Dim DateVar as Date

    End Function
```

Because the variables *ABC* and *DateVar* were declared inside calcValue() with Dim statements, they cannot be used outside the calcValue() function.

You cannot make a variable declared *within a function* available to other functions by using the Public statement. The statement

```
    Public XYZ as Long
```

is a valid statement, but not inside of a function. Variables declared with the Public statement, to be available to any function (or Sub) in a module must be declared in the "Declarations" section of a module.

8.6   WORKING WITH FORMS

VBA also allows you to create forms (e.g., *dialog boxes*). Forms can contain standard programming objects, such as the following:

- *Labels* (display text, not for data entry)
- *Text Fields* (for display and data entry)
- *List Boxes*
- *Check Boxes*
- *Option Buttons* (a.k.a. radio buttons)
- *Frames* (for collecting objects, especially option buttons)
- *Button* (a.k.a. command buttons)

You can tie program code to the objects to make the form do what you want. VBA gives you a lot of programming power, and using forms can be very useful for some tasks. Think of all of the routine tasks that Microsoft has made simpler through the use of wizards. Wizards are forms that someone has programmed to make a multistep task easier and less error prone.

EXAMPLE 8.2

Choosing a Gas Constant Form

As an example of what you can do with forms, and the various objects that you can include on forms, we will develop a form for selecting the units on a gas constant.

With this form, every time the user chooses a new unit, the value of the gas constant is updated at the bottom of the form. When the unit selection is complete, the user clicks, the Return R button and the displayed gas-constant value is returned to the active cell in the spreadsheet.

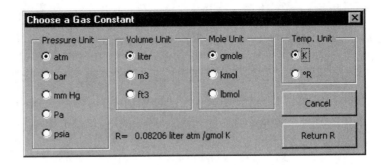

The objects on the Choose a Gas Constant form include the following:

- Four frames
  - o   Pressure Unit
  - o   Volume Unit
  - o   Mole Unit
  - o   Temp. Unit
- Several option buttons per frame
- Two labels
  - o   One label caption contains "R=", and never changes.
  - o   One label caption contains "0.08206 liter atm / gmol K" and is updated each time a new unit is selected.
- Two buttons
  - o   One Cancel button that simply hides the form from view.
  - o   One Return R button that sets the value of the active cell to the current R value, then hides the form.

The development process for this form follows such steps as the following sequence:

1. Decide what you want the form to do.
2. Develop a general layout for the form.
3. Insert a user form and a module into the current project.
4. Build the form by dragging and dropping control objects from the controls toolbox onto the form.
5. Declare variables that should be public (visible) to all of the control objects' functions in the Declarations section of a module.
6. For those variables that require initial values, set them, using the User-Form_Initialize() sub. (VBA creates this sub when the form is created. You don't need to create it; simply add code to it.)
7. Add code to the option buttons to change the R value when an option is selected.
8. Add code to the option buttons to change the display of the R value when an option is selected.
9. Add code to the cancel button to hide the form.
10. Add code to the Return R button to set the value of the active cell to the current R value and then hide the form.
11. Create a sub that shows the form.
12. Tie the sub that shows the form to a shortcut key, so that the form can be displayed from an Excel spreadsheet.

Steps 1, 2. Figure Out What You Want

The final form has already been displayed, so these steps are now apparent. The form is to collect user inputs (desired units), display the current R value, and provide a way (a button) to send the R value back to a spreadsheet.

Step 3. Insert a User Form and a Module into the Current Project

From the Visual Basic editor, use Insert/UserForm to add a new form to the project.

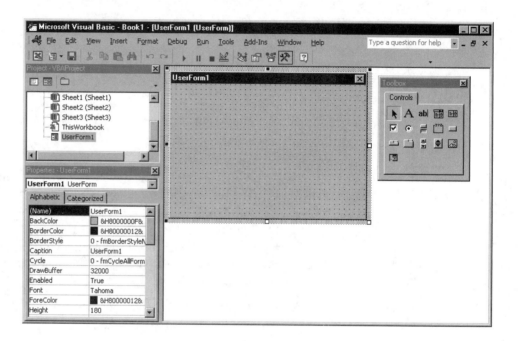

Grab and drag any of the handles along the edges to resize the form. Properties of the form are listed in the Properties panel, shown at the bottom left of the editor window. The following properties of the form were changed:

PROPERTY	CHANGED TO . . .
(Name)	Rform
Caption	Choose a Gas Constant

The caption was changed to make the heading that appears on the form more instructive.

A module provides a place to store program code. You insert a module by using Insert/Module. A module named Module1 is added to the project.

Step 4. Add Controls to the Form

The available controls are displayed on the Controls panel of the Toolbox, shown at the right side of the editor window in the previous figure. To add a control, simply select it on the toolbox, then use the mouse to indicate where and how big the control should be on the form.

Add the First Frame (for the Pressure Units)

Select the frame icon on the toolbox, then indicate where the frame should be placed on the form. You don't have to be too precise; the frame can be moved and resized later.

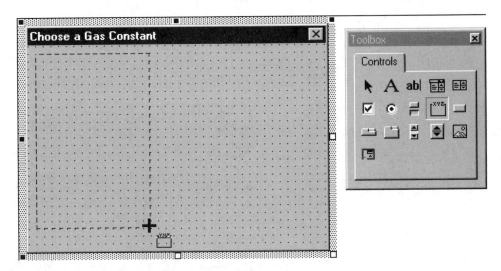

When you release the mouse button, the frame is added to the form:

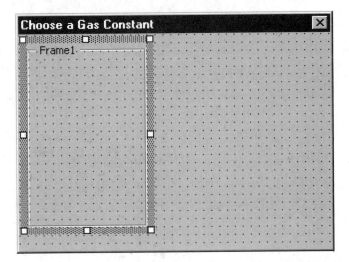

Note that the frame is left selected. (The dark border around the frame indicates it is selected.) This is so that you can change properties of the frame and add objects (such as option buttons) to the frame. The following properties of the frame were changed:

PROPERTY	CHANGED TO ...
(Name)	freP
Caption	Pressure Unit

Try to select names you can remember (they come in handy when you need to refer to an object while writing a program statement). The (Name) was changed to freP: *fre* as a reminder that this object is a frame, and *P* for pressure. The caption is the title that is displayed on the top line of the frame and should tell the user either (a) what the frame contains, or (b) what to do with the frame.

Add Option Buttons to the Frame

While the frame is selected, begin adding option buttons to the frame. In this example, we will add five option buttons to this frame.

First, one option button is added to the selected frame:

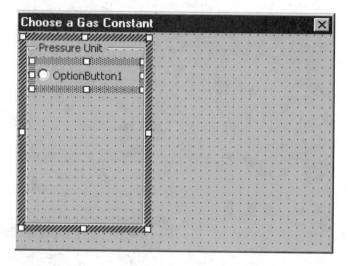

Four more option buttons can be added, or the first option button can be copied and pasted into the frame four times. Then the caption property of each option button is changed to display the unit abbreviation:

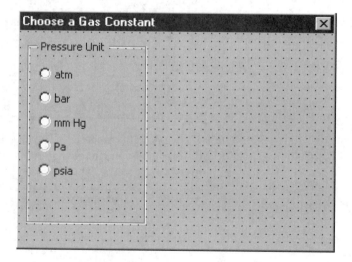

The size of the option buttons and the frame can now be reduced to fit the length of the unit names. Also, one of the options must be selected. This is done by setting the Value property of one option button to True.

OBJECT	PROPERTY	CHANGED TO...
OptionButton1	Caption	atm
	Value	True
OptionButton2	Caption	bar
OptionButton3	Caption	mm Hg
OptionButton13	Caption	Pa
OptionButton4	Caption	psia

Note that the option button labeled Pa (pascals) is out of sequence. It was added later. It makes no difference to the operation of the buttons.

The frame is vital to the correct operation of the option buttons. Only one option button per frame may be selected at any time. At run time, when one button is clicked, its Value property is set to true, and all other option buttons in the frame have their Value property set to False. The frame is what makes the option buttons a collection.

Repeat the Process for the Other Frames and Option Buttons

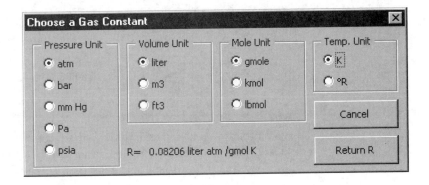

VOLUME UNIT FRAME	PROPERTY	CHANGED TO...
Frame	(Name)	freV
	Caption	Volume Unit
OptionButton5	Caption	liter
	Value	True
OptionButton6	Caption	m3
OptionButton7	Caption	ft3

MOLE UNIT FRAME	PROPERTY	CHANGED TO...
Frame	(Name)	freN
	Caption	Mole Unit
OptionButton8	Caption	gmole
	Value	True
OptionButton9	Caption	kmol
OptionButton10	Caption	lbmol

TEMPERATURE UNIT FRAME	PROPERTY	CHANGED TO...
Frame	(Name)	freT
	Caption	Temp. Unit
OptionButton11	Caption	K
	Value	True
OptionButton12	Caption	°R

Note: The degree symbol is entered as [Alt 0 1 7 6]: The [Alt] key is held down as the numbers 0, 1, 7, and 6 are pressed on the numeric keypad.

Add the Label Fields to the Form

Two label fields are used to display the current value of R. One contains just "$R=$" and never changes. It is added to the form as Label1. The caption field is changed to $R=$, but it will never be changed, so the name was left as Label1.

The second label is called Label2 when installed, but the (Name) property was changed to lblR: *lbl* as a reminder that it is a label, R, because that is what will be displayed. The caption was changed to "0.08206 liter atm/gmole K" so that the label would display the default value of R when the form is first displayed.

PROPERTY	CHANGED TO...
(Name)	lblR
Caption	0.08206 liter atm/gmole K

Add Two Buttons to the Form

Two buttons are added to the form so that users will have two ways out: one that saves the displayed value of R on their spreadsheet (the Return R button), and one that lets them leave the form without changing their spreadsheet (the Cancel button).

CANCEL BUTTON	PROPERTY	CHANGED TO...
btnCancel	(Name)	btnCancel
	Caption	Cancel
btnReturn	(Name)	btnReturn
	Caption	Return R

Step 5. Declare Public Variables

Any variables that need to be available to all of the procedures (functions and subs) used by the objects on the form must be declared in the Declarations section of the module. In this example, the following variables are declared as Public to all procedures in the module:

```
Public R As Double

Public Pvalue As Double
Public Vvalue As Double
Public Nvalue As Double
Public Tvalue As Double

Public Plabel As String
Public Vlabel As String
Public Nlabel As String
Public Tlabel As String
```

The value of *R* is stored in variable *R*. The *Pvalue, Vvalue*, and so on, variables will hold the standard values of the pressure, the volume, and so on. These values are used to calculate *R* each time a unit is changed. The *Plabel, Vlabel*, and so on, variables will hold the text descriptions of the units in use and are used to display the units in the lblR label field.

Step 6. Initialize Variables

Most computer programs require some variables to receive *initial values*. To do this with a form, you place the program statements that initialize the variables in a procedure that is run when the form loads. In VBA, it is called `Private Sub UserForm_-Initialize()`. It is run only when the form is loaded, so the variables are given these values once, but only once.

```
Private Sub UserForm_Initialize()
     R = 0.08206          ' liter atm / gmol K

     Pvalue = 1           ' atm
     Vvalue = 22.4138     ' liter
     Tvalue = 273.15      ' K
     Nvalue = 1           ' gmol

     Plabel = "atm /"
     Vlabel = "liter "
     Nlabel = "gmol "
     Tlabel = "K"
End Sub
```

This Sub is created by VBA when the form is created. To access it, you double click on some blank space on the form, then select `UserForm_Initialize` from the (long) list of empty procedures that were created in the module when the form was created.

Steps 7, 8. Add Code to the Option Buttons

This is the most tedious step in the process, but it can be expedited by using a lot of judicious cutting and pasting, because each option button's behavior should be very similar.

We begin by considering OptionButton2, the option button that is to change the pressure unit to bars. The *R* values are calculated as

$$R = \frac{PV}{NT}.$$

Initially, the pressure was 1 atm; temperature, 273.15 K; moles, 1 gmole; and volume, 22.4138 liters. These are standard conditions. Any time a unit is changed, we simply change the value in such a way as to maintain equivalency. So, when atmospheres are changed to bars, 1 atm = 1.01325 bars. Pvalue is then changed to 1.01325 and the value of *R* is recalculated:

```
Private Sub OptionButton2_Click()
     Pvalue = 1.01325
     R = Pvalue * Vvalue / (Nvalue * Tvalue)
     Plabel = "bar /"
     LblR.caption = Format(R, "###0.00000") & " " & Vlabel & Plabel
          & Nlabel & Tlabel
End Sub
```

The rest of the function is there to allow the current value and units of R to be displayed. The *Plabel* variable is changed to bar (with the divide symbol, as well), and the caption of the lblR label is pieced together by using the R value (formatted) and the current pressure, volume, mole, and temperature unit labels.

The Format() function is used when you want to control the way values are displayed. The "#" means "display a digit or nothing," and the "0" means "always display a digit, even if it is zero." The Format() function used here will cause the R value to be displayed with five decimal places:

```
Format(R, "###0.00000")
```

All of the other option buttons require similar code. Only a few of those *Subs* are reproduced here:

```
Private Sub OptionButton1_Click()

  Pvalue = 1
  R = Pvalue * Vvalue / (Nvalue * Tvalue)
  Plabel = "atm /"

  lblR = Format(R, "###0.00000") & " " & Vlabel & Plabel & Nlabel & Tlabel

End Sub

Private Sub OptionButton4_Click()

  Pvalue = 14.696
  R = Pvalue * Vvalue / (Nvalue * Tvalue)
  Plabel = "psia /"

  lblR = Format(R, "###0.00000") & " " & Vlabel & Plabel & Nlabel & Tlabel

End Sub

Private Sub OptionButton5_Click()

  Vvalue = 22.4138
  R = Pvalue * Vvalue / (Nvalue * Tvalue)
  Llabel = "liter "

  lblR = Format(R, "###0.00000") & " " & Vlabel & Plabel & Nlabel & Tlabel

End Sub

Private Sub OptionButton6_Click()

  Vvalue = 0.0224138
  R = Pvalue * Vvalue / (Nvalue * Tvalue)
  Llabel = "m3 "

  lblR = Format(R, "###0.00000") & " " & Vlabel & Plabel & Nlabel & Tlabel

End Sub

Private Sub OptionButton9_Click()

  Nvalue = 0.001
  R = Pvalue * Vvalue / (Nvalue * Tvalue)
  Nlabel = "kmol "
```

```
    lblR = Format(R, "###0.00000") & " " & Vlabel & Plabel & Nlabel & Tlabel
End Sub

Private Sub OptionButton12_Click()

  Tvalue = 459.67 + 32
  R = Pvalue * Vvalue / (Nvalue * Tvalue)
  Tlabel = "˚R"

  lblR = Format(R, "###0.00000") & " " & Vlabel & Plabel & Nlabel & Tlabel

End Sub
```

Steps 9, 10. Add Code to the *Cancel* and *Return R* Buttons

The Cancel button simply needs to close the form and do nothing. Technically, it hides the form (leaving it in memory so that it can be used again quickly.) The code that runs when the button is clicked goes in the "Click" Sub for the button that was created by VBA when the button was created:

```
    Private Sub btnCancel_Click()

        Rform.Hide

    End Sub
```

The Return R button needs to do something with the value of R before hiding the form. In the following code, the value of the active cell is given the value of R; then, the form is hidden:

```
    Private Sub btnReturn_Click()

        ActiveCell.Value = R
        Rform.Hide

    End Sub
```

Step 11. Create a Sub that Shows the Form

If you want the form to pop up from a spreadsheet, you need to create a Sub that shows the form. The reason for this is that Excel automatically considers all VBA Subs that take no arguments to be macros—and you can start macros from the spreadsheet. The following Sub, called getR(), was created by inserting a Sub procedure into the module:

```
    Public Sub getR()

        Rform.Show

    End Sub
```

To insert the procedure, make sure the module is displayed in the VBA editor, and then use Insert/Procedure. Make sure to insert a Sub, not a function, and do not give the Sub any arguments (or it will not be considered a macro by Excel).

Step 12. Tie the getR Macro to a Shortcut Key

Finally, back in Excel, change the options on the getR macro (the macro list was obtained from the Excel spreadsheet using Tools/Macro/Macros . . .) to add a shortcut key:

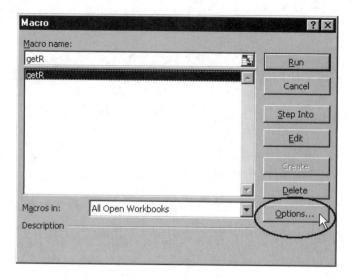

Here, a capital R was entered in the shortcut-key field, so Excel displays the shortcut key as Ctrl + Shift + R:

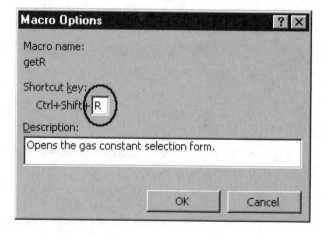

Press OK to close the Macro Options dialog, and Cancel to close the Macro list; then ... ... we're finally ready to use the form from the spreadsheet.

8.6.1 Saving Your Project

Before we go too much farther, it might be a good time to save all of the program code. Pressing the save button on the Visual Basic editor's toolbar (or using File/Save) saves the project, including the form and the program code, as part of the workbook.

8.6.2 Using the Form from the Spreadsheet

Back in the spreadsheet, you simply press [Ctrl-Shift-R] from the cell you want to contain the gas constant, and the form will be displayed:

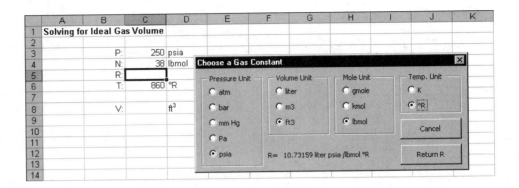

Select the desired units, and then click the Return *R* button to place the R value in the active cell (cell C5 in this example):

	A	B	C	D	E
1	**Solving for Ideal Gas Volume**				
2					
3		P:	250	psia	
4		N:	38	lbmol	
5		R:	10.73159		
6		T:	860	°R	
7					
8		V:		ft^3	
9					

The *R* value is inserted in the active cell as a number, so it can be used in subsequent calculations:

C8 f_x =(C4*C5*C6)/C3

	A	B	C	D	E
1	**Solving for Ideal Gas Volume**				
2					
3		P:	250	psia	
4		N:	38	lbmol	
5		R:	10.73159	ft^3 atm / lbmol °R	
6		T:	860	°R	
7					
8		V:	1402.833	ft^3	
9					

KEY TERMS

Arguments
Assigning a default value
Button
Check box
Collection
Conditional execution
Data type
Decision tree
Development area
Dialog box
Flowchart
Form
Form objects
Frame

Functions
Initial values
Input sources
Label
Loop structures
Macro virus
Module
Operators
Operator precedence rules
Option button
Output
Parameters
Passing by reference
Passing by value

Private (scope)
Project panel
Properties panel
Public (scope)
Return value
Reynold's number
Scope
Sheets
Subprograms
Text field
VB Project
Visual Basic Editor
Visual Basic for Applications
Workbook

SUMMARY

Starting the Visual Basic Editor   Tools/Macro/Visual Basic Editor

Inserting a Module into the Visual Basic Project—to hold Functions
From the Visual Basic Editor: Insert/Module

Function Summary

- Insert a function into a module—use Insert/Procedure ... and the Add Procedure dialog.

- Function Scope: Public—visible to entire workbook (and other workbooks); Private—visible to module only.

- Variable Scope: Public—visible to entire workbook (and other workbooks); *Private*—visible to function only.

  o   Store public variables in a module.

- Passing values—through an argument list

- Data types for variables—Single, Double, String, Boolean, Variant (default)

- Returning a Value—Use the function name as a variable holding the function's return value.

Form Summary

Standard Form Objects

- Labels (display text, not for data entry)
- Text Fields (for display and data entry)
- List Boxes
- Check Boxes
- Option Buttons (a.k.a. radio buttons)
- Frames (for collecting objects, especially option buttons)
- Buttons (a.k.a. command buttons)

General Development Process

1. Decide what you want the form to do.
2. Develop a general layout.
3. Insert a user form and a module into the current project.
4. Build the form by dragging and dropping control objects from the controls toolbox onto the form.

5. Declare public variables in the Declarations section of a module.
6. Set initial values by using UserForm_Initialize().
7. Add code to the controls on the form.
8. Set any return value(s), usually through code attached to an OK button.
9. Create a Sub that shows the form.
10. Tie the Sub that shows the form to a shortcut key, so that the form can be displayed from an Excel spreadsheet.

Problems

Ideal-Gas Volume

1. Write a VBA function that receives the absolute temperature, absolute pressure, number of moles, and ideal gas constant (optional) and then returns the calculated ideal-gas volume. Use the equation

$$PV = nRT$$

Use an IF() statement to see whether the function received a zero for the gas constant:

```
IF (R = 0) THEN
        R = 0.08206    ' use your default R value here
END IF
```

If R is equal to zero, use your favorite default value from this list:

Ideal Gas Constants	
8.314	$\dfrac{m^3\,Pa}{gmol\,K}$
0.08314	$\dfrac{L\,bar}{gmol\,K}$
0.08206	$\dfrac{L\,atm}{gmol\,K}$
0.7302	$\dfrac{ft^3\,atm}{lbmol°R}$
10.73	$\dfrac{ft^3\,psia}{lbmol\,°R}$
8.314	$\dfrac{J}{gmol\,K}$
1.987	$\dfrac{BTU}{lbmol\,°R}$

Absolute Pressure from Gauge Pressure

2. Write a VBA function that receives a gauge pressure and the barometric pressure and returns the absolute pressure. Have the function check whether the barometric pressure is zero, indicating that the program should use a default value of 1 atmosphere as the barometric pressure. If the barometric pressure is zero, add, 1 atmosphere (or 14.696 psia) to the gauge pressure to calculate the absolute pressure.

Ideal-Gas Solver

3. Write a VBA function that receives a variable name plus four known values and then solves the ideal-gas equation for a specified variable. Use a SELECT CASE statement to select the correct equation to solve. The first two options (solve for P and V) are listed next. You will need to add additional cases (for N, R, and T) ahead of the Case Else statement to complete the function.

    Note: PVNRT is declared as a Variant type in the next line so that it can return either a value (the solution) or a text string (the error message from the Case Else statement).

```
Public Function PVNRT(SolveFor As String, V1 As Single, V2 As Single,
V3 As Single, V4 As Single) As Variant
    Dim P As Single
    Dim V As Single
    Dim N As Single
    Dim R As Single
    Dim T As Single

    Select Case SolveFor

        Case "P"
            V = V1
            N = V2
            R = V3
            T = V4

            PVNRT = N * R * T / V
        Case "V"
            P = V1
            N = V2
            R = V3
            T = V4

            PVNRT = N * R * T / P
        Case Else
        PVNRT = "Ouch! Could not recognize the variable to solve for."
    End Select

End Function
```

Computing Gas Heat Capacity II

4. There are standard equations for calculating the heat capacity (specific heat) of a gas at a specified temperature. One common form of a heat-capacity equation is a simple third-order polynomial in T:[1]

$$C_P = a + bT + cT^2 + dT^3$$

    If the coefficients a through d are known for a particular gas, you can calculate the heat capacity of the gas at any T (within an allowable range).

---

[1]From *Elementary Principles of Chemical Processes*, 3d ed., R. M. Felder and R. W. Rousseau, New York: Wiley, 2000.

The coefficients for a few common gases are listed here:

			Heat Capacity Coefficients			
GAS	A	B	C	D	UNITS ON T	VALID RANGE
Air	28.94×10^{-3}	0.4147×10^{-5}	0.3191×10^{-8}	-1.965×10^{-12}	°C	0–1500°C
CO_2	36.11×10^{-3}	4.233×10^{-5}	-2.887×10^{-8}	7.464×10^{-12}	°C	0–1500°C
CH_4	34.31×10^{-3}	5.469×10^{-5}	0.3661×10^{-8}	-11.00×10^{-12}	°C	0–1200°C
H_2O	33.46×10^{-3}	0.6880×10^{-5}	0.7607×10^{-8}	-3.593×10^{-12}	°C	0–1500°C

The units on the heat-capacity values computed by using this equation are kJ/gmol °C.

a. Write a VBA function that receives a temperature and four coefficients and then returns the computed heat capacity at the temperature.

b. Use your macro to find the heat capacity of water vapor at 300°, 500°, and 750°C.

c. Does the heat capacity of steam change significantly with temperature?

Note: Because of the very small values of coefficients c *and* d, *you should use double-precision variables in this problem.*

Resolving Forces into Components

5. Write two functions, xComp() and yComp(), that will receive a force and an angle in degrees from horizontal and return the horizontal or vertical component, respectively. Test your functions with the following data:

A force $F = 1000$ N, acting at 30° (0.5235 rad) from the horizontal, resolves into the horizontal component 866 N and the vertical component 500 N:

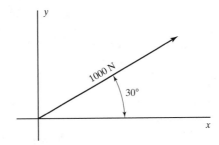

Use the functions you have written to find the horizontal and vertical components of the following forces:

a.

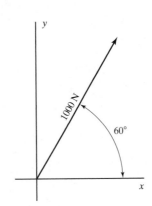

b.

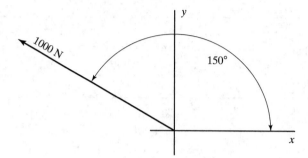

c.

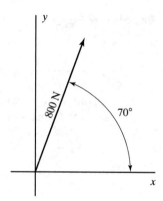

d.

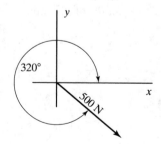

Calculating Magnitude and Angle of a Resultant Force

6. Write two functions, `resForce()` and `resAngle()`; each is to receive horizontal and vertical force components. One returns the magnitude, the other the angle, of the resultant force (in degrees from horizontal), respectively. Test your functions on this example: A horizontal component, 866 N, and a vertical component, 500 N, produce as a resultant force 1000 N acting at 30° from horizontal.

   Solve for the resultant force and angle for the following:

   a. $f_H = 500$ N, $f_V = 500$ N
   b. $f_H = 200$ N, $f_V = 600$ N
   c. $f_H = -200$ N, $f_V = 300$ N
   d. $f_H = 750$ N, $f_V = -150$ N

   Combine the following horizontal and vertical components, and then solve for the magnitude and angle of the resultant force:

   $$f_{H1} = 200 \text{ N}, f_{H2} = -120 \text{ N}, f_{H3} = 700 \text{ N}$$
   $$f_{V1} = 600 \text{ N}, f_{V2} = 400 \text{ N}, f_{V3} = -200 \text{ N}$$

9

Sharing Excel Information with Other Programs

9.1 INTRODUCTION TO SHARING EXCEL INFORMATION WITH OTHER PROGRAMS

The original method for exchanging information between programs was to save one program's results in a file (usually a text file or database file) and have the second program read the file. This is still done, and it is very effective for many situations. Excel can read the data with or without maintaining a link to the source file (called a *data source*). If the link is maintained, Excel can reread the data source file periodically to update the data in the workbook.

The *copy and paste operations* that use the Windows® *clipboard* are available to Windows programs. Most current software packages try to help out with this by analyzing the contents of the copied object and acting in "appropriate" ways: Numbers are pasted into cells as values in Excel, Excel cell contents become tables in Word®, an image is displayed as a picture, and so on. Much of this is seamless and intuitive to the user, but some information on how the process works can help if you want to control how two programs share information. The added control is usually called *Object Linking and Embedding (OLE)*®. In computer terms, an *object* is a collection of data and the procedures and methods required to handle the data. Objects can contain other objects. For example, an Excel workbook is an object, but so is a graph within the workbook, and so is the legend displayed on the graph.

9.2 USING COPY AND PASTE TO MOVE INFORMATION

The simplest way to move information is to copy it from the source program and paste it into the destination program. Using a simple copy-and-paste operation, we can transfer information (as *objects*, in computer terms) from one program

OBJECTIVES

After reading this chapter, you will know

- How to share information between Excel and Word and between Excel and Mathcad
- How to exchange information between Excel and the Web
- How to use external data sources with Excel

to another, and the objects are not *linked*. The objects may or might not be *embedded* (it depends on the type of object).

- When an object is said to be *embedded*, it means that it is stored with the file in which it is embedded and that the connection, or link, to the program that created the object is severed. For example, an Excel graph object can be embedded in a Word document. After it is embedded in the document file, it will be stored with the Word document. No connection to the workbook used to create the graph is maintained, but the graph object is still recognizable as an Excel graph. Double clicking on the graph object in Word will start Excel so that the object can be edited by using Excel.

- When an object is *linked*, a connection between the object and the source file used to create the object is maintained. Any changes to the source file will cause the object to be changed in the destination file as well. For example, if an Excel graph object were linked to a Word document, any changes to the data values in the Excel workbook would cause the graph to be updated in both the Excel file and the Word file.

9.2.1   Moving Information Between Excel and Word

Copying Cell Contents from Excel to Word   A range of cells in Excel becomes a table in Word. This is one of the easiest ways to create a table in Word.

To copy the range of cells in Excel to the clipboard,

1.   select the cells containing the information to be copied;
2.   copy the cell contents to the Windows clipboard, using Edit/Copy from Excel.

To paste the information into Word as a table,

1.   position the cursor (also called the "I-beam pointer") in Word at the location where the table should be created;
2.   paste the contents of the Windows clipboard by selecting Edit/Paste from within Word.

Example: Temperature-versus-time data used in the following polynomial-regression example can be pasted into Word as a table:

Step 1. Select the Cells Containing the Information to be Copied

Step 2. Copy the Selected Cells to the Windows Clipboard

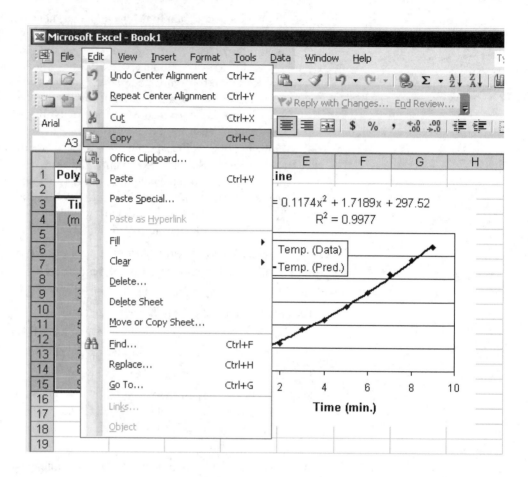

Step 3. Position the Cursor in Word at the Location Desired for the Table

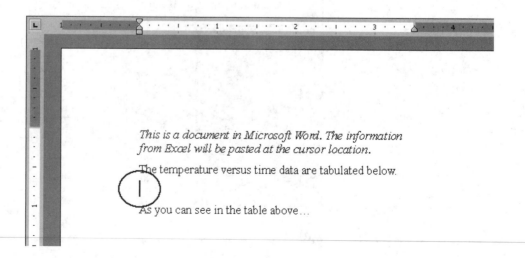

Step 4. Paste the Information By Selecting Edit/Paste from within Word

This is a document in Microsoft Word. The information from Excel will be pasted at the cursor location.

The temperature versus time data are tabulated below.

Time (min.)	Temp. (K)
0	298
1	299
2	301
3	304
4	306
5	309
6	312
7	316
8	319
9	322

As you can see in the table above...

At this point, the information from Excel has been copied to, and is stored in, Word. There is no link between the table in Word and the data in Excel. If the Excel cells contained labels or values, the cells of the Word table will display text (words or numbers). If the Excel cells contained formulas, the cells of the Word table will display the results of the formulas as text. To format the table, use Word's formatting options.

Copying a Word Table into Excel   The process of copying the contents of a Word table to Excel is basically the same.

To copy the table in Word to the clipboard,

1.   select the cells of the table containing the information to be copied;
2.   copy the cell contents to the Windows clipboard by selecting Edit/Copy within Word.

To paste the information into Excel cells,

1.   click on the cell that will contain the top left corner of the pasted range of cells;
2.   paste the contents of the Windows clipboard by selecting Edit/Paste from within Excel.

If the cells of the Word table contained text, the Excel cells will contain labels. If the Word table contained numbers, the Excel cells will contain values.

Copying an Excel Graph to Word   Creating a graph in Excel and then pasting it into Word is a very easy way to get a graph into a report. To copy a graph from Excel to Word,

1.   In Excel, click on the whitespace around the plot area to select the entire graph. Drag handles will appear on the corners and edges of the graph.

   Note: If you click on one of the objects that make up a graph (axis, series, titles, etc.) that object will be selected rather than the entire graph. By clicking in the whitespace around the plot area, you will select the entire graph, not just one element of the graph.

2.   Copy it to the Windows clipboard by selecting Edit/Copy within Excel.

3. Position the Word cursor at the location where the table should be pasted.

4. Paste the contents of the Windows clipboard by selecting Edit/Paste from within Word.

An Excel graph object is pasted into Word as an embedded object. No link is maintained with the original file, but double clicking on the graph in Word allows you to edit the graph just as you would in Excel. In the following figure, the trendline object has been selected for editing (in Word) (double clicking on the trendline again would bring up the Format Trendline dialog from Excel):

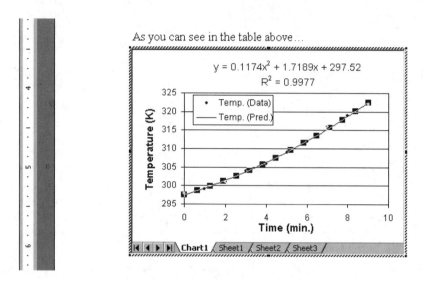

9.2.2 Moving Information between Excel and Mathcad

Excel and Mathcad® are both commonly used math packages, and each has particular strengths and weaknesses. It is very handy to be able to move information back and forth to make use of the best features of each program. Most objects copy to and from Mathcad as graphic images—just graphical "snapshots" of the Mathcad objects. An exception is an array of values, which can be moved to and from Excel.

Copying Excel Values to Mathcad Mathcad can handle arrays of numbers, but it is easier to enter columns of values in Excel. Once you have a column of values in Excel, you can copy and paste them to Mathcad. The procedure is as follows:

1. Enter the values in Excel.

2. Select the range of cells containing the values.

   Note: If there are any labels or other nonnumbers in the column of values, Mathcad will interpret the information to be pasted as an image and place a picture of the Excel values on the Mathcad worksheet.

3. Copy the values to the clipboard from Excel (Edit/Copy in Excel).

4. Enter the variable name that will be used with the matrix in Mathcad and then the *define as equal to* symbol, := (entered by pressing the [: (colon)] key).

5. Click on the placeholder that Mathcad creates on the right side of the *define as equal to* symbol.

6. Paste the values into the placeholder, using Edit/Paste from Mathcad.

EXAMPLE 9.1

Copying the Time Values to Mathcad

Steps 1 and 2. Enter the Values in Excel, and Select the Range of Cells to Be Copied

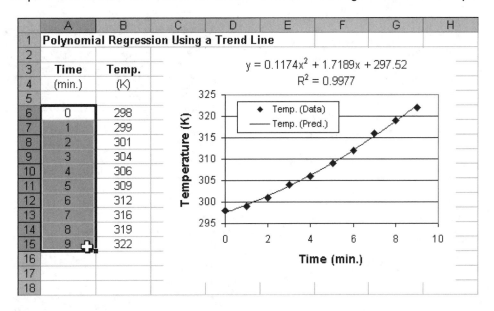

Step 3. Copy the Cell Values to the Clipboard

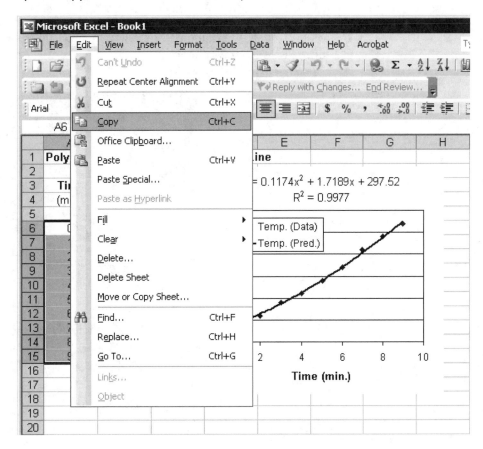

Steps 4 and 5. Create the Matrix Variable Definition in Mathcad that will Receive the Values.

The variable name `time` has been used in this Example:

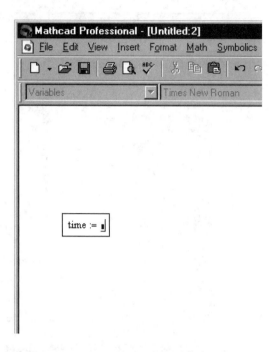

Step 6. Paste the contents of the Clipboard into the Placeholder by Selecting Edit/Paste from Mathcad

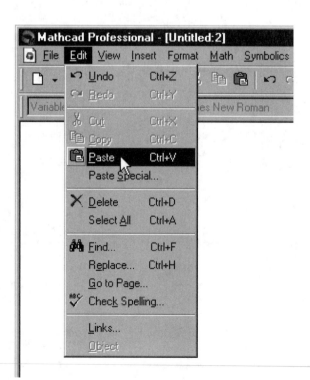

The values from Excel are used to create the `time` matrix:

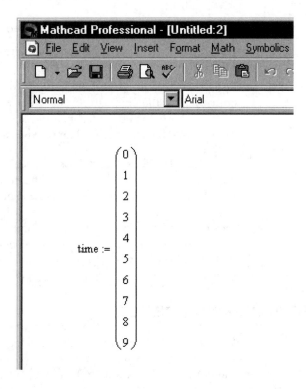

There is no link from the time matrix in Mathcad back to the Excel spreadsheet, and this is not an embedded Excel object on the Mathcad worksheet. This is simply a one-way transfer of information from Excel to Mathcad.

Copying an Array from Mathcad to Excel The process is very similar, but you must be certain to select only the values in the Mathcad array, not the entire definition (the entire *equation region*, in terms of Mathcad's terminology). When the values in the array have been selected, the selection bars surround the array, but not the "define as equal to" symbol, as shown here:

$$
\text{time} :=
\begin{pmatrix}
0 \\
1 \\
2 \\
3 \\
4 \\
5 \\
6 \\
7 \\
8 \\
9
\end{pmatrix}
$$

Once the array has been selected, use Edit/Copy from Mathcad to copy the values to the clipboard. Then use Edit/Paste from Excel to paste the values into cells.

9.2.3 Moving Information between Excel and the World Wide Web

Copying Tabular Data from HTML Pages to Excel Tabular data from HTML pages on the World Wide Web can be copied and pasted into Excel. To do so, select the cell values you want to copy (you must copy entire rows in HTML files) and copy them to the clipboard by selecting Edit/Copy from a browser. Then, in Excel, select Edit/Paste to paste the data into Excel. Excel tries to reproduce the formatting used by the HTML page's table when the data are pasted.

Creating HTML Pages from Excel This section is included here for the sake of completeness, but you do not use the copy-and-paste process to create HTML pages from Excel. Instead, use Save as Web Page . . . from the File menu to create an HTML version of your Excel spreadsheet. You will have an option during the save process to indicate whether you want the entire workbook saved as a Web page (with tabs for the various sheets) or just the current spreadsheet. An HTML document is prepared, along with a folder with the same file name. The folder is used to house any images that the page needs, such as the graph on the polynomial-regression page.

9.3 EMBEDDED AND LINKED OBJECTS

Whenever information (actually, an object) is shared between two programs, there is a *source* program in which the object was created and a *destination* that receives either the object itself or a link to the object in the other program. When the destination program receives and stores a copy of the object, the object is said to be embedded in the destination program. When the object continues to be stored with the source program and the destination program only knows how to find the object, the object is linked.

- If you double click on a **linked object** in a destination file, the original source file in which the object was created will be opened by the program used to create the object. Any changes you make to the linked object will be made to the source file and displayed in both the source and destination file.

  Note: The object to be linked must be in a saved file before you create the link. If OLE doesn't know where the object is stored, it cannot update the link.

- If you double click on an **embedded object** in a destination file, the program used to create the object will be started—but only the copy of the object that was embedded will be displayed, not the original. Any changes you make to the embedded object will be made to the copy of the object stored with the destination file.

The copy-and-paste operations just described occasionally create embedded objects (e.g., embedding an Excel graph in Word), but more often, you use the *paste special* operation to embed or link objects.

9.3.1 Embedded and Linked Objects in Word and Excel

Using Paste Special . . . to Paste an Excel Graph into Word If the polynomial-regression graph were copied to the clipboard by using Excel, we could use Edit/Paste Special . . . :

Then we would have several options on how to paste the contents of the clipboard (the Excel graph). Edit/Paste Special ... opens the Paste Special dialog box. Word's default for Excel chart objects causes the Excel graph to be pasted (embedded) as a Microsoft Excel Chart Object:

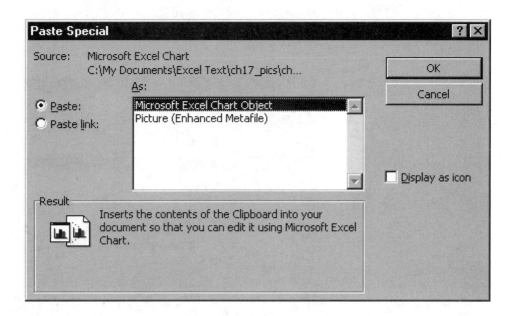

This is exactly how the Excel graph was pasted into Word with the simple copy-and-paste operation described.

The other option in the As: list (i.e., "Paste As:") is to paste the graph as a picture. This would place a picture of the Excel graph in the Word document, but it would no longer be an Excel object, so double clicking on the picture would not start Excel.

Alternatively, you could select the Paste link: option:

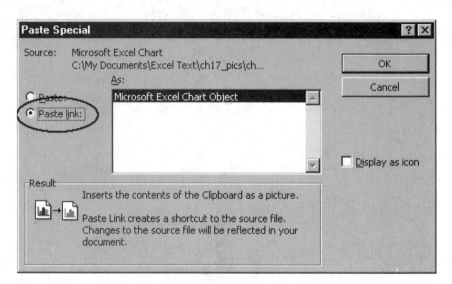

This option is used to create a link from the Word document to the graph stored in the Excel spreadsheet. The usual reason for doing this is to allow the graph displayed in the Word document (but stored in the Excel spreadsheet) to be updated automatically, any time the graph in the Excel spreadsheet is changed.

If the graph were pasted as a linked object, double clicking on the graph in Word would cause the Excel spreadsheet used to create the graph to be opened.

Using Paste Special ... to Paste Excel Cell Values as a Table in Word

If you copy a range of Excel cell values to the clipboard, such as the table of temperature and time values used earlier, you have a number of options available under the Paste Special dialog box:

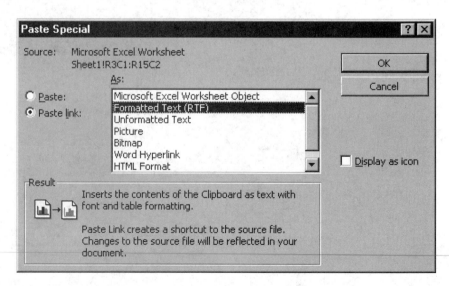

The possible ways to paste the cell values listed in the As: list are nearly the same for both the Paste: and Paste link: options. The difference is that the Paste link: option will cause the table displayed in Word to be updated any time the cell values in Excel are changed. The Paste: option will place the current clipboard contents into the Word document and not maintain a link to the Excel spreadsheet.

The various ways the clipboard contents can be pasted into the Word document are listed in the As: box. Formatted Text (RTF) is a good choice for most tables, because it allows the table to be formatted by using Word's table-formatting options, which are quite extensive. Picture and Bitmap are graphic image formats and will not allow formatting of the table contents. The Word Hyperlink method (Paste Link: only) creates a table of hyperlinks back to the original Excel data. The hyperlinks display the contents of the cells when the table is created and do not update if the Excel values are changed.

9.3.2 Embedded and Linked Objects in Mathcad and Excel

It is possible to insert an embedded or linked Mathcad object in Excel, or an Excel object in Mathcad. This effectively makes it possible to use the capabilities of Mathcad with Excel.

Embedding a Mathcad Object in Excel To insert a Mathcad object into Excel, there are a few requirements:

- You must have Mathcad and Excel running on the same computer.
- The Mathcad add-in in Excel should be activated. Select Tools/Add-Ins . . . from Excel to open the list of available add-ins for Excel. Check the Mathcad box to activate the Mathcad add-in:

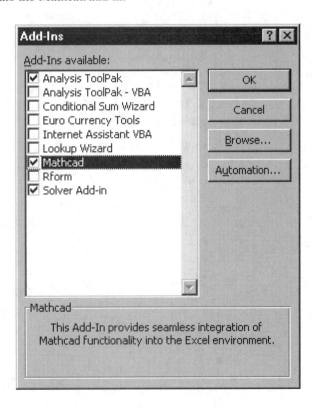

If Mathcad does not appear in the Add-Ins list, it must be installed from a Mathcad installation CD.
- The object to be linked must exist *in a saved file* before you create the link.

As an example, we'll use Mathcad's *cubic spline interpolation* functions, `cspline()` and `interp()`, to add a cubic-spline fit to the temperature–time data. We will have Mathcad calculate temperature values at half-minute intervals, using a cubic-spline interpolation, and save the additional time and interpolated temperature values back on the Excel spreadsheet. The steps required are as follows:

1. Enter the data in Excel (same as previous examples).
2. Insert an embedded Mathcad object in the Excel spreadsheet.
3. Set the properties of the Mathcad object to assign the time and temperature values to Mathcad variables.

Mathcad uses preassigned variable names for the input and output variables associated with linked and embedded objects. Input variables are called in0, in1, ..., and output variables are out0, out1, ... These Mathcad variable names are associated with Excel cell ranges through the properties of the embedded or linked object.

4. Add the Mathcad equations needed to compute the spline fit, and assign the results to output variables. The results will be transferred to Excel by OLE.
5. Use Mathcad's computed results to plot the spline curve on the temperature-versus-time graph in Excel.

Step 1. Enter the Original Data

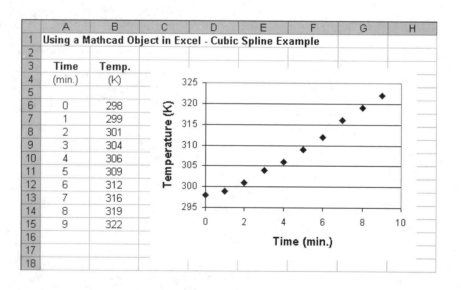

Step 2. Set the Properties of the Mathcad Object

a. Display the Mathcad Toolbar in Excel by selecting View/Toolbars/Mathcad:

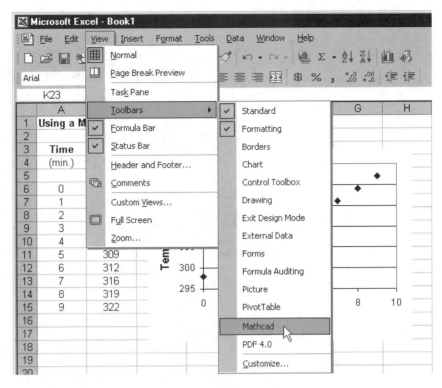

If Mathcad does not appear on the list of available toolbars, the Mathcad Add-in needs to be activated in Excel by selecting Tools/Add-Ins ... and checking the Mathcad item in the Add-Ins list.

b. Add the Mathcad object, using the Add Mathcad Object button on the Mathcad toolbar:

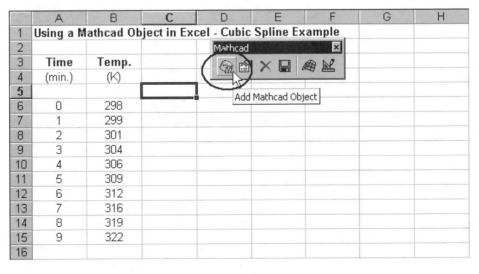

The new Mathcad object will (eventually) be displayed as a rectangular region on the Excel spreadsheet, with the top-left corner at the location of the active cell (cell C5 in our figure). Note that the existing graph has been moved out of the way. It will be brought back later a bit lower on the spreadsheet.

c. Indicate whether the Mathcad object should be embedded or linked on the Add Mathcad Object dialog.

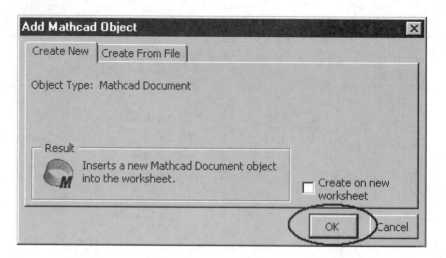

If you use the Create New panel and do not check the Create on New Worksheet box, the Mathcad object will be embedded (and stored as part of the Excel workbook). If you use the Create from File, you have the option of linking to an existing file to create a linked object.

d. Press OK on the Add Mathcad Object dialog box to insert the embedded Mathcad object on the Excel worksheet:

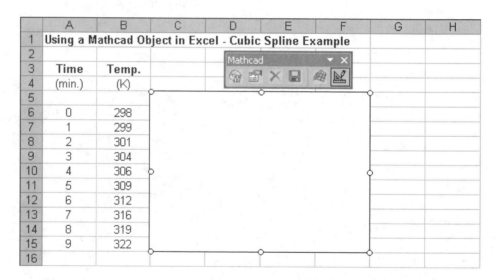

The grab handles around the Mathcad box indicate that the Mathcad object is selected.

Step 3. Set the Properties of the Mathcad Object to Assign Input and Output Variables

a. Click on the Set Mathcad Object Properties button on the Mathcad toolbar to bring up the Mathcad Object Properties dialog box:

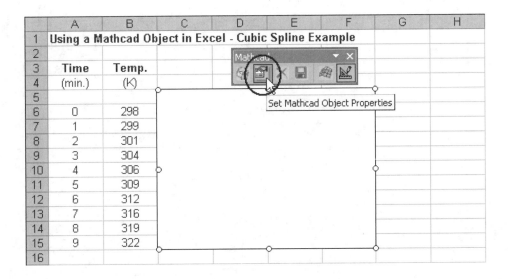

b. Indicate where the Mathcad object should find the time values.

The time values are in cells A6 through A15. If you click on the 0 under Input on the left side of the Mathcad Object Properties dialog box, you can then use the mouse to indicate the cell range containing the time values:

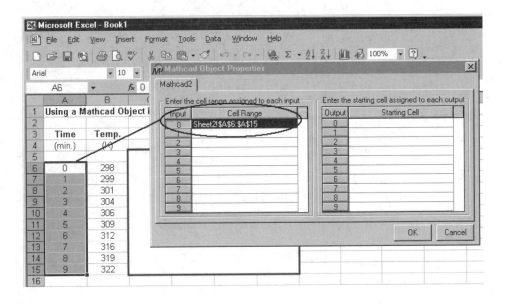

c. Indicate where the Mathcad object should find the temperature values (cells B6:B15):

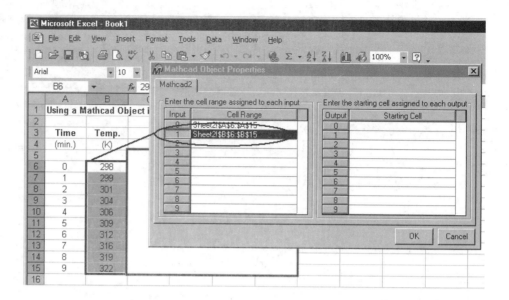

d. Indicate where the Mathcad object should put the new time and interpolated temperature values:

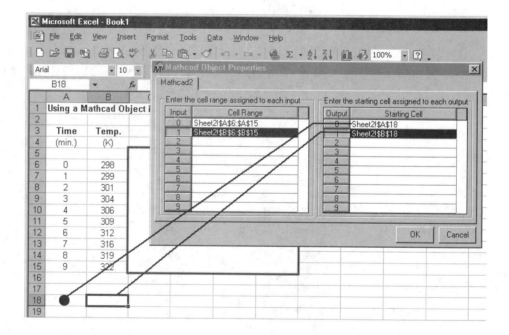

e. Press OK to close the Mathcad Object Properties dialog box.

At this point, the Mathcad object knows the time and temperature values. They are available to Mathcad as variables in0 and in1. In the next illustration, we have Mathcad display the values to be sure Mathcad is using the right values. To do so, we double click the Mathcad object and enter the Mathcad equations:

in0 =

in1 =

This causes Mathcad to display the contents of the in0 and in1 variables:

	A	B	C	D	E	F	G	H
1	Using a Mathcad Object in Excel - Cubic Spline Example							
2								
3	Time	Temp.						
4	(min.)	(K)						
5								
6	0	298						
7	1	299						
8	2	301						
9	3	304						
10	4	306						
11	5	309						
12	6	312						
13	7	316						
14	8	319						
15	9	322						
16								
17								
18								
19								
20								
21								
22								
23								
24								
25								

Within the embedded Mathcad object:

$$\text{in0} = \begin{pmatrix} 0 \\ 1 \\ 2 \\ 3 \\ 4 \\ 5 \\ 6 \\ 7 \\ 8 \\ 9 \end{pmatrix} \qquad \text{in1} = \begin{pmatrix} 298 \\ 299 \\ 301 \\ 304 \\ 306 \\ 309 \\ 312 \\ 316 \\ 319 \\ 322 \end{pmatrix}$$

Note: Displaying these arrays in the Mathcad object is not necessary to performing the cubic-spline interpolation.

Step 4. Add the Mathcad Equations to Compute the Spline Fit

	A	B	C	D	E	F	G	H
1	Using a Mathcad Object in Excel - Cubic Spline Example							
2								
3	Time	Temp.						
4	(min.)	(K)						
5								
6	0	298						
7	1	299						
8	2	301						
9	3	304						
10	4	306						
11	5	309						
12	6	312						
13	7	316						
14	8	319						
15	9	322						
16								
17								
18								
19								
20								
21								

Within the embedded Mathcad object:

$$\text{time} := \text{in0}$$

$$\text{Temp} := \text{in1}$$

$$\text{vs} := \text{cspline}(\text{time}, \text{Temp})$$

$$j := 0 .. 18$$

$$\text{newtimes}_j := \frac{j}{2}$$

$$\text{newTemps} := \text{interp}(\text{vs}, \text{time}, \text{Temp}, \text{newtimes})$$

$$\text{out0} := \text{newtimes}$$

$$\text{out1} := \text{newTemps}$$

Assigning the array called `in0` to variable `time` and array `in1` to variable `Temp` is not necessary, but it makes the equations that follow a little bit easier to understand.

The Mathcad equation

```
vs := cspline (time, Temp)
```

calculates a vector of second derivatives using a cubic-spline method and stores the values with variable name vs. These second derivatives will be used by the interp() function to interpolate new temperature values.

The equation

```
j := 0..18
```

is a Mathcad statement that creates a *range variable*. This effectively causes variable j to take on sequential values 0 through 18 each time it is used. Any equation that uses j will be evaluated 19 times, with a different value of j each time. The equation that uses j is

```
newtimesⱼ := j / 2
```

This creates a 19-element array of new time values starting at 0 (because the first value assigned to j is 0) and increasing by 0.5 each time j is incremented.

The new temperature values are calculated by the interp () function:

```
NewTemps := interp (vs, time, Temp, newtimes)
```

This statement causes new temperature values to be interpolated from the original time and temperature values by using the second-derivative values calculated by cspline().

Finally, the new time and temperature values are assigned to the standard output variables for Mathcad objects: out0 and out1. Again, these variable names could have been used directly, but newtimes and newTemps were used to try to make the equations easier to understand.

The new time and temperature values are sent back to the Excel spreadsheet at the locations defined on the Mathcad Object Properties dialog box. They will be displayed the first time that Excel recalculates the spreadsheet. Variable out0 was assigned to cell A18, and out1 was assigned to cell B18. These cell addresses are the starting addresses for the arrays returned by the Mathcad object. Only the first portion of the returned time and temperature values are displayed in the next example; they will be shown in their entirety in the next step.

	A	B	C	D	E	F	G	H
1	Using a Mathcad Object in Excel - Cubic Spline Example							
2								
3	**Time**	**Temp.**						
4	(min.)	(K)						
5								
6	0	298		$time := in0$				
7	1	299		$Temp := in1$				
8	2	301						
9	3	304		$vs := cspline(time, Temp)$				
10	4	306		$j := 0..18$				
11	5	309						
12	6	312		$newtimes_j := \dfrac{j}{2}$				
13	7	316						
14	8	319		$newTemps := interp(vs, time, Temp, newtimes)$				
15	9	322						
16								
17	**newTime**	**newTemp**		$out0 := newtimes$				
18	0.0	298.0		$out1 := newTemps$				
19	0.5	298.4						
20	1.0	299.0						
21	1.5	299.8						

Step 5. Use Mathcad's Computed Results to Plot the Spline Curve in Excel

	A	B	C	D	E	F	G	H
17	newTime	newTemp						
18	0.0	298.0						
19	0.5	298.4						
20	1.0	299.0						
21	1.5	299.8						
22	2.0	301.0						
23	2.5	302.5						
24	3.0	304.0						
25	3.5	305.0						
26	4.0	306.0						
27	4.5	307.4						
28	5.0	309.0						
29	5.5	310.4						
30	6.0	312.0						
31	6.5	314.0						
32	7.0	316.0						
33	7.5	317.6						
34	8.0	319.0						
35	8.5	320.4						
36	9.0	322.0						
37								

The open squares in this graph are the original data. The small markers were used to show interpolated values.

This example shows how a Mathcad object can be embedded in Excel. It is also possible to embed an Excel object in Mathcad.

Inserting a Linked Excel Object in Mathcad   A linked Excel object in Mathcad is an image of an Excel spreadsheet. Linked Excel objects in Mathcad do not give you access to the numbers on the Excel spreadsheet.

Before you can insert a linked object, the object must exist in a saved file. To insert a linked Excel object, do the following:

1. Create the object in Excel (the spreadsheet containing the temperature and time data will be used as an example).
2. From Mathcad, use Insert/Object and select the file containing the object.

Step 1. Create the Object in Excel

The time and temperature data used throughout this chapter will be used again in this example:

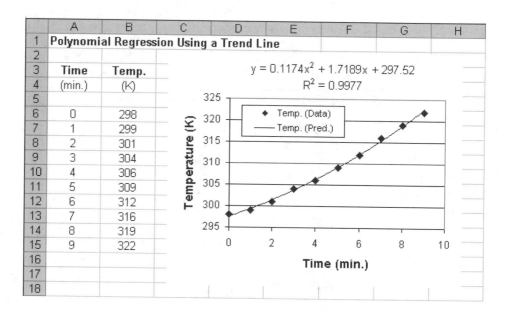

Step 2. From Mathcad, Use Inert/Object... to Insert the Excel Object

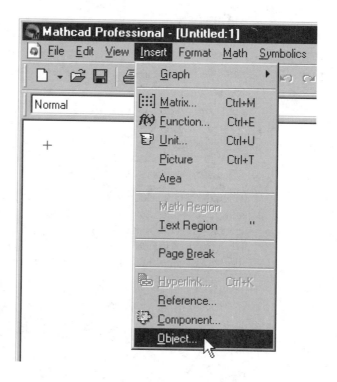

The Insert Object dialog box is displayed. Select Create from File (if you create a new object, it will be embedded, not linked); browse for the Excel spreadsheet you want to display; and then check the Link box. Press OK to close the Insert Object dialog box, and insert the linked Excel spreadsheet into the Mathcad worksheet:

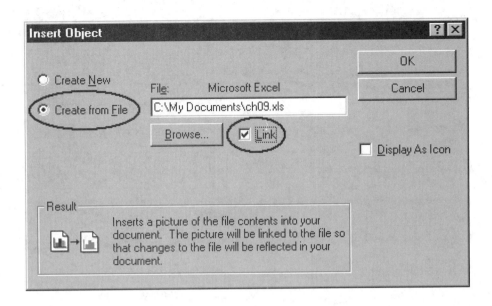

The Excel spreadsheet will be displayed on the Mathcad worksheet:

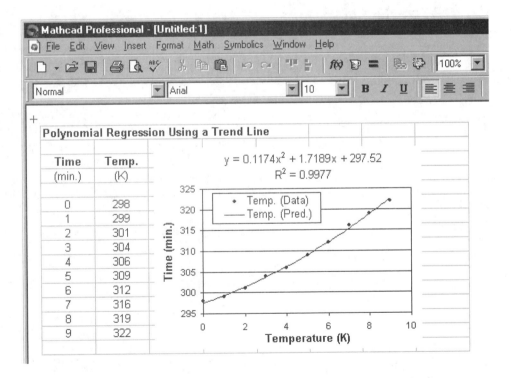

Because it is a linked object, changes to the Excel spreadsheet will be shown on the Mathcad worksheet also. For example, if the temperature at 2 minutes is (arbitrarily) changed in the Excel spreadsheet to 310 K, that shows up in the image on the Mathcad worksheet:

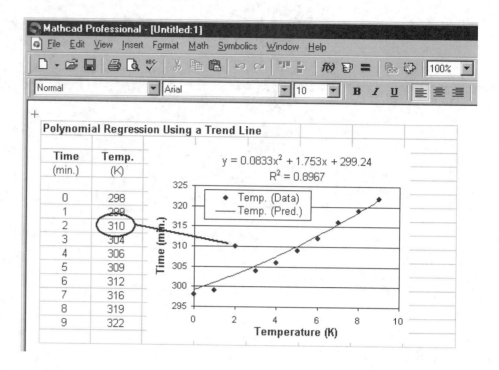

9.4 EXTERNAL DATA SOURCES

An external data source is a data file used as input to an Excel spreadsheet. (Similarly, an Excel spreadsheet can be used as a data source for another program—even another Excel spreadsheet.) The typical use of a data source is to provide automatic updating of input for a program. If the Excel spreadsheet creates, for example, a report, using an external data source, the report is automatically updated whenever the data changes.

It is common for industrial plants to keep a record of all routine plant measurements (e.g., temperatures, pressures, flow rates, and power consumptions) in a data source available for all plant technical staff to refer to and use. Three common data sources are text files, database files, and Excel files. For example, some data from a fictitious industrial plant have been saved as an Excel spreadsheet, a text file, and a Microsoft Access database table. The data in each file are the same:

Title:	Plant Standard Data Sheet
Date:	7/5/2003
Time:	12:02 AM
T102:	342.4 °C
T104:	131.6 °C
T118:	22.1 °C
T231:	121.2 °C
Q104:	43.3 GPM
Q320:	457.1 GPM
P102:	89.8 PSI
P104:	124.6 PSI
P112:	63.1 PSI
W104:	112.4 KW

The variable names are typical cryptic plant names. T102, for example, would be a temperature measurement from unit 102. Q refers to flow rate, P to pressure, and W to power. We will use these files as external data sources for a very simple Excel spreadsheet that monitors the status of unit 104, a pump. The spreadsheet checks temperature (T104 < 150°C is OK), flow rate (Q104 > 45 gpm is OK), pressure (P104 < 180 psi is OK), and power consumption (W104 < 170 kW is OK). If all are within normal operating limits, the status reads "OK," but, if not, a warning message is displayed.

Note: The external data source should be saved in a file before the reference to the data source is added to the Excel spreadsheet.

9.4.1   Text Files as Data Sources for Excel

Text files are very easy-to-use data sources for Excel. To set up a text file as an external data source, use Data/Import External Data/Import Data . . . :

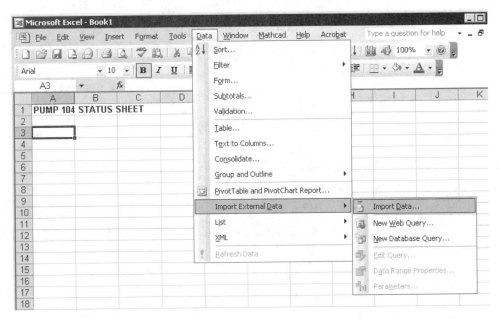

Then select the text file that contains the data you want to import. Here, it is in a file called plantData.txt in a folder named DataSource:

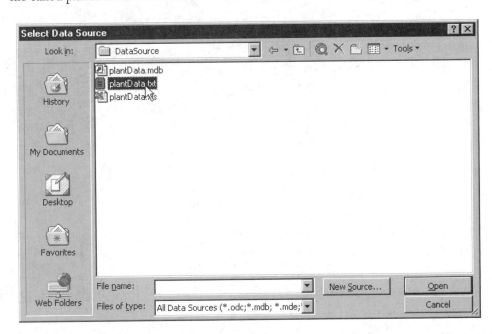

Once Excel knows which file to open, it starts the Text Import Wizard. In Step 1, you indicate the file type. This text file is tab delimited (there is a tab character between values on each row):

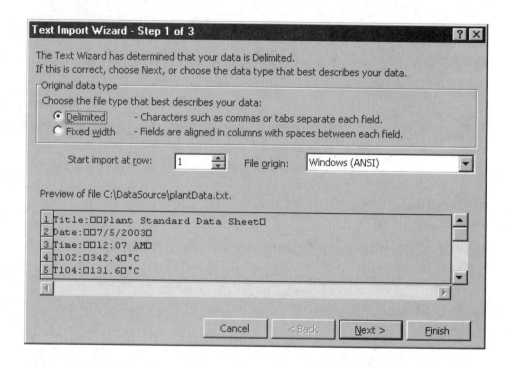

Step 2 of the wizard shows how the data will be imported and gives you a chance to change the delimiters. For this file, no change was needed:

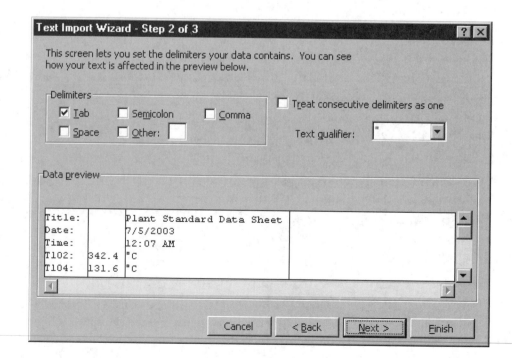

Step 3 of the wizard allows you to specify the formats used for the imported data. General format is fine, so no changes are needed:

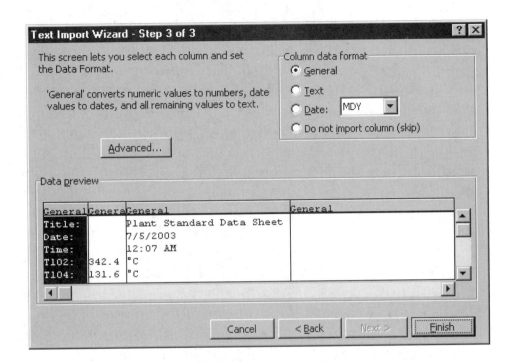

When you click the Finish button on the Text Import Wizard, the Import Data dialog box is displayed. First indicate where the imported data should go on the worksheet (cell A3 was used here), and then click the Properties button to open the External Data Range Properties dialog box:

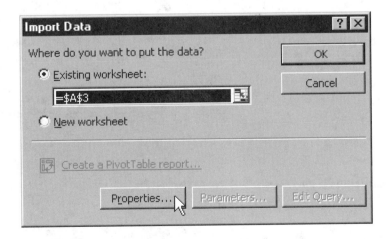

This dialog gives you a lot of control over how (and how often) the data should be refreshed. In this example, the data are to be refreshed every five minutes and whenever the Excel file is opened. You must select Save query definition if you want the data to be updated automatically.

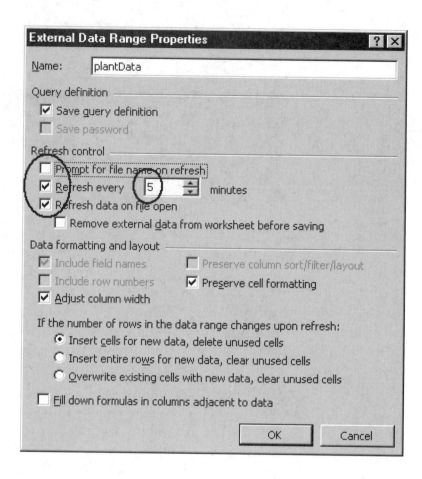

When you click the OK button on the External Data Range Properties dialog box, you are returned to the Import Data dialog box. Click OK again to finish importing the data:

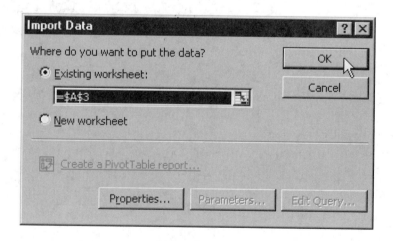

The plantData.txt file is read, and the contents are inserted on the worksheet:

	A	B	C	D	E	F	G
1	**PUMP 104 STATUS SHEET**						
2							
3	Title:		Plant Standard Data Sheet				
4	Date:		7/5/03				
5	Time:		12:07 AM				
6	T102:	342.4	°C				
7	T104:	131.6	°C				
8	T118:	22.1	°C				
9	T231:	121.2	°C				
10	Q104:	43.3	GPM				
11	Q320:	457.1	GPM				
12	P102:	89.8	PSI				
13	P104:	174.6	PSI				
14	P112:	63.1	PSI				
15	W104:	112.4	kW				
16							

Finally, the status-checking statements are added, a box is drawn around the imported data (as a reminder not to edit that portion of the spreadsheet), and the spreadsheet is complete:

E10		f_x =IF(B10>45,"OK","TOO LOW")			
	A	B	C	D	E	F
1	**PUMP 104 STATUS SHEET**					
2					**STATUS**	
3	Title:		Plant Standard Data Sheet			
4	Date:		7/5/03			
5	Time:		12:07 AM			
6	T102:	342.4	°C			
7	T104:	131.6	°C		OK	
8	T118:	22.1	°C			
9	T231:	121.2	°C			
10	Q104:	43.3	GPM		TOO LOW	
11	Q320:	457.1	GPM			
12	P102:	89.8	PSI			
13	P104:	174.6	PSI		OK	
14	P112:	63.1	PSI			
15	W104:	112.4	kW		OK	
16						

The status-checking statements in column E are as follows:

```
E7:    =IF(B7<150, "OK", "TOO HIGH")
E10:   =IF(B10>45, "OK", "TOO LOW")
E13:   =IF(B13<180, "OK", "TOO HIGH")
E15:   =IF(B15<170, "OK", "TOO HIGH")
```

Five minutes later, the data are refreshed. The flow rate and pressure recorded in the plantData.txt file have gone up, so the status of the pump has changed:

E13		▼		f_x	=IF(B13<180,"OK","TOO HIGH")	
	A	B	C	D	E	F
1	PUMP 104 STATUS SHEET					
2					**STATUS**	
3	Title:		Plant Standard Data Sheet			
4	Date:		7/5/03			
5	Time:		12:07 AM			
6	T102:	342.4	°C			
7	T104:	131.6	°C		OK	
8	T118:	22.1	°C			
9	T231:	121.2	°C			
10	Q104:	47.2	GPM		OK	
11	Q320:	457.1	GPM			
12	P102:	89.8	PSI			
13	P104:	184.6	PSI		TOO HIGH	
14	P112:	63.1	PSI			
15	W104:	112.4	kW		OK	
16						

9.4.2 Database Files as Data Sources for Excel

Database programs use tables to store information, and these tables have rows and columns, much like a spreadsheet. It's not surprising, then, that you can use database tables as data sources for Excel. The major difference between spreadsheet files, database tables, and text files is the format used to store the data, but, once the translation algorithms have been written (and many are included with Excel), the various data sources are easy to work with.

The process for using a database table as an external data source is very similar to that for using a text file. You start by defining a database query by selecting Data/Import External Data/New Database Query:

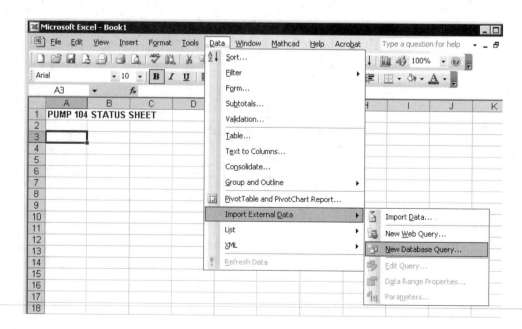

First, you select the type of database you will connect to (Microsoft Access in this example):

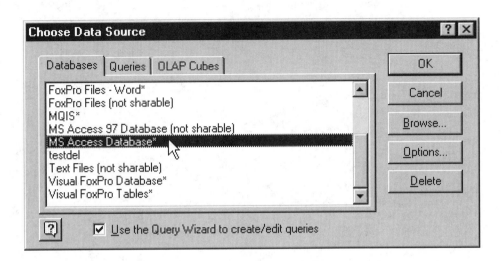

Then, find the database on your computer's hard drive or a network drive:

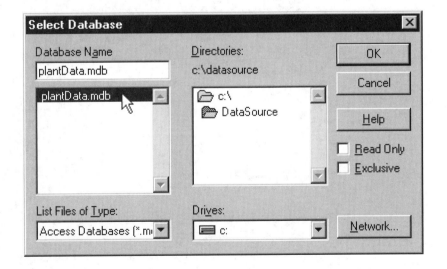

Once the database has been identified, Excel starts the Query Wizard. Within the Query Wizard, you first select the tables and columns of data you want to import. The plantData.mdb database contains only one table, called PlantDataTable, and we want to import all of the columns of data. To import all of the columns in a table, select the table in the left panel and press the [>] button to move all of the selected columns to the right panel:

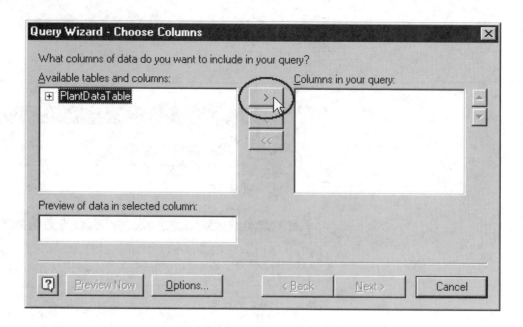

The columns to be imported are shown in the right panel. Click on the Next button to move to the next step:

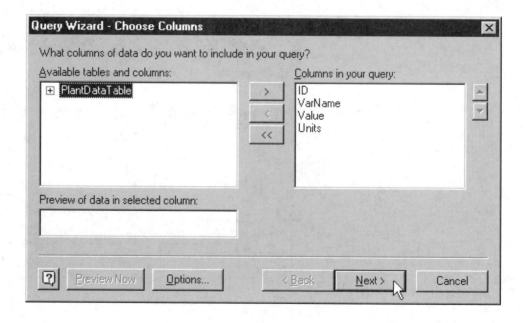

There are several steps where you can limit and rearrange the data that will be imported. Most are not needed here. We will not filter the data, so click Next:

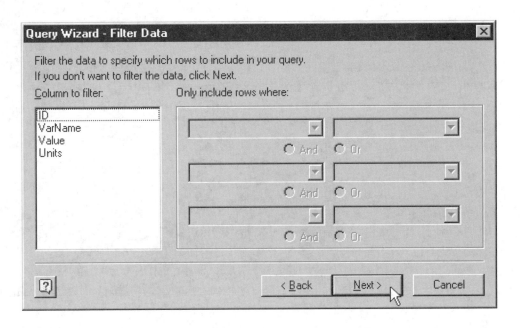

As an example of rearranging the data, we will sort the data by ID number, in ascending order:

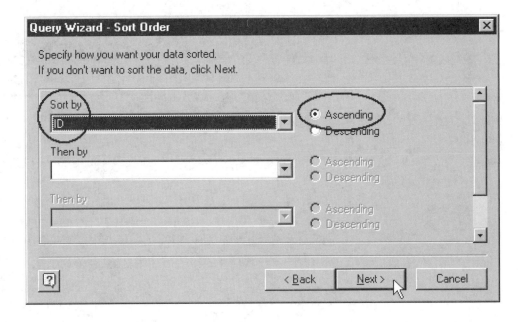

Before you finish with the Query Wizard, be sure to save the query; this is necessary if you want the data to be refreshed automatically:

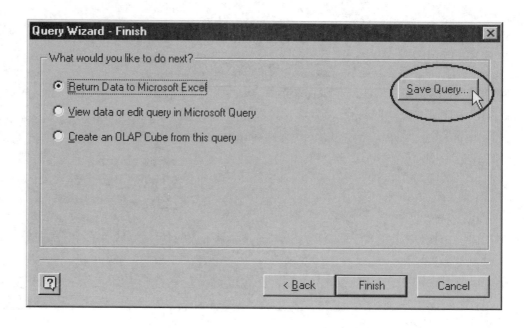

Indicate a name for the query (PlantDataQueryAccess.dqy has been used here):

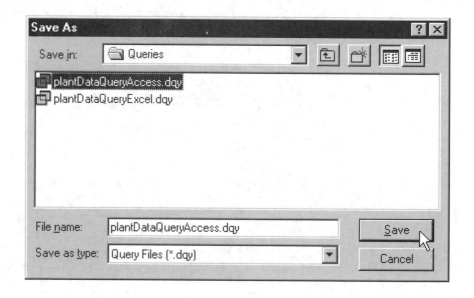

Once the query has been saved, you are returned to the Finish page of the Query Wizard. Click the Finish button:

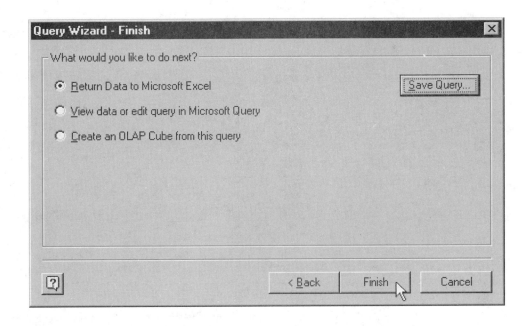

Now you need to tell Excel where to put the data on the spreadsheet. Again, cell A3 has been used here. Be sure to click the Properties button on the Import Data dialog box—that's where you set up automatic refreshing of the data:

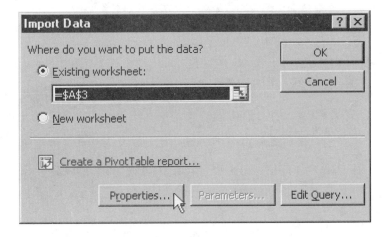

On the External Data Range Properties dialog box, Excel has been instructed to refresh the data (1) in the background, (2) every five minutes, and (3) each time the spreadsheet is opened:

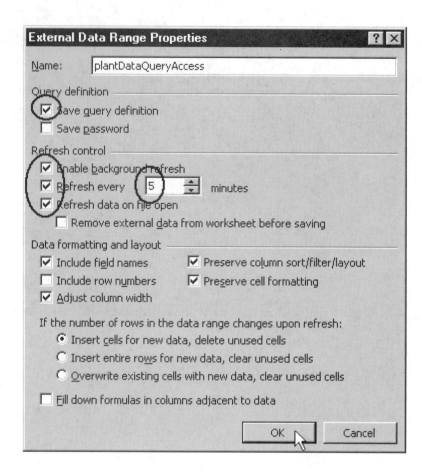

When you click OK on the External Data Range Properties dialog box, you are returned to the Import Data dialog box. Click OK to finish importing data from the database table:

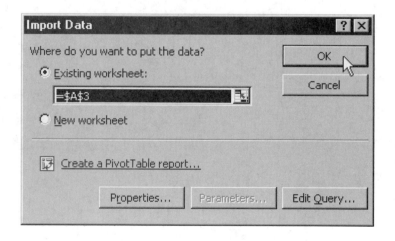

Finally, the data are available in the spreadsheet:

	A	B	C	D	E	F
1	**PUMP 104 STATUS SHEET**					
2						
3	ID	VarName	Value	Units		
4	1	Title:		Plant Standard Data Sheet		
5	2	Date:		2/3/2003		
6	3	Time:		12:02 AM		
7	4	T102:	342.4	°C		
8	5	T104:	131.6	°C		
9	6	T118:	22.1	°C		
10	7	T231:	121.2	°C		
11	8	Q104:	43.3	GPM		
12	9	Q320:	457.1	GPM		
13	10	P102:	89.8	PSI		
14	11	P104:	124.6	PSI		
15	12	P112:	63.1	PSI		
16	13	W104:	112.4	kW		
17						

After the same changes that we made last time (i.e., a box placed around the external data and status formulas put in column E), the Pump 104 Status Sheet is complete:

	E11	▼		f_x =IF(C11>45,"OK","TOO HIGH")		
	A	B	C	D	E	F
1	**PUMP 104 STATUS SHEET**					
2					**STATUS**	
3	ID	VarName	Value	Units		
4	1	Title:		Plant Standard Data Sheet		
5	2	Date:		2/3/2003		
6	3	Time:		12:02 AM		
7	4	T102:	342.4	°C		
8	5	T104:	131.6	°C	OK	
9	6	T118:	22.1	°C		
10	7	T231:	121.2	°C		
11	8	Q104:	43.3	GPM	TOO HIGH	
12	9	Q320:	457.1	GPM		
13	10	P102:	89.8	PSI		
14	11	P104:	124.6	PSI	OK	
15	12	P112:	63.1	PSI		
16	13	W104:	112.4	kW	OK	
17						

The status-checking statements in column E are as follows:

```
E8:     =IF(C8<150, "OK", "TOO HIGH")
E11:    =IF(C11>45, "OK", "TOO LOW")
E14:    =IF(C14<180, "OK", "TOO HIGH")
E16:    =IF(C16<170, "OK", "TOO HIGH")
```

9.4.3   Excel Files as Data Sources for Excel Spreadsheets

The process used to set up an Excel file as an external data source for another Excel spreadsheet is very nearly the same as that used for a database. You still start the process by selecting Data/Get External Data/New Database Query..., but, for the database type, select Excel Files:

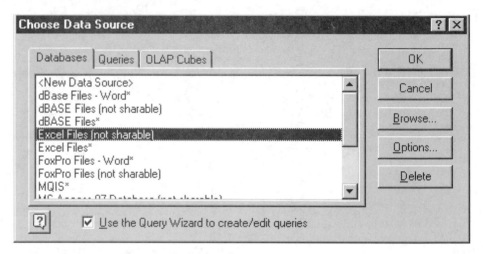

There is, however, one requirement for the data source file that must be completed before the Excel file can be used as an external data source: each range of cells to be imported must be named. The name of the cell range is used in the same way as the name of a database table. To name the range of cells to be imported, do the following:

1. Select the cells to be imported.
2. Click on the name box at the left side of the formula bar, and enter a name for the range of cells.
3. Press enter to finish naming the range of cells.

In the following example, column labels have been added to the plant data, and the range of cells is being named dataTable:

	A	B	C	D	E
	dataTable	▼	f_x VarName		
1	VarName	Value	Units		
2		Title:	Plant Standard Data Sheet		
3		Date:	2/3/2003		
4		Time:	12:02 AM		
5		T102:	342.4 °C		
6		T104:	131.6 °C		
7		T118:	22.1 °C		
8		T231:	121.2 °C		
9		Q104:	43.3 GPM		
10		Q320:	457.1 GPM		
11		P102:	89.8 PSI		
12		P104:	124.6 PSI		
13		P112:	63.1 PSI		
14		W104:	112.4 kW		
15					

Once the range of cells to be imported has been named, follow the process for importing data from a database, but set the type of data source to Excel File.

Clipboard	Embedded object	Paste special
Copy and paste operations	Linked object	Range variable
Data source	Object	Source
Destination	OLE: Object Linking and Embedding	

SUMMARY

Embedded and Linked Objects

- **Linked object**: stored with the source program; all the destination program knows is how to find the object.
  Note: The object to be linked must be in a saved file before the creation of the link.
- **Embedded object**: the destination program receives and stores a *copy* of the object.

You typically use the paste special operation to embed or link objects.

Moving Information between Excel and Word
Copying Cell Contents From Excel to Word

1. Select the cells.
2. Copy the cell contents by selecting Edit/Copy from Excel.

To paste the information into Word as a table,

1. Position the cursor at the location where the table should be created.
2. Paste the contents of the Windows clipboard by selecting Edit/Paste from within Word.

Copying an Excel Graph to Word

1. In Excel, click on the graph.
2. Copy it by selecting Edit/Copy within Excel.
3. Position the Word cursor at the location where the graph should be pasted.
4. Paste the graph by selecting Edit/Paste from within Word.

Moving Information between Excel and Mathcad
Copying Excel Values to Mathcad

1. Enter the values in Excel.
2. Select the range of cells containing the values.
3. Copy the values by selecting Edit/Copy in Excel.
4. In Mathcad, enter the variable name that will be used with the matrix and the *define as equal to* symbol, :=
5. Click on the placeholder that Mathcad creates.
6. Paste the values into the placeholder, using Edit/Paste from Mathcad.

Copying an Array from Mathcad to Excel
Select only the values in the Mathcad array, not the entire definition (the entire *equation region*, using Mathcad's terminology). When the values in the array have been selected, the selection bars surround the array, but not the define as equal to symbol, as shown here:

$$time := \begin{pmatrix} 0 \\ 1 \\ 2 \\ 3 \\ 4 \\ 5 \\ 6 \\ 7 \\ 8 \\ 9 \end{pmatrix}$$

Once the array has been selected, choose Edit/Copy from Mathcad, and then Edit/Paste from Excel, to paste the values into cells.

Moving Information between Excel and the World Wide Web

Copying Tabular Data from HTML Pages to Excel

Select the cell values you want to copy (you must copy entire rows in HTML tables) and copy them by selecting Edit/Copy from a browser. Then, in Excel, select Edit/Paste to paste the data into Excel.

Creating HTML Pages from Excel

Use Save as Web Page … from the File menu to create an HTML version of your Excel spreadsheet.

External Data Sources

Text Files

Choose Data/Get External Data/Import Text File. The text import wizard will start.

Database Files

1. Select Data/Get External Data/New Database Query . . . .
2. Select the type of database you will connect to.
3. Browse for the database file.
4. The Query Wizard will start when the database file has been identified.

Excel Files

Requirement: Cells to be imported must be in named ranges.

1. Select Data/Get External Data/New Database Query . . . .
2. Select *Excel Spreadsheet* as the type of database you will connect to.
3. Browse for the Excel file.
4. The Query Wizard will start when the database file has been identified.

10

Time Value of Money with Excel

10.1 TIME, MONEY, AND CASH FLOWS

Making decisions that are based upon economic considerations is a common task for an engineer. The decision may be as simple as deciding which vendor to use for the parts needed for a construction project. The cost of the parts will be a large part of the decision-making process, but other nonmonetary considerations also come into play, such as the availability of the parts and the reliability of the vendor at delivering the parts on time. These nonmonetary questions can be significant parts of the decision-making process; they are termed *intangibles*. Here we will consider only the monetary aspects of making decisions.

Other common decisions include

1. Choosing between competing technologies.
2. Making the best use of limited resources.

Examples are as follows:

1. An electronic switching device could use a mechanical relay or a solid-state relay. The mechanical relay would be less expensive, but slower to open and close, and it would wear out and have to be replaced periodically. The solid-state relay would be more reliable and operate more quickly, but would be more expensive than a mechanical relay. Which type should be specified?
2. Two engineers working for the same company each have an idea that will save money. Ben's idea will cost $10,000 to implement, but will save the company $200 a day after the project is implemented. Anna's idea will cost $2.3 million to build and will actually increase operating costs by $360,000 per month during a six-month startup period. However, after the startup period, the savings will be $520,000 per month forever. Where should the company put its money?

OBJECTIVES

After reading this chapter, you will know

- The principles of the time value of money and engineering economic decisions
- How to move money through time at specified interest rates
- How to use built-in Excel functions for present values, future values, and regular payments
- How to use time-value-of-money concepts to make economic decisions
- The terminology used with time-value-of-money calculation

In the first example, the speed and reliability of the solid-state relays might make them the better choice if the additional cost is not excessive. If they are a lot more expensive, the costs and hassle associated with replacing the mechanical relays would need to be considered.

In the second example, the company might decide to implement both ideas. Or, the company might not have $2.3 million available or be unable to handle the increased operating costs for the six-month startup period, so Anna's project might not get funded, even if it is a great idea.

It is important to know how to make these economic decisions.

10.1.1 Cash Flows

A big return on an investment in two years will not help you pay your rent next month. For your personal finances or your company's finances, the cash must be available when it is needed. To keep track of the availability of cash over time, we look at the *cash flow*. In the second example, Anna's idea required a large initial outlay of cash ($2.3 million) and regular cash payments for six months ($360,000 per month) to take care of the increased operating costs. The company would need a cash flow sufficient to handle all of these costs; otherwise, the company would go bankrupt before it began reaping the benefit of the $520,000 per month cost savings.

Cash flows can be diagrammed in a couple of ways. A common method of illustrating expense and income payments over time is to use arrows on a *timeline*. Arrows pointing towards the timeline indicate out-of-pocket *expenses* (from you, or from your company). Arrows pointing away indicate *incomes*. The timeline for the first 24 months of Anna's idea would look like this:

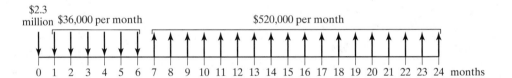

An alternative is to scale the arrows to the size of the expense or income to keep track of the total funds available to the company over time. For example, if the company had $3.5 million in their capital fund when they began implementing Anna's idea, the capital fund balance as a cash flow diagram (neglecting the time value of money) might look like this:

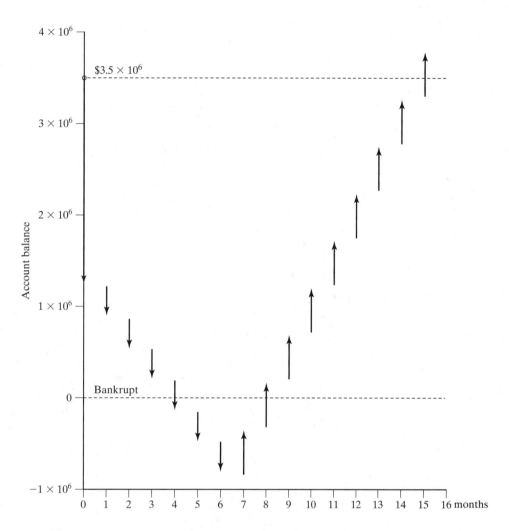

Trying to implement Anna's idea with $3.5 million in the capital fund would bankrupt the fund in four months. The diagram shows that (ignoring interest rates for a moment) nearly another million dollars is required to cover the operating costs in the first six months. If the company had $4.5 million available for the project (at time zero), then Anna's idea would pay back the capital fund in 15 months, and afterwards the $520,000 per month savings would be pure profit.

10.1.2 Types of Expenses

Expenses are categorized by when they occur and by whether they recur. Single payments, such as the $2.3 million expense at time zero in the last example, occur frequently in time value of money problems. When these expenses occur at time zero, they are called *present values* and are given the symbol PV. Single payments at known later times are called *future values*, or FV.

Regularly recurring, constant-amount expenses, such as the $360,000 operating cost in the last example, are also common, and time-value-of-money formulas have been developed to handle this type of payment. A classic type of regularly occurring expense is an annual payment into an investment. This annual payment is called an *annuity payment*, and the symbol *A* (for annuity) is sometimes used for any regularly recurring payment, even if the interval between payments is not one year. In this text, regular payments will by labeled with the symbol Pmt. Regularly occurring expenses with differing amounts are handled as a series of single payments.

Expense payments may be made at the beginning of the time period (common for rents and tuition) or at the end of the period (common for provided services, such as utility and telephone bills). Excel's time-value-of-money functions allow you to specify whether the payments are made at the beginning or end of the time periods.

10.1.3 Types of Income

Types of income are also categorized by when they occur and by whether they recur. Single payments at time zero are called present values (PV) or, in the context of a loan, the *principal*. Incomes resulting from the sale of used equipment at the end of a time period (the *service life* of the equipment) is called a *salvage value* and is a type of future value (FV). Incomes could also be received at regular intervals and are included in Excel's time-value-of-money functions as regular payments (to you) (Pmt). If the amount of the periodic income is not constant, the incomes are handled as a series of single payments.

10.2   INTEREST RATES AND COMPOUNDING

Money left in a bank grows with time because of interest payments. However, if the nation's *inflation rate* is greater than the bank's *interest rate*, the buying power, or value, of your banked money decreases with time. The inflation rate is effectively a negative interest rate.

The value of money over time depends on the interest rate. There is no single interest rate used. Banks give their customers one rate for saving accounts, but often charge a different rate for car loans and yet another rate for home loans. Your company might use an effective interest rate when deciding which projects to fund. Before you can work with money over time, you need to know the applicable rate(s).

The *compounding period* is also significant. Bank loans typically require monthly payments and calculate the interest you owe on a monthly basis (monthly compounding). Many banks compound daily on savings accounts, although many credit unions use quarterly compounding. Excel's time-value-of-money functions use *periodic interest rates* to handle any of these options.

There are several ways of expressing interest rates:

- *Periodic interest rates* are the interest rates used over a single compounding period. If a bank pays interest daily, the interest rate they use to compute the amount of the interest payment is the daily (periodic) interest rate.
- *Nominal interest rates* are the most common, and least useful, way of expressing interest rates. A nominal interest rate is an annual rate that ignores the effect of compounding. If you have a periodic interest rate of 1% per month, the nominal annual rate would be 12% compounded monthly. The "compounded monthly" is an important part of the nominal-interest-rate specification; you must know the compounding interval in order to use a nominal annual interest rate, and you use a nominal rate by converting it into a periodic interest rate.
- *Effective annual interest rates* are calculated annual interest rates that yield the same results as periodic interest rates with nonannual compounding.
- *Annual Percentage Rates (APR)* are the way lending institutions in the United States are required to present interest rate information on funds you borrow. They are not the same as an effective annual interest rate for two reasons. First, any up-front fees that are part of the total cost of borrowing are included when calculating the APR. Second, the APR tells you what the total cost of

borrowing would be with annual compounding, but that does not mean that annual compounding will be used. Most loans that quote an APR rate use monthly compounding. The actual interest cost is therefore greater than what you calculate using the APR and a compounding period of one year. If there are no up-front fees, the APR can be used as a nominal interest rate if the compounding period is known.

- *Annual Percentage Yields (APY)* are the way savings institutions in the United States are required to present interest rate information on interest-bearing accounts, such as savings accounts or certificates of deposit (CDs). The APY is a nominal interest rate calculated by assuming daily compounding.

10.2.1 Periodic Interest Rates (i_P)

Nonannual compounding periods are very common. The interest rate per compounding period (e.g., percent per day) is called a periodic interest rate. Excel's time-value-of-money functions all use periodic interest rates.

Note: You must be careful that the periodic interest rate and the number of payments are on the same time basis. If your interest rate is expressed as percent per month, then the number of compounding periods used in the function must also be in months.

10.2.2 Nominal Interest Rates (i_{nom})

Nominal interest rates are a type of annual interest rate, but usually with nonannual compounding. You must know the compounding period in order to compute a periodic interest rate from the nominal rate. Banks are required to report APR and APY, but most still report nominal interest rates, as well as APR and APY. The following table is typical of how a bank might present their CD rates:

CERTIFICATE OF DEPOSIT		
COMPOUNDED DAILY: 365 BASIS	**RATE**	**APY**
6 Month	5.85%	6.02%
1 Year	5.20%	5.34%
1 1/2 Year	5.30%	5.44%
3 Year	5.60%	5.76%
5 Year	5.70%	5.87%

The rate column shows a nominal interest rate. The compounding period is stated in the table, "Compounded Daily: 365 Basis." This means that the bank is assuming a 365-day year when calculating the annual percentage yield (APY) from the nominal rate. (Banks are required to use a 365-day basis when calculating APY.) We can use this example to learn how to convert the nominal interest rate to a periodic rate and an effective annual rate.

Converting a Nominal Interest Rate (i_{nom}) to a Periodic Interest Rate (i_p)

To convert a nominal interest rate to a periodic interest rate, simply divide the nominal rate by the number of compounding periods per year. For the six-month CD shown in the preceding table, the nominal rate is

$$i_{nom} = 5.85\%, \text{ or } 0.0585. \tag{10.1}$$

The table states that compounding is daily and uses a 365-day-year basis. The periodic rate (daily interest rate) is found by dividing the nominal rate by 365 days per year:

$$i_p = i_{\text{nom}} / 365$$

$$= 5.85\% / 365 = 0.0160\% \text{ per day}$$

$$= 0.0585 / 365 = 0.000160 \text{ fractional interest rate per day.} \quad (10.2)$$

10.2.3 Effective Annual Interest Rates (i_e)

It is sometimes convenient to compute the effective annual interest rate for nonannual periodic interest payments. The effective annual interest rate is simply the rate that produces the same end-of-year value with a single interest payment (at the end of the year) as would be produced by more frequent periodic interest payments.

For example, $100.00 invested at a monthly interest rate of 1% would be worth $112.68 twelve months later. The interest rate required to generate the same result with a single interest payment is 12.68%. So, the effective annual interest rate equivalent to a 1% monthly rate is 12.68%.

Calculating the Effective Annual Interest Rate (i_e)   Because the effective annual interest rate is designed to produce the same future (end-of-year) value as the periodic interest rate, we begin by finding out how much an investment made at time zero would be worth one year later using the periodic interest rate (i_p). To do this, we use the equation for calculating a future value (FV) when given a present value (PV) and a periodic interest rate (i_P):

$$FV = PV \cdot (1 + i_p)^{N_{\text{PPY}}}. \quad (10.3)$$

Here,

> i_P is the periodic interest rate,
> N_{PPY} is the number of compounding periods per year,
> FV is the future value, and
> PV is the present value.

We need to get the same future value by using the effective annual interest rate (i_e), and a compounding period of one year. With the effective annual interest rate, the future-value equation becomes

$$FV = PV \cdot (i + i_e)^1. \quad (10.4)$$

Because the future value must be the same regardless of whether the periodic interest rate or the effective annual interest rate is used, we can set the future-value equations equal to each other to solve for i_e:

$$i_e = (1 + i_p)^{N_{\text{PPY}}} - 1. \quad (10.5)$$

The periodic interest rate is simply the nominal rate divided by the number of periods per year, so the equation for (i_e) in terms of the nominal interest rate can be written as

$$i_e = \left(1 + \frac{i_{\text{nom}}}{N_P}\right)^{N_{\text{PPY}}} - 1. \quad (10.6)$$

At the beginning of this section, it was stated that the effective annual interest rate equivalent to a periodic rate of 1% per month was 12.68%. We can use this fact to test the conversion equation we just derived:

$$i_e = (1 + 0.01)^{12} - 1 = 0.1268 \qquad (10.7)$$

or

$$12.68\%.$$

EXAMPLE 10.1

If the nominal rate for the six-month CD in the table on page 393 is 5.85% and the number of periods per year is 365, then

$$i_e = \left(1 + \frac{5.85\%}{365}\right)^{365} - 1 = 6.02\%. \qquad (10.8)$$

The effective annual interest rate (and APY) is 6.02%, as reported in the bank's table of CD yields.

Excel provides a function for calculating effective annual interest rates from nominal rates. The EFFECT() function has the syntax

```
=EFFECT (Nominal_Rate, N_PPY)
```

where

Nominal_Rate is the nominal (annual) interest rate (required) and
N_{PPY} is the number of periods per year (required)

The nominal rate may be entered as a percentage (e.g., 5.85%) or as a fractional rate (0.0585):

A6		f_x =EFFECT(A3,A4)		
	A	B	C	D	E
1	EFFECT Calculation				
2					
3	5.85%	Nominal Rate, percent (or fraction)			
4	365	Periods per year			
5					
6	6.02%	Effective annual interest rate			
7					

The corresponding function to convert an effective annual interest rate to a nominal annual interest rate is the NOMINAL() function:

A14		f_x =NOMINAL(A11,A12)		
	A	B	C	D	E
9	NOMINAL Calculation				
10					
11	6.02%	Effective annual interest rate			
12	365	Periods per year			
13					
14	5.85%	Nominal annual interest rate			
15					

The syntax of the nominal function is

```
=NOMINAL (Eff_Ann_Rate, N_PPY)
```

where

Eff_Ann_Rate is the effective annual interest rate (required) and
N_{PPY} is the number of periods per year (required)

The effective annual rate may be entered as a percentage (e.g., 6.02%) or as a fractional rate (0.0602).

10.3 MOVING AMOUNTS THROUGH TIME

Excel's time-value-of-money functions allow you to calculate the effect of interest rates on money as it moves through time. It is a standard part of an engineer's job to calculate what a proposed project will cost, how much a possible improvement might be worth, or how long it will take for a new piece of equipment to pay for itself. All of these are time-value-of-money problems.

10.3.1 Present and Future Values

One hundred dollars invested for one year at 1% per month will be worth $112.68. There is an equality relating those two amounts at different points in time; it is the equation for calculating a future value when given a present value, periodic interest rate, and number of compounding periods:

$$FV = PV \cdot (1 + i_p)^{N_{per}} \tag{10.9}$$

In this equation,

i_p is the periodic interest rate,
N_{per} is the number of compounding periods between the present and future,
FV   is the future value, and
PV   is the present value.

Note: The N_{per} *(number of compounding periods, over the term of the loan for example) in this equation is not the same as the* N_{PPY} *(number of compounding periods per year) that is used in calculating the effective annual interest rate.*

One hundred dollars today and the $112.68 one year in the future are considered equivalent when making economic decisions (if the interest rate is 1% per month). The numerical values are not equivalent, but the two values *at their respective times* are equivalent. This is an example of the *time value of money*, and illustrates an important point: To compare numerical values, the values must be at the same point in time. Moving values through time (at specified interest rates) is the subject here.

Calculating Future Values from Present Values   If you have $1,000 today, what will it be worth in 20 years if

1.   you put it in a savings account at 5% APR?
2.   you invest it in the stock market with an average annual rate of 12.3%?
3.   you hide it in the cookie jar and the effective annual inflation rate is 2.7% per year?

The "$1,000 today" is a present value. The value at the end of the specified period (20 years) is a future value. Excel provides a function, called FV(), that can be used to calculate future values when given present values, periodic interest rates, and the number of compounding periods. The syntax of the function is

```
=FV(Rate, Nper, Pmt, PV, Type)
```

where

Rate	is the interest rate per compounding period (required),
Nper	is the number of periods (required),
Pmt	is the amount of any *regular payment* each period (required—if there is no regular payment, include a 0 for this argument),
PV	is the amount of any present value (optional), and
Type	indicates whether payments occur at the beginning of the period (Type=1 or at the end of the period (Type=0). If Type is omitted, payments at the end of the period are assumed.

Note: Excel, like most financial software programs, uses a sign convention to indicate which direction money is moving. Out-of-pocket expenses are negative, and incomes are positive.

We can use the FV() function to answer the questions just listed.

Question 1 If you have $1,000 today, what will it be worth in 20 years if you put it in a savings account at 5% APR?

$1,000 invested at 5% APR for 20 years would be worth $2653.30 (assuming annual compounding). The use of the FV() function to calculate this result is illustrated in the following spreadsheet:

	B11	▼	fx	=FV(B5,B6,B7,B8,B9)				
	A	B	C	D	E	F	G	H
1	**FV Calculations**							
2								
3	$1000 invested at 5% APR for 20 years.							
4								
5	Rate:	5%	per year					
6	Nper:	20	years					
7	Pmt:	0	(no regular payments)					
8	PV:	-$1,000	(out-of-pocket expense)					
9	Type:	0	(end of year payments)					
10								
11	FV:	$2,653.30						
12								

Note the following:

1. *In Excel, 5% and 0.05 are equivalent. You can enter interest rates either as percentages (including the percent sign) or as decimal. In this example, the rate was entered as a percentage (5%).*

2. *The $1,000 investment is an expense and was entered into the spreadsheet as a negative value. The $2,653.30 is income, so the result was positive.*

Question 2 If you have $1,000 today, what will it be worth in 20 years if you put it in the stock market with an average annual rate of return of 12.3%?

If the $1,000 were invested in the stock market with an average annual rate of return of 12.3%, it would be worth $10,176. (But stock market rates are not guaranteed. Historical rates make no promises for the future.)

B11	▼	f_x =FV(B5,B6,B7,B8,B9)					
A	B	C	D	E	F	G	H
1 FV Calculations							
2							
3 $1000 invested at 12.3% (effective annual rate) for 20 years.							
4							
5 Rate:	12.3%	per year					
6 Nper:	20	years					
7 Pmt:	0	(no regular payments)					
8 PV:	-$1,000	(out-of-pocket expense)					
9 Type:	0	(end of year payments)					
10							
11 FV:	$10,176.42						
12							

Question 3   If you have $1,000 today, what will it be worth in 20 years if you hide it in the cookie jar and the effective annual inflation rate is 2.7% per year?

If the $1,000 were left in a cookie jar at time zero, the buying power would go down because of inflation. After 20 years of 2.7% inflation, you would still have $1,000, but the buying power would be equivalent to $578 in today's dollars.

B11	▼	f_x =FV(B5,B6,B7,B8,B9)					
A	B	C	D	E	F	G	H
1 FV Calculations							
2							
3 $1000 subject to inflation of 2.7% (effective annual rate) for 20 years.							
4							
5 Rate:	-2.7%	per year					
6 Nper:	20	years					
7 Pmt:	0	(no regular payments)					
8 PV:	-$1,000	(out-of-pocket expense)					
9 Type:	0	(end of year payments)					
10							
11 FV:	$578.44						
12							

Calculating Present Values from Future Values   Excel also provides a function for calculating present values from future values, the PV() function. It's syntax is similar to that of the FV() function, namely,

```
=PV(Rate, Nper, Pmt, FV, Type)
```

where

Rate   is the interest rate per period (required),

Nper   is the number of periods (required),

Pmt   is the amount of any regular payment each period (required—
       if there is no regular payment, include a 0 for this argument),

FV   is the amount of any future value (optional), and

Type indicates whether payments occur at the beginning of the period (Type=1) or end of the period (Type=0). If Type is omitted, payments at the end of the period are assumed.

Again, out-of-pocket expenses should be entered as negative values, and incomes are treated as positive.

The PV() function can be used to answer questions such as the following:

1. How much do I need to invest today at 5% APR in order to have $1,000 available in one year?

2. How much do we need to reserve from this year's budget surplus to buy a $30,000 piece of equipment when it becomes available in 15 months? (Assume that your company makes an effective annual rate of 7% on reserved monies.)

A spreadsheet capable of answering the first question might look like the following:

	B11		fx =PV(B5,B6,B7,B8,B9)					
	A	B	C	D	E	F	G	H
1	**PV Calculations**							
2								
3	$1000 from an investment at 5% APR for 1 year.							
4								
5	Rate:	5%	per year					
6	Nper:	1	years					
7	Pmt:	0	(no regular payments)					
8	FV:	$1,000	(out-of-pocket expense)					
9	Type:	0	(end of year payments)					
10								
11	PV:	-$952.38	(expense)					
12								

An investment of $952.38 (shown in the spreadsheet as a negative value since it is an expense) at 5% APR would be worth $1,000 one year later.

An additional calculation is required in order to answer the second question, because the number of periods is expressed in months, while the interest rate is on an annual basis. We must convert the 15 months to years. This has been done in cell B6 in the following spreadsheet:

	B11		fx =PV(B5,B6,B7,B8,B9)					
	A	B	C	D	E	F	G	H
1	**PV Calculations**							
2								
3	$30,000 from reserved funds invested at 7% (effective annual rate) for 15 months.							
4								
5	Rate:	7%	per year					
6	Nper:	1.25	years					
7	Pmt:	0	(no regular payments)					
8	FV:	$30,000	(out-of-pocket expense)					
9	Type:	0	(end of year payments)					
10								
11	PV:	-$27,567.13	(expense)					
12								

APPLICATIONS: ECONOMICS

Preparing for Retirement: part 1

It is fairly common to try to predict the amount of money you will need to live comfortably after retirement. Some of the first questions to be considered are

- How much income do I need to live comfortably today?
- With inflation, how much income will I need when I retire?

All economic predictions are full of assumptions. In this example, we will have to assume an inflation rate, the number of years you have left before retirement, and the amount of income required today to live comfortably. A common assumption made when deciding what income level is required to live comfortably is that, by the time you retire, most of your major bills will be paid off, such as your home mortgage and the education expenses for your children.[1] Here are the assumptions we'll use in this calculation:

1. Income required to live comfortably today (without major bill payments): $25,000 per year,
2. Inflation rate: 4% per year,
3. Years left before retirement: 25.

With inflation, how much income will I need when I retire?

To answer this question, we need to find the future value that is equivalent to $25,000 today. We've already developed a spreadsheet for calculating future values from present values. Here it is again, modified for this application:

B11		fx =FV(B5,B6,B7,B8,B9)					
	A	B	C	D	E	F	G	H
1	**FV Calculations**						
2							
3	$25,000 today is equivalent to what amount 25 years in the future with 4% inflation?						
4							
5	Rate:	4%	per year				
6	Nper:	25	years				
7	Pmt:	0	(no regular payments)				
8	PV:	-$25,000	(out-of-pocket expense)				
9	Type:	0	(end of year payments)				
10							
11	FV:	$66,646					
12							

You will need an annual income of over $66,000 to have the same buying power as $25,000 today. This, of course, assumes that the inflation rate will be 4% for the next 25 years. If the inflation rate is higher during some of those years, the necessary future income would also be higher, but the interest rates on the investments used to generate that income tend to go up when inflation goes up. So, you may need more income after periods of high inflation, but your investments also tend to yield higher returns during those high-inflation years.

In another application, we will look at how much money you will need to have available to generate an income of over $66,000 per year.

[1] The biggest uncertainty tied to this assumption is the cost of health care during your retirement years. This analysis leaves it out. Including several hundred dollars a month (in today's dollars) for health insurance changes the required annual income considerably.

10.3.2 Regular Payments

Loan payments, periodic maintenance payments, rental payments, and even salaries are regular, periodic payments. Because regular payments are very common, Excel provides functions for working with these payments over time. The FV() and PV() functions described before can be used with payments, and there is a function PMT() that calculates payment amounts from present and future values.

The PMT(), FV(), and PV() functions can be used to answer such questions as the following:

1. What will be the monthly payment on a four-year car loan at 9% (nominal) interest, compounded monthly, if the loan amount is $15,000?

2. If your company earns 5% (effective annual rate) on unspent balances, how much money do you need to have in the annual budget to cover the $600 per month rental rate (payments at the end of each month) on a piece of lab equipment?

3. If a relative gives you $10,000 and you invest it at 7% nominal rate, compounded monthly, how much can you spend each month and have the money last five years?

The PMT() function uses the syntax

```
=PMT(Rate, Nper, PV, FV, Type)
```

where

Rate   is the interest rate per period (required),

Nper   is the number of periods (required),

PV   is the amount of any future value (required—if there is no present value, include a 0 for this argument),

FV   is the amount of any future value (optional), and

Type   indicates whether payments occur at the beginning of the period (Type=1) or end of the period (Type=0). If Type is omitted, payments at the end of the period are assumed.

Again, out-of-pocket expenses should be entered as negative values, incomes as positive values.

Question 1   To calculate the required payment on a four-year, $15,000 car loan at 9% nominal interest, compounded monthly, the PMT() function is used, as in the following spreadsheet:

	B11	▾	fx =PMT(B5,B6,B7,B8,B9)					
	A	B	C	D	E	F	G	H
1	PMT Calculations							
2								
3	What is the payment on a 4-year, $15,000 loan at 9% APR (monthly compounding)?							
4								
5	Rate:	0.75%	per month					
6	Nper:	48	months					
7	PV:	$15,000	(principal)					
8	FV:	0						
9	Type:	0						
10								
11	PMT:	-$373.28	(out-of-pocket expense)					
12								

Because the interest rate on the loan was a nominal annual rate, the periodic interest rate corresponding to the compounding period (one month) had to be computed. This was done in cell B5, with the following formula:

    B5:   =9%/12

The total number of compounding periods was computed in cell B6 as

    B6:   =4*12   (four years, twelve months per year)

The $373.28 payment is shown as a negative number because it represents an out-of-pocket expense.

Question 2   If your company earns 5% (effective annual rate) on unspent balances, how much money do you need to have in the annual budget to cover the $600 per month rental rate (payments at the end of each month) on a piece of lab equipment?

In this question, the amount of the regular payment is stated, and you are asked to work out how much needs to be allocated in the budget at time zero to cover all of the rent payments. The PV() function will be used to answer this question.

The effective annual rate, 5%, needs to be converted to a periodic (monthly) interest rate to correspond to the monthly rent payments. The equation for converting an effective annual rate to a periodic rate is

$$i_p = \left[\sqrt[N_{PPY}]{i_e + 1} \right] - 1, \tag{10.10}$$

where

i_e is the effective annual interest rate,
i_p is the periodic interest rate, and
N_{PPY} is the number of compounding periods per year.

An effective annual interest rate of 5% is equivalent to a monthly interest rate of 0.4074%:

$$i_p = \left[\sqrt[12]{0.05 + 1} \right] - 1 = 0.004074, \tag{10.11}$$

or

$$0.4074\%, \text{ per month.}$$

The amount required in the annual budget to cover the rent payments is found by using the PV() function:

	B11	▼	f_x =PV(B5,B6,B7,B8,B9)					
	A	B	C	D	E	F	G	H
1	PV Calculations							
2								
3	How much should be budgeted to pay $600 rent each month?							
4								
5	Rate:	0.4074%	per month					
6	Nper:	12	months					
7	Pmt:	-$600	(expense)					
8	FV:	0						
9	Type:	0	(end of month payments)					
10								
11	PV:	$7,012.91						
12								

Because the unspent funds are invested, slightly over $7,000 is enough to pay all 12 rental payments for the year.

Question 3   If a relative gives you $10,000 and you invest it at 7% nominal rate, compounded monthly, how much can you spend each month and have the money last five years? This time you are looking for a monthly payment, so you would use the PMT() function to answer this question.

First, the nominal rate of 7% compounded monthly needs to be converted to a periodic rate via the equation

$$\text{B5:} \quad = 7\%/12$$

in cell B5. The number of monthly payments over the five-year period is computed in cell B6 as

$$\text{B6:} \quad = 5*12$$

If you assume payments at the end of the month (Type $= 0$), the result is $198.01:

	B11	▼	f_x =PMT(B5,B6,B7,B8,B9)					
	A	B	C	D	E	F	G	H
1	**PMT Calculations**							
2	How much can you spend each month to make $10,000 invested at 7% APR,							
3	compounded monthly, last 5 years?							
4								
5	Rate:	0.583%	per month					
6	Nper:	60	months					
7	PV:	-$10,000	(investment expense)					
8	FV:	0						
9	Type:	0						
10								
11	PMT:	$198.01	(income)					
12								

What if you want the income to last forever? You can't put infinity into the PMT() function's Nper argument. You can put large numbers into cell B4 and keep increasing the number of months until the calculated payment stops changing. But what you are doing is calculating the monthly interest payment on a $15,000 investment at 0.583% per month. If you spend only the interest, the monthly payments will last forever.

The interest payment on a $15,000 investment at 0.583% per month is $87.50:

$$0.00583 \times \$15,000 = \$87.50. \tag{10.12}$$

Preparing for Retirement: part 2

In the first part of this application example, we found that an income of $66,000 per year would be required 25 years from now to provide buying power equivalent to $25,000 today. If you stop working 25 years from now, you will need enough money in a retirement fund to generate that $66,000 per year.

Question 1 How much money must be in the fund (at 7% effective annual rate) to generate an annual interest payment of $66,000? (Remember that, if you spend only the interest, the income can continue forever.)

We have

$$\$66,000 = 0.07 \times \text{PV},$$

$$\text{PV} = \frac{\$66,000}{0.07} = \$943,000. \tag{10.13}$$

If you don't plan to live forever, you can reduce the amount required by spending down the principal over time.

Question 2 How much money must be in the fund (at 7% effective annual rate) to generate payments of $66,000 per year for 20 years? (The principal will be gone after 20 years.)

To answer this question, we need to find the present value (in an account at 7%) that is equivalent to a series of $66,000 payments over 20 years. This can be calculated using Excel's PV() function:

```
=PV(7%, 20, 66000)
```

B11	▾	*fx* =PV(B5,B6,B7,B8,B9)					
	A	B	C	D	E	F	G	H
1	PV Calculations							
2	What amount, at 7% eff. ann. rate, will generate annual payments of $66,000 for							
3	a period of 20 years?							
4								
5	Rate:	7%	per year					
6	Nper:	20	years					
7	Pmt:	$66,000						
8	FV:	0						
9	Type:	0						
10								
11	PV:	-$699,205						
12								

The answer is $699,000.

Question 3 How much do you need to put into the retirement account (at 12% APR compounded monthly) each month for 25 years to generate a fund of $699,000? You can use Excel's PMT() function to answer this question:

```
=PMT(1%, 300, 0, 699000)
```

B11	▼		f_x =PMT(B5,B6,B7,B8,B9)					
	A	B	C	D	E	F	G	H
1	**PMT Calculations**							
2	How much should you pay into a retirement account (12% APR) each month for 25							
3	years to accumulate $699,000?							
4								
5	Rate:	1.00%	per month					
6	Nper:	300	months					
7	PV:	0						
8	FV:	$699,000						
9	Type:	0						
10								
11	FV:	-$372.04	(out-of-pocket expense)					
12								

This answer is over $370.

10.3.3 Discounting-Factor Tables and Nomenclature

Long before Excel existed, engineers used tables of *discounting factors* to solve time-value-of-money problems. A typical table showing factors for computing future values of a $1 investment at various interest rates is shown here:

SINGLE-PAYMENT COMPOUND AMOUNT FACTORS (FV|PV)

$(FV|PV, i_p, N_{per})$

INTEREST RATE: N_{per}	3%	4%	5%	6%	7%	8%	9%	10%
0	1.0000	1.0000	1.0000	1.0000	1.0000	1.0000	1.0000	1.0000
1	1.0300	1.0400	1.0500	1.0600	1.0700	1.0800	1.0900	1.1000
2	1.0609	1.0816	1.1025	1.1236	1.1449	1.1664	1.1881	1.2100
3	1.0927	1.1249	1.1576	1.1910	1.2250	1.2597	1.2950	1.3310
4	1.1255	1.1699	1.2155	1.2625	1.3108	1.3605	1.4116	1.4641
5	1.1593	1.2167	1.2763	1.3382	1.4026	1.4693	1.5386	1.6105
6	1.1941	1.2653	1.3401	1.4185	1.5007	1.5869	1.6771	1.7716
7	1.2299	1.3159	1.4071	1.5036	1.6058	1.7138	1.8280	1.9487
8	1.2668	1.3686	1.4775	1.5938	1.7182	1.8509	1.9926	2.1436
9	1.3048	1.4233	1.5513	1.6895	1.8385	1.9990	2.1719	2.3579
10	1.3439	1.4802	1.6289	1.7908	1.9672	2.1589	2.3674	2.5937
15	1.5580	1.8009	2.0789	2.3966	2.7590	3.1722	3.6425	4.1772
20	1.8061	2.1911	2.6533	3.2071	3.8697	4.6610	5.6044	6.7275
25	2.0938	2.6658	3.3864	4.2919	5.4274	6.8485	8.6231	10.8347
30	2.4273	3.2434	4.3219	5.7435	7.6123	10.0627	13.2677	17.4494
35	2.8139	3.9461	5.5160	7.6861	10.6766	14.7853	20.4140	28.1024
40	3.2620	4.8010	7.0400	10.2857	14.9745	21.7245	31.4094	45.2593
45	3.7816	5.8412	8.9850	13.7646	21.0025	31.9204	48.3273	72.8905
50	4.3839	7.1067	11.4674	18.4202	29.4570	46.9016	74.3575	117.3909

The interest rate indicated in the table is in percent per period (i_p). The table indicates that $1 invested at 3% per year for one year would be worth $1.03, but after 50 years, that $1 would be worth $4.38. This table was created by using the FV() function in Excel. Each of the various types of discounting factors has a name, and there is a common shorthand notation describing each factor. The factors used to compute future values from present values are called *single-payment compound amount factors* and are given the shorthand

$$(FV|PV, i_p, N_{per}), \tag{10.14}$$

which is read as "Find FV given ($|$) PV, i_p and N."

To see how discounting-factor tables are used, let's consider the following example: If you have $1,000 today, what will it amount to in 20 years if you put it in a savings account at 5% APR?

Assuming annual compounding, we get

$$PV = \$1,000,$$

$$i_p = 5\% \text{ per year,}$$

$$N_{per} = 20 \text{ years.} \tag{10.15}$$

Then, using the single payment *compound amount factor* table, we get

$$(FV|\$1, 5\%, 20) = 2.6533. \tag{10.16}$$

The future value is computed by using the actual present value multiplied by the factor, so

$$FV = PV \times (FV|\$1, 5\%, 20)$$

$$= \$1000 \times (2.6533)$$

$$= \$2653.30. \tag{10.17}$$

The equivalent calculation using Excel's FV ( ) function is

$$= FV(i_p, N_{per}, Pmt, PV, Type)$$

$$= FV(5\%, 20, 0, \$1000, 0)$$

$$= \$2653.30, \tag{10.18}$$

or, in the spreadsheet developed earlier,

B11		f_x =FV(B5,B6,B7,B8,B9)					
	A	B	C	D	E	F	G	H
1	FV Calculations							
2								
3	$1000 invested at 5% APR (compounded annually) for 20 years.							
4								
5	Rate:	5%	per year					
6	Nper:	20	years					
7	Pmt:	0	(no regular payments)					
8	PV:	-$1,000	(out-of-pocket expense)					
9	Type:	0	(end of year payments)					
10								
11	FV:	$2,653.30						
12								

Summary of Common Discounting Factors   There are six common discounting factors. The following table summarizes the factors, showing each historical name, the computational equation for the factor, a shorthand notation, and the Excel function that could be used to calculate the factor:

SUMMARY OF COMMON DISCOUNTING FACTORS

SHORTHAND NOTATION	FACTOR NAME	EQUATION FOR FACTOR	EXCEL FUNCTION
$(FV \vert PV, i_p, N_{per})$	Single Payment Compound Amount Factor	$(1 + i_p)^{N_{per}}$	=FV(iP, Nper, 0, $1, 0)
$(PV \vert FV, i_p, N_{per})$	Single Payment Present Worth Factor	$(1 + i_p)^{-N_{per}}$	=PV(iP, Nper, 0, $1, 0)
$(Pmt \vert FV, i_p, N_{per})$	Uniform Series Sinking Fund Factor	$\dfrac{1}{(1 + i_p)^{N_{per}} - 1}$	=PMT(iP, Nper, 0, $1, 0)
$(Pmt \vert PV, i_p, N_{per})$	Capital Recovery Factor	$\dfrac{i_p(1 + i_p)^{N_{per}}}{(1 + i_p)^{N_{per}} - 1}$	=PMT(iP, Nper, $1, 0, 0)
$(FV \vert Pmt, i_p, N_{per})$	Uniform Series Compound Amount Factor	$\dfrac{(1 + i_p)^{N_{per}} - 1}{i_p}$	=FV(iP, Nper, $1, 0, 0)
$(PV \vert Pmt, i_p, N_{per})$	Uniform Series Present Worth Factor	$\dfrac{(1 + i_p)^{N_{per}} - 1}{i_p(1 + i_p)^{N_{per}}}$	=PV(iP, Nper, $1, 0, 0)

Be aware that the equations shown in the table are for the discounting factors; that's why the "$1" amounts appear in the Excel function calls. To skip the calculation of the discount factor and calculate the final result directly, use the equations and function calls listed in the following table:

SUMMARY OF TIME VALUE OF MONEY CALCULATIONS

SHORTHAND NOTATION	EQUATION	EXCEL FUNCTION
$(FV \vert PV, i_p, N_{per})$	$-PV \times (1 + i_p)^{N_{per}}$	=FV(iP, Nper, 0, PV, 0)
$(PV \vert FV, i_p, N_{per})$	$-FV \times (1 + i_p)^{-N_{per}}$	=PV(iP, Nper, 0, FV, 0)
$(Pmt \vert FV, i_p, N_{per})$	$-FV \times \dfrac{1}{(1 + i_p)^{N_{per}} - 1}$	=PMT(iP, Nper, 0, FV, 0)
$(Pmt \vert PV, i_p, N_{per})$	$-PV \times \dfrac{i_p(1 + i_p)^{N_{per}}}{(1 + i_p)^{N_{per}} - 1}$	=PMT(iP, Nper, PV, 0, 0)
$(FV \vert Pmt, i_p, N_{per})$	$-Pmt \times \dfrac{(1 + i_p)^{N_{per}} - 1}{i_p}$	=FV(iP, Nper, Pmt, 0, 0)
$(PV \vert Pmt, i_p, N_{per})$	$-Pmt \times \dfrac{(1 + i_p)^{N_{per}} - 1}{i_p(1 + i_p)^{N_{per}}}$	=PV(iP, Nper, Pmt, 0, 0)

Note: These equations begin with a minus sign to make them agree with Excel's sign convention: Expenses are negative, and incomes are positive.

10.4 ECONOMIC ALTERNATIVES: MAKING DECISIONS

One of the most common reasons for doing time-value-of-money calculations is to make a decision between two or more *economic alternatives*. Selecting between alternatives is a routine part of an engineer's job. These decisions can include the following:

- Which is the better deal: The expensive compressor with lower annual maintenance costs and longer life or the cheaper compressor with higher annual maintenance costs and shorter life?
- Which is the better deal when financing a new truck: 1.5% APR on the loan or $1500 cash back to apply towards the down payment?

To answer questions such as these, it is necessary to move all of the money to the same point in time, usually a present value.

APPLICATIONS: EVALUATING ALTERNATIVES

Choosing the Best Compressor

Two compressor options are available for a new process design:

- Option *A* is a $60,000 compressor with a 10-year life expectancy, annual maintenance costs of $2,200, and a salvage value of $12,000 when it is removed from service.
- Option *B* is a $32,000 compressor with a 5-year life expectancy, annual maintenance costs of $2,600, and no salvage value.

Which is the better deal?

Before you can make a decision, you need to get all of the dollars moved to the same point in time (usually a present value) by using a specified interest rate. Here, we will assume an effective annual rate of 7%. You also need to get the total time span for each option to be the same, in this case by purchasing a second compressor in year 5 for Option *B*. (Alternatively, you can calculate an equivalent cost per year for each option to make the decision.)

The cash flow diagrams for each option are as follows:

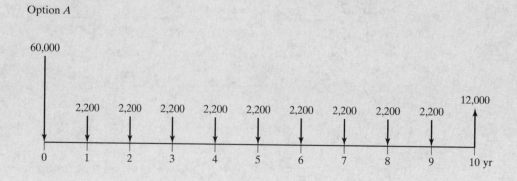

Option B

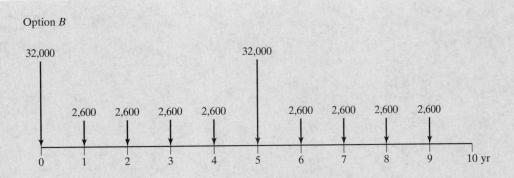

The present value of Option A is approximately $68,000, as calculated in the following spreadsheet:

	A	B	C	D	E	F	G
1	**Option A**						
2	**PV**	**Item**	**iP**	**Nper**	**Pmt**	**FV**	**Type**
3			(% per year)	(years)			
4							
5	$60,000	Initial purchase cost (at time zero)					
6	$14,334	Annual maintenance costs	7%	9	-$2,200	0	0
7	-$6,100	Salvage value	7%	10	0	$12,000	0
8							
9	$68,233	**Total PV**					
10							

Cells C6 through G7 contain the values shown. The values or formulas in column A are as follows:

A5: $60,00   (a value)
A6:   =PV(C6, D6, E6, F6, G6)   (The "F6" and "G6" are optional, and could be omitted.)
A7:   =PV(C7, D7, E7, F7, G7)   (The "G7" could be omitted.)

Calculating the present value of Option B requires a few more steps. One way to simplify the calculations is to take $2,600 from the purchase price of the second pump in year 5 ($32,000 − $2,600 = $29,400) and use it to create a uniform series of $2,600 payments in years 1 through 9, as shown on the following cash-flow diagram:

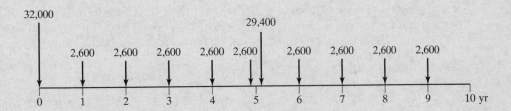

Now, the present value of Option B can be calculated:

	A	B	C	D	E	F	G
13	**Option B**						
14	**PV**	**Item**	**iP**	**Nper**	**Pmt**	**FV**	**Type**
15			(% per year)	(years)			
16							
17	$32,000	Purchase cost, Comp. 1 (at time zero)					
18	$20,962	Purchase cost, Comp. 2 (year 5)	7%	5	0	-$29,400	0
19	$16,940	Annual maintenance costs	7%	9	-$2,600	0	0
20	$0	Salvage value	7%	10	0	0	0
21							
22	$69,901	**Total PV**					
23							

The values or formulas in column A are as follows:

```
A17:    $32,000
A18:    =PV(C18, D18, E18, F18, G18)
A19:    =PV(C19, D19, E19, F19, G19)
```

The present value of Option B is approximately $70,000, almost $2000 higher than Option A. Option A is the way to go (at least, if you think the compressor will be needed for more than five years).

EXAMPLE 10.2

Selecting the Best Car-Financing Deal

For years, advertisements for new cars have offered low interest rates as incentives. Those ads typically offer either a low interest rate or cash back (usually applied to the down payment), but not both. The cash-back option is provided for individuals who can pay cash for the new car, since they will have no interest in a low APR on a loan.

If you will finance the purchase, you get to choose between the following options:

1. Pay the full purchase price, but get a reduced interest rate on the loan.
2. Pay a standard interest rate (e.g., 9% APR, compounded monthly), but get "cash back" to lower the effective purchase price.

Which is the better deal? You can use time value of money calculations to find out. Specifically, you can use Excel's PMT ( ) function to calculate which option results in the lower monthly payment.

A new pickup truck sells for $20,000. You plan to offer a down payment of $5,000 and finance the rest. The company's latest advertisement offers either 1.5% APR for up to 48 months or $1,500 cash back. The $1,500 cash back would effectively reduce the purchase price to $18,500. After your $5,000 down payment, the remaining $13,500 would be financed with a 9% APR loan for 48 months. Which is the better deal for you?

Find the monthly payment for each loan option:

- 1.5% APR (cell B3: =1.5%/12, or 0.125% per month), 48 Months, $15,000 → Pmt = $322.16:

B11	▼	f_x =PMT(B5,B6,B7,B8,B9)					
	A	B	C	D	E	F	G	H

	A	B	C	D	E	F	G	H
1	**PMT Calculations**							
2								
3	What is the monthly payment on a $15,000 loan at 1.5% APR for 4 years?							
4								
5	Rate:	0.125%	per month					
6	Nper:	48	months					
7	PV:	$15,000						
8	FV:	0						
9	Type:	0						
10								
11	FV:	-$322.16	(payment)					
12								

- 9% APR (cell B3: =9%/12, or 0.750% per month), 48 Months, $13,500 → Pmt = $335.95:

B11	▼	f_x =PMT(B5,B6,B7,B8,B9)					
	A	B	C	D	E	F	G	H

	A	B	C	D	E	F	G	H
1	**PMT Calculations**							
2								
3	What is the monthly payment on a $13,500 loan at 9% APR for 4 years?							
4								
5	Rate:	0.750%	per month					
6	Nper:	48	months					
7	PV:	$13,500						
8	FV:	0						
9	Type:	0						
10								
11	FV:	-$335.95	(payment)					
12								

The 1.5% APR loan offers the lowest payments, although the difference is very small. For this example, that's the better deal; but the situation can change if you have a large down payment. If you put $15,000 down and finance $5,000 at 1.5% APR the payment will be $107 a month. If you accept the $1,500 cash back, and finance $3,500 at 9% APR the monthly payment amount will be $87. If you can make a large down payment, the cash-back option becomes the better deal.

KEY TERMS

Annual percentage rate (APR)	Expenses	Present value
Annual percentage yield (APY)	Future value	Principal
Annuity payment	Incomes	Regular payment
Cash flow	Inflation rate	Salvage value
Compounding period	Intangibles	Service life
Compound amount factor	Interest rate	Timeline
Discounting factor	Nominal interest rate	
Effective annual interest rate	Periodic interest rate	

SUMMARY

There are several ways of expressing interest rates:

- Periodic interest rates are the interest rates used over a single compounding period.
- Nominal interest rates are annual rates that ignore the effect of compounding. You use a nominal rate by converting it into a periodic interest rate.
- Effective annual interest rates are calculated annual interest rates that yield the same results as periodic interest rates with nonannual compounding.
- Annual Percentage Rates (APR) include any up-front fees that are part of the total cost of borrowing. If there are no up-front fees, the APR can be used as a nominal interest rate if the compounding period is known.
- Annual Percentage Yields (APY) are nominal interest rates calculated on the assumption of daily compounding.

Excel's Time-Value-of-Money Functions

Nomenclature

Rate	is the interest rate per period.
Nper	is the number of periods.
Pmt	is the amount of any regular payment each period.
PV	is the amount of any present value.
Type	indicates whether payments occur at the beginning of the period (Type=1) or end of the period (Type=0). If Type is omitted, payments at theend of the period are assumed.

Present Value Function

=PV (Rate, Nper, Pmt, *FV*, *Type*)

Future Value Function

=FV (Rate, Nper, Pmt, *PV*, *Type*)

Payment Function

=PMT (Rate, Nper, PV, *FV*, *Type*)

Summary Table

SUMMARY OF TIME-VALUE-OF-MONEY CALCULATIONS

SHORTHAND NOTATION	EQUATION	EXCEL FUNCTION
$(FV|PV, i_p, N_{per})$	$-PV \times (1 + i_p)^{N_{per}}$	$=FV(i_p, N_{per}, 0, PV, 0)$
$(PV|FV, i_p, N_{per})$	$-FV \times (1 + i_p)^{-N_{per}}$	$=PV(i_p, N_{per}, 0, FV, 0)$
$(Pmt|FV, i_p, N_{per})$	$-FV \times \dfrac{1}{(1 + i_p)^{N_{per}} - 1}$	$=PMT(i_p, N_{per}, 0, FV, 0)$
$(Pmt|PV, i_p, N_{per})$	$-PV \times \dfrac{i_p(1 + i_p)^{N_{per}}}{(1 + i_p)^{N_{per}} - 1}$	$=PMT(i_p, N_{per}, PV, 0, 0)$
$(FV|Pmt, i_p, i_p, N)$	$-Pmt \times \dfrac{(1 + i_p)^{N_{per}} - 1}{i_p}$	$=FV(i_p, N_{per}, Pmt, 0, 0)$
$(PV|Pmt, i_p, N_{per})$	$-Pmt \times \dfrac{(1 + i_p)^{N_{per}} - 1}{i_p(1 + i_p)^{N_{per}}}$	$=PV(i_p, N_{per}, Pmt, 0, 0)$

Evaluating Options

Principles

- Compare values at the same point in time (present values, future values, or equivalent annual cost).
- Projects should have the same duration before comparison, or use equivalent annual cost.

Problems

Loan Calculations I

1. Terry wants to purchase a used car for $8,000. The dealer has offered to finance the deal at 8.5% APR for four years with a 10% down payment. Terry's bank will provide a loan at 7.8% for three years, but wants a 20% down payment.

    a. Calculate the required down payment and monthly payment for each loan option. (Assume monthly compounding and monthly loan payments.)

    b. The present value for both options (including the down payments) is the same, $8,000. What additional factors should Terry consider when deciding which is the best deal for her?

Loan Calculations II

2. It is not uncommon to borrow $100,000 to purchase a home. Banks typically offer 15-, 20-, or 30-year loans.

    a. Calculate the monthly loan payment to repay a $100,000 home loan at 7% APR over 30 years.

    b. How much more must be paid each month to pay off the loan in 15 years?

Loan Calculations III

3. John has estimated that trading in his current car and adding the money he has saved will allow him to pay $8,000 in cash when he buys a new car. He also feels that his budget can handle a $350 payment each month, but doesn't want to make payments for more than three years. His credit union will lend him money at 6.8% APR (monthly compounding). What is the maximum-priced car he can afford?

Loan Calculations IV

4. Some of the loans college students get for school start charging interest immediately, even if repayment is not required until the student leaves school. The unpaid interest while the student is in school is added to the principal.

    If a student borrows $2,300 at 8% APR compounded monthly when he or she starts school, how much will be owed 5.5 years later when loan repayment begins? (The 5.5-year figure assumes five years to graduate plus six months after graduation to begin loan payments.)

Loan Calculations V

5. Patricia and Bill want to buy a home, and they feel that they can handle a $1,000-per-month payment. If they make $1,000 payments for 30 years on a 7% APR loan, how much can they borrow?

Time-Value-of-Money Problems I

6. Fill in the missing values in the following table:

PRESENT VALUE	FUTURE VALUE	REG. PAYMENT	INTEREST RATE	COMPOUNDING	YEARS
$50,000		—	5% APR	Annual	12
$50,000		—	5% APR	Weekly	12
	$250,000	—	17.3% (eff. ann. rate)	Annual	7
—	$250,000		12.4% (eff. ann. rate)	Annual	30
$50,000	—		5% APR	Weekly	12

Time-Value-of-Money Problems II

7. Compute the present value of the following cash flows if the interest rate is 4% APR compounded monthly:

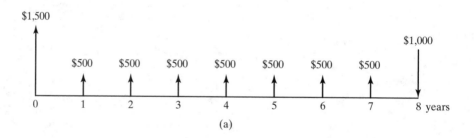

(a)

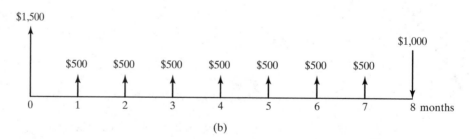

(b)

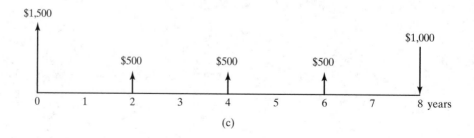

(c)

Evaluating Options I

8. Consumers today have a choice of buying or leasing new cars. Compare the net present value of the following options:

Three-Year-Life Calculations

- Option *A*: Lease a new $26,000 sport utility vehicle for 36 months with $380 monthly payments and $1,750 required up-front. At the end of the lease, you return the car and get no cash back.

- Option *B*: Purchase a new $26,000 sport utility vehicle with a 60-month loan at 9% APR and $1,750 down payment. At the end of three years, sell the vehicle for $15,000 and use the money to pay off the rest of the loan and generate some cash.

Nine-Year-Life Calculations

- Option *C*: Use three consecutive three-year leases to cover the nine-year time span. Assume 4% annual inflation rate for the monthly payments and up-front costs each time a new vehicle is leased.

- Option *D*: Purchase a new $26,000 sport utility vehicle with a 60-month loan at 9% APR and $1,750 down payment. Keep the vehicle for nine years. Assume no maintenance costs the first three years (warranty coverage) and $400 maintenance in the fourth year, with maintenance costs increasing by $200 each year, thereafter. After nine years, sell the vehicle for $5,000.

Sketch the cash-flow diagram for each option, and calculate the net present value for each option, using an interest rate of 6% APR. Which seems like the best deal?

Note: Why a 6% rate when the bank loan is 9%? The 9% is what the bank owners want to get for their money. The 6% is a rate that the person buying the car could readily get for his or her own money if it was not tied up in the vehicle.

11 Financial Calculations with Excel

11.1 INTRODUCTION

This chapter introduces a few financial calculations that are not always covered in engineering curricula, but often are needed on the job. These calculations are often required to determine the projected return on investment of proposed projects. This typically is a significant factor for determining whether the proposed projects are funded or not.

11.2 INTERNAL RATE OF RETURN AND NET PRESENT VALUE

An *internal rate of return* is a common financial calculation. Effectively, it is the interest rate (rate of return) that "balances" the income and expenses for a project over a specified period. More precisely, the internal rate of return is the periodic interest rate that produces a net present value of zero for a time series of incomes and expenses. For a typical engineering cash flow (big expense up front, smaller incomes annually for several years) the internal rate of return is the rate of return required just to break even.

Note: Not all cash flows fit the description of a "typical" engineering cash flow of large initial investments with smaller incomes over a long period. Large payments at the end of a project's life can have a dramatic impact on the internal rate of return. This is the subject of Problem 11.3.

OBJECTIVES

After reading this chapter, you will know

- How to use Excel to calculate internal rate of return
- How to use methods for asset depreciation

EXAMPLE 11.1

A friend asks to borrow $1,000 and promises to pay back $1,200 after five years. You can use time-value-of-money principles to see what the present value of the $1,200 would be with various interest rates. For example, with a 5% annual rate, the present value of the $1,200 repayment is

$$\begin{aligned} PV &= FV \times (1 + i_P)^{-N_{per}} \\ &= \$1200 \times (1 + 0.05)^{-5} \\ &= \$940. \end{aligned}$$

That doesn't look like such a good deal. You're paying him $1,000 (−$1,000, since it is an expense) and he is paying back an amount equivalent to $940 in today's dollars. You lose $60 (if you use an interest rate of 5%). What happens if you use an interest rate of 3%?

$$\begin{aligned} PV &= \$1200 \times (1 + 0.03)^{-5} \\ &= \$1035. \end{aligned}$$

That looks like a better deal, since you come out ahead by $35 (if you use an interest rate of 3%).

What is the break-even interest rate that makes the present values of your payment (−$1,000) and his repayment ($1,200 after five years) equal (but opposite in sign)?

The answer is 3.7138%.

The interest rate that makes the present values of the expenses and incomes balance is the internal rate of return. For this simple example, that rate is 3.7138%. If you want your money to earn more than 3.7138%, you won't make the loan. Similarly, at work, if your proposed project's expenses and incomes have an internal rate of return of 3.7% and your company wants a higher rate of return on their investments, they won't fund your project. But if your project's projected internal rate of return is 37%, you'll have your employer's attention.

11.2.1 Net Present Value

The internal rate of return of a series of incomes and expenses is the interest rate that causes the net present value to be zero, so let's learn to calculate net present values.

EXAMPLE 11.2

A new heat exchanger will cost $40,000, but is expected to generate an income of $7,000 per year for the first two years and then $5,500 per year for another five years. If the interest rate is 5% per year, what is the net present value?

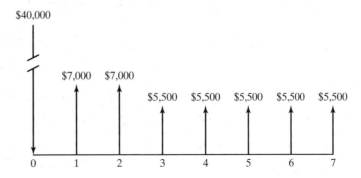

First, we set up a spreadsheet that calculates the present value of each income and expense for the specified interest rate, using the usual sign convention of negative values for expenses and positive values for incomes:

	C5		f_x =PV($C$2,A5,0,-B5)		
	A	B	C	D	E
1	Net Present Value				
2		Rate:	5%	per year	
3					
4	Year	Income	PV		
5	0	-$40,000	-$40,000		
6	1	$7,000	$6,667		
7	2	$7,000	$6,349		
8	3	$5,500	$4,751		
9	4	$5,500	$4,525		
10	5	$5,500	$4,309		
11	6	$5,500	$4,104		
12	7	$5,500	$3,909		
13					

The net present value is simply the sum of the present values of all of the incomes and expenses:

	C14		f_x =SUM(C5:C12)	
	A	B	C	D
1	Net Present Value			
2		Rate:	5%	per year
3				
4	Year	Income	PV	
5	0	-$40,000	-$40,000	
6	1	$7,000	$6,667	
7	2	$7,000	$6,349	
8	3	$5,500	$4,751	
9	4	$5,500	$4,525	
10	5	$5,500	$4,309	
11	6	$5,500	$4,104	
12	7	$5,500	$3,909	
13				
14			NPV:	-$5,385.83
15				

Excel's Net Present Value Function, *NPV()* Excel provides a function, NPV(), that will calculate net present values directly from a stream of incomes. The syntax of the NPV() function is

$$=NPV(i_P, \text{ Incomes1, Incomes2, ...})$$

where

i_P	is the interest rate per period for the cash flow (required),
Incomes1	is a value, cell address, or range of cells containing one or more incomes, with expenses as negative incomes (required), and
Incomes2, ...	are used to include a series of individual incomes (up to 29) if the incomes are not stored as a range of cells

When the NPV() function is used for the heat exchanger example, it isn't necessary to calculate the PV values. Instead, we simply send the column of incomes directly to the NPV() function:

	C14	▼	f_x =NPV(C2,B6:B12)+B5	
	A	B	C	D	E
1	**Net Present Value**				
2		**Rate:**	**5%**	per year	
3					
4	**Year**	**Income**			
5	0	-$40,000			
6	1	$7,000			
7	2	$7,000			
8	3	$5,500			
9	4	$5,500			
10	5	$5,500			
11	6	$5,500			
12	7	$5,500			
13					
14		**NPV:**	-$5,385.83		
15					

Because the initial expense is not at the end of a period, it was not included in the values sent to the NPV() function. But it is already at time zero, so it can simply be added on to the result returned by NPV():

```
C14:=NPV(C2,B6:B12)+B5
```

Note the following:

- *The NPV() function works with the incomes, not the present values of the incomes. It calculates the present values internally.*
- *The incomes sent to the NPV() function must be* equally spaced in time *and occur at the* end of each interval. *If you have both an expense and an income at the same point in time, send NPV() the net income at that time.*
- *Do not include a payment (or income) at time zero in the values sent to the NPV() function, because time zero is not the end of an interval. If there is an amount at time zero, it should be added on to the result returned by NPV().*

APPLICATIONS: ECONOMICS

Payout Period

The "typical" engineering project requires a source of cash to purchase equipment and get things started. Then, income starts coming in, and, eventually (hopefully), income exceeds expenses, and the project starts making money. The moment when income from the project has paid back the startup expenses and initial operating costs marks the end of the *payout period*.

The net present value can be used to figure out the payout period for a cash flow. You simply start at the beginning of the cash flow and calculate the net present value at time zero. Then you include income and expenses for one time internal and recalculate the net present value. You continue to include more time intervals until the net present value becomes positive. That's the end of the payout period.

Considering the heat-exchanger example again, we see that the $40,000 expense at time zero creates a net present value of $-\$40,000$ at time zero if all other incomes and expenses are ignored. At the end of the first year, the income of $7,000 changes the net present value to $-\$33,000$ (if the time value of money is ignored by setting $i_P = 0$). Building the incomes over time into a spreadsheet, we find that it is easy to calculate the net present values by using the NPV() function, and it is clear that the heat-exchanger project (with $i_P = 0$) has a payout period of seven years:

	C6	▼		f_x =NPV($C$2,$B$6:B6)+$C$5	
	A	B	C	D	E
1	**Payout Period**				
2		i_P:	0%		
3					
4	**Year**	**Income**	**NPV**		
5	0	-$40,000	-$40,000		
6	1	$7,000	-$33,000		
7	2	$7,000	-$26,000		
8	3	$5,500	-$20,500		
9	4	$5,500	-$15,000		
10	5	$5,500	-$9,500		
11	6	$5,500	-$4,000		
12	7	$5,500	$1,500		
13					

```
C5: = B5
C6: = NPV($C$2, $B$6:B6)+$C$5
```

The net present value in cell C6 was calculated from the range $B$6:B6, which contains only one value ($7,000). This unusual range (with the dollar signs on only the starting cell) was used so that the equation in cell C6 could be copied down column C to calculate the net present values in cells C7 through C12. As the equation is copied, the range expands. In year 3, for example, the copied equation reads

```
C8: =NPV($C$2, $B$6:B8)+$C$5
```

and includes the $-\$40,000$, $7,000$, $7,000$, and $5,500$ in the calculation of the net present value.

Note: It is possible to account for the time value of money when computing the payout period. Simply use an interest rate greater than zero when calculating the net present value.

11.2.2   Internal Rate of Return

Because the internal rate of return is defined as the interest rate that causes the net present value to be zero, you can find the internal rate of return by changing the interest rate used in the spreadsheet until the calculated NPV equals zero:

	A	B	C	D
1	**Net Present Value**			
2		**Rate:**	**5%**	per year
3				
4	**Year**	**Income**		
5	0	-$40,000		
6	1	$7,000		
7	2	$7,000		
8	3	$5,500		
9	4	$5,500		
10	5	$5,500		
11	6	$5,500		
12	7	$5,500		
13				
14		**NPV:**	-$5,385.83	
15				

The rate 5% returned a negative NPV. There is more than $40,000 of income in this example, but, at an interest rate of 5%, the future amounts are reduced (*discounted*) enough to cause the project not to pay for itself in seven years.

We can try lower interest rates to see what rate makes the NPV zero. Try 4% per year:

	A	B	C	D
1	**Net Present Value**			
2		**Rate:**	**4%**	per year
3				
4	**Year**	**Income**		
5	0	-$40,000		
6	1	$7,000		
7	2	$7,000		
8	3	$5,500		
9	4	$5,500		
10	5	$5,500		
11	6	$5,500		
12	7	$5,500		
13				
14		**NPV:**	-$4,159.56	
15				

The NPV is closer to zero, but still negative.

With a little trial and error, we find that a rate of 0.9737% makes the NPV nearly zero:

	A	B	C	D
1	Net Present Value			
2		Rate:	0.9737%	per year
3				
4	Year	Income		
5	0	-$40,000		
6	1	$7,000		
7	2	$7,000		
8	3	$5,500		
9	4	$5,500		
10	5	$5,500		
11	6	$5,500		
12	7	$5,500		
13				
14		NPV:	$0.01	
15				

This is the internal rate of return for the heat exchanger. This is a pretty unfavorable internal rate of return; the company could earn nearly this amount simply by putting their $40,000 in a savings account!

PRACTICE!

Would the internal rate of return increase or decrease if the exchanger could be sold for $20,000 at the end of seven years? Why?

Excel's Internal Rate of Return Function: IRR()   In the last example, the interest rate in cell C2 was changed by hand until a net present value close to zero was obtained. This works, but Excel provides the IRR() function to calculate internal rates of return directly.

The syntax of the IRR() function is

    =IRR(Incomes, irrGuess)

where

Incomes is a range of cells containing the incomes, with expenses as negative incomes (required) and irrGuess is a starting value, or guess value, for the internal rate of return. (This is optional, and Excel uses 10% if it is omitted.)

C2		▾	f_x =IRR(B5:B12,5%)		
	A	B	C	D	E
1	Internal Rate of Return				
2		IRR:	0.9737%	per year	
3					
4	Year	Income			
5	0	-$40,000			
6	1	$7,000			
7	2	$7,000			
8	3	$5,500			
9	4	$5,500			
10	5	$5,500			
11	6	$5,500			
12	7	$5,500			
13					

Note the following:

- *The* IRR() *function works with the incomes, not the present values of the incomes. It calculates the present values internally.*

- *The incomes sent to the* IRR() *function must be equally spaced in time and, with the exception of the first value (an amount at time zero), must occur at the end of the interval. If you have both an expense and an income at the same point in time, send* IRR() *the net income at that time.*

- *The rate returned by the* IRR() *function is the rate per time interval (% per year, % per month, etc.).*

- *The* IRR() *function uses an iterative approach to solving for IRR. If the solution does not converge on a value within 20 iterations,* IRR() *returns a #NUM! error. A closer* irrGuess *value may help in finding the correct internal rate of return.*

11.3 DEPRECIATION OF ASSETS

Taxes are a very significant cost of doing business and must be considered when estimating the profit or loss from a proposed business venture. Depreciation is an important factor in tax planning, and numerous methods have been used over the years to calculate depreciation amounts. Only a few of the more common methods will be presented here. Excel provides built-in functions for each of these depreciation methods.

11.3.1 Methods of Depreciation

U.S. tax codes subject most capital expenditures to *depreciation. Capital expenditures* are substantial expenditures for *assets* (things like equipment purchases) that will be used for more than one year. (That is, the impact of the equipment purchase is felt over multiple taxation periods.) Expenses can be used to offset income before the calculation of taxes, but capital expenditures impact multiple taxation periods, so the amount of the capital expenditure that may be used to offset income is also spread over multiple taxation periods. The amount of the capital expenditure that may be deducted each year may be computed by several depreciation methods. The U.S. tax laws require businesses to use a depreciation system called *Modified Accelerated Cost Recovery System (MACRS)*, which includes the following depreciation methods (the actual methods used depend on the asset type and service life):

- Straight-line depreciation method
- *Declining-balance depreciation* methods
    - 200% declining-balance, or double-declining-balance, method
    - 150% declining-balance method

The MACRS system starts out by using declining-balance methods for most capital expenditures. These methods allow larger depreciation amounts in the years immediately after the capital expenditure than does the straight-line method. Under the MACRS system, the deductible amount is calculated by using declining-balance methods for the early years and then switches to the *straight-line depreciation* method when the allowable deduction calculated by that method exceeds the amount calculated by the declining-balance method.

Sounds complicated? It is, but we'll explain it all in the sections that follow.

Straight-Line Depreciation Method   If you buy a heat exchanger for $40,000, and it has an expected service life of 10 years and an expected salvage value of $2,000, how much can you deduct each year?

Using the straight-line depreciation method, we see that the amount that can be deducted each year, D, is the initial cost minus the salvage value, divided by the expected service life, so

$$D = \frac{C_{init} - S}{N_{SL}} = \frac{\$40,000 - \$2,000}{10} = \$3,800, \tag{11.1}$$

where

D	is the depreciation amount (same each year using the straight-line method),
C_{init}	is the initial cost of the asset,
S	is the salvage value at the end of the service life, and
N_{SL}	is the service life (also called the *recovery period*) of the asset.

The difference between the initial cost and the salvage value (the numerator in the depreciation equation) is called the *depreciation basis* of the asset. (Depreciation is calculated based on this amount.) Some depreciation systems (including MACRS) do not take the salvage value into account when calculating depreciation amounts.

Because the capital expense of an asset must be depreciated over its service life, you must keep track of the current value of the asset throughout its service life. The current value of an asset is called its *book value*, B. For straight-line depreciation, the book value in the current year is last year's book value minus the depreciation amount, so

$$B_j = B_{j-1} - D, \tag{11.2}$$

where j represents the current year and j-1 represents the previous year. B_0 is equivalent to the initial cost of the asset.

Note: Depreciation is an accounting practice used to spread the cost of assets over several tax years. The book value used for tax purposes simply keeps track of the amount of the original purchase cost that has yet to be deducted. It is not necessarily related to the actual value of a piece of equipment.

The book value of an asset subject to straight-line depreciation decreases linearly over the service life of the asset (that's why it is called straight-line depreciation), as is shown in the following figure:

200% or Double-Declining-Balance Depreciation   *Declining-balance methods* calculate the allowable depreciation amount, based on the current book value of an asset, the declining depreciation factor, F_{DB}, rate R_{DB}, and the salvage value s of the asset. Thus,

$$D_j = (B_{j-1} - S)F_{DB}, \tag{11.3}$$

or, if the salvage value is not considered (as in MACRS),

$$D_j = B_{j-1}F_{DB}, \tag{11.4}$$

where

D_j	is the allowable depreciation amount in the current year (year j),
B_{j-1}	is the book value at the end of the previous year ($B_0 = C_{init}$),
N_{SL}	is the service life of the asset,
F_{DB}	is the declining-balance depreciation factor, and
R_{DB}	is the declining-balance percentage. For the double-declining-balance method, $R_{DB} = 200\%$, or 2.

The declining-balance depreciation factor is computed for the double-declining method as

$$F_{DB} = \frac{200\%}{N_{SL}}. \tag{11.5}$$

Considering the heater example again ($C_{init} = \$40,000$, $S = \$2,000$, $N_{SL} = 10$ years), we see that the double-declining method of depreciation produces the following declining-balance rate:

C8		f_x =200%/C7						
A	B	C	D	E	F	G	H	I
1	**Double-Declining Balance Depreciation (Salvage Value Included)**							
2	(Half-Year Convention Not Included)							
3								
4	Initial Cost:	$40,000						
5	Salvage Value:	$2,000						
6	Depreciation Basis:	$38,000						
7	Service Life:	10	years					
8	F_{DB}:	20%						
9								

The depreciation amounts are

	B13	▼		*fx* =(C12-$C$5)*$C$8				
	A	B	C	D	E	F	G	H	I

	A	B	C	D
1	**Double-Declining Balance Depreciation (Salvage Value Included)**			
2	(Half-Year Convention Not Included)			
3				
4	Initial Cost:		$40,000	
5	Salvage Value:		$2,000	
6	Depreciation Basis:		$38,000	
7	Service Life:		10	years
8	F_{DB}:		20%	
9				
10		**Depreciation**		
11	**Year**	**Amount**	**Book Value**	
12	0		$40,000	
13	1	$7,600	$32,400	
14	2	$6,080	$26,320	
15	3	$4,864	$21,456	
16	4	$3,891	$17,565	
17	5	$3,113	$14,452	
18	6	$2,490	$11,961	
19	7	$1,992	$9,969	
20	8	$1,594	$8,375	
21	9	$1,275	$7,100	
22	10	$1,020	$6,080	
23				
24				

Notice that the depreciation amounts are much larger in the first few years under the double-declining-balance method. This allows significantly larger tax deductions in the first years after capital expenditures.

MACRS Depreciation System   MACRS uses declining-balance methods as long as the depreciation amounts are greater than those calculated by the straight-line method. When the depreciation amounts from the straight-line method are greater, MACRS switches to straight-line depreciation. (This occurs in the latter years of an asset's service life.) MACRS does not consider salvage value when calculating depreciation amounts.

With MACRS, assets are categorized, and the service lives are established as 3, 5, 7, 10, 15, or 20 years. 200% declining-balance methods are used for service lives of 10 years or less, 150% declining-balance methods for service lives of 15 or 20 years.

MACRS also uses a *half-year convention* that assumes that assets are placed in service in the middle of the tax year and calculates a half-year of depreciation in the first and last years of service. When the remaining service life falls below one year, the straight-line depreciation factor for that tax year is 1.00 (100%), which means that the entire remaining book value may be deducted.

We can use the MACRS depreciation system on the heat-exchanger example. Because this is a complex calculation, we will proceed step by step, first showing the equations and then the Excel spreadsheet implementation.

Note: Because this is a complex calculation, MACRS depreciation is usually calculated by using tabulated depreciation factors. Those factors will be presented after this example.

Year 1

- The first year's depreciation is actually for six months, because of the half-year convention. The years remaining (needed for the straight-line (SL) method) are still 10. The SL depreciation factor for the first year is calculated by dividing one by the number of years remaining and then by 2, because of the half-year convention:

$$f_{\text{SL}_1} = \frac{1}{n_{\text{remain}}} \Big/ 2 = \frac{1}{10} \Big/ 2 = 0.050 \quad (\text{or } 5\%).\tag{11.6}$$

	D10	▾	f_x =1/C10/2						
	A	B	C	D	E	F	G	H	I
1	**MACRS Depreciation**								
2									
3		Initial Cost:	$40,000						
4		Service Life:	10	years					
5		F_DB:	200%						
6									
7		**Beginning**	**SL**	**SL**	**DDB**			**End**	
8		**of Year**	**Years**	**Depreciation**	**Depreciation**	**Method**	**Depreciation**	**of Year**	
9	**Year**	**Book Value**	**Remaining**	**Factor**	**Factor**	**Used**	**Amount**	**Book Value**	
10	1	$40,000	10	0.050					
11	2								
12	3								

The double-declining-balance depreciation factor for the first year is

$$F_{\text{DDB}_1} = \frac{F_{\text{DB}}}{N_{\text{SL}}} \Big/ 2 = \frac{200\%}{10} \Big/ 2 = 0.010 \quad (\text{or } 10\%).\tag{11.7}$$

Note: For full years, $f_{\text{DDB}} = F_{\text{DB}}$, but, although F_{DB} is a simple constant, f_{DDB} must be calculated each year, because of the half-year convention in the first and last years of depreciation.

	E10	▾	f_x =(C5/C4)/2						
	A	B	C	D	E	F	G	H	I
1	**MACRS Depreciation**								
2									
3		Initial Cost:	$40,000						
4		Service Life:	10	years					
5		F_DB:	200%						
6									
7		**Beginning**	**SL**	**SL**	**DDB**			**End**	
8		**of Year**	**Years**	**Depreciation**	**Depreciation**	**Method**	**Depreciation**	**of Year**	
9	**Year**	**Book Value**	**Remaining**	**Factor**	**Factor**	**Used**	**Amount**	**Book Value**	
10	1	$40,000	10	0.050	0.100				
11	2								
12	3								

Cell F10 contains an `IF()` statement and was included just to show which depreciation method will actually be used to compute the depreciation amount. The method that is used is the method that produces the largest depreciation factor. The `IF()` statement checks this and then displays either "SL" or "DDB":

```
F10: =IF(D10>E10, "SL", "DDB")
```

The `IF()` statement is read as, "If the contents of cell D10 are larger than the contents of cell E10, then display "SL"; otherwise, display "DDB".

Cell G10 uses another `IF()` statement to calculate the depreciation amount, using the larger depreciation factor:

G10		f_x =IF(D10>E10,D10*B10,E10*B10)						
	A	B	C	D	E	F	G	H	I
1	MACRS Depreciation								
2									
3		Initial Cost:	$40,000						
4		Service Life:	10	years					
5		F$_{DB}$:	200%						
6									
7		Beginning	SL	SL	DDB			End	
8		of Year	Years	Depreciation	Depreciation	Method	Depreciation	of Year	
9	Year	Book Value	Remaining	Factor	Factor	Used	Amount	Book Value	
10	1	$40,000	10	0.050	0.100	DDB	$4,000		
11	2								
12	3								

Finally, the end-of-year book value is calculated by subtracting the depreciation amount from the beginning-of-year book value:

$$B_j = B_{j-1} - D_j, \quad \text{where } j = 1 \text{ for the first year.} \tag{11.8}$$

H10		f_x =B10-G10						
	A	B	C	D	E	F	G	H	I
1	MACRS Depreciation								
2									
3		Initial Cost:	$40,000						
4		Service Life:	10	years					
5		F$_{DB}$:	200%						
6									
7		Beginning	SL	SL	DDB			End	
8		of Year	Years	Depreciation	Depreciation	Method	Depreciation	of Year	
9	Year	Book Value	Remaining	Factor	Factor	Used	Amount	Book Value	
10	1	$40,000	10	0.050	0.100	DDB	$4,000	$36,000	
11	2								
12	3								

Year 2

The book value at the beginning of Year 2 is simply the end-of-year book value from the previous year:

B11		f_x =H10						
	A	B	C	D	E	F	G	H	I
1	MACRS Depreciation								
2									
3		Initial Cost:	$40,000						
4		Service Life:	10	years					
5		F$_{DB}$:	200%						
6									
7		Beginning	SL	SL	DDB			End	
8		of Year	Years	Depreciation	Depreciation	Method	Depreciation	of Year	
9	Year	Book Value	Remaining	Factor	Factor	Used	Amount	Book Value	
10	1	$40,000	10	0.050	0.100	DDB	$4,000	$36,000	
11	2	$36,000							
12	3								

Because of the half-year convention, the years remaining in cell C11 is 9.5, not 9:

C11		f_x =C10-0.5						
	A	B	C	D	E	F	G	H	I
1	MACRS Depreciation								
2									
3		Initial Cost:	$40,000						
4		Service Life:	10	years					
5		F$_{DB}$:	200%						
6									
7		Beginning	SL	SL	DDB			End	
8		of Year	Years	Depreciation	Depreciation	Method	Depreciation	of Year	
9	Year	Book Value	Remaining	Factor	Factor	Used	Amount	Book Value	
10	1	$40,000	10	0.050	0.100	DDB	$4,000	$36,000	
11	2	$36,000	9.5						
12	3								

The straight-line depreciation factor is calculated by dividing 1.0 by the years remaining:

$$f_{SL_2} = \frac{1}{n_{remain}} = \frac{1}{9.5} = 0.105 \text{ (or } 10.5\%\text{)}. \tag{11.9}$$

	D11		f_x =1/C11						
	A	B	C	D	E	F	G	H	I
1	MACRS Depreciation								
2									
3		Initial Cost:	$40,000						
4		Service Life:	10	years					
5		F_{DB}:	200%						
6									
7		Beginning	SL	SL	DDB			End	
8		of Year	Years	Depreciation	Depreciation	Method	Depreciation	of Year	
9	Year	Book Value	Remaining	Factor	Factor	Used	Amount	Book Value	
10	1	$40,000	10	0.050	0.100	DDB	$4,000	$36,000	
11	2	$36,000	9.5	0.105					
12	3								

The double-declining-balance depreciation factor is simply the declining-balance rate, R_{DB}, divided by the service life of the asset, N_{SL}:

$$f_{DDB_1} = \frac{R_{DB}}{N_{SL}} = \frac{200\%}{10} = 0.200 \text{ (or } 20\%\text{)}. \tag{11.10}$$

	E11		f_x =$C$5/$C$4						
	A	B	C	D	E	F	G	H	I
1	MACRS Depreciation								
2									
3		Initial Cost:	$40,000						
4		Service Life:	10	years					
5		F_{DB}:	200%						
6									
7		Beginning	SL	SL	DDB			End	
8		of Year	Years	Depreciation	Depreciation	Method	Depreciation	of Year	
9	Year	Book Value	Remaining	Factor	Factor	Used	Amount	Book Value	
10	1	$40,000	10	0.050	0.100	DDB	$4,000	$36,000	
11	2	$36,000	9.5	0.105	0.200				
12	3								

Simply copy the formulas in cells F10 through H10 down one row to complete the calculations for Year 2:

	A	B	C	D	E	F	G	H	I
1	MACRS Depreciation								
2									
3		Initial Cost:	$40,000						
4		Service Life:	10	years					
5		F_{DB}:	200%						
6									
7		Beginning	SL	SL	DDB			End	
8		of Year	Years	Depreciation	Depreciation	Method	Depreciation	of Year	
9	Year	Book Value	Remaining	Factor	Factor	Used	Amount	Book Value	
10	1	$40,000	10	0.050	0.100	DDB	$4,000	$36,000	
11	2	$36,000	9.5	0.105	0.200	DDB	$7,200	$28,800	
12	3	$28,800							
13	4								

Year 3

Again, the book value at the beginning of the year is simply the end-of-year book value from the previous year (shown in the preceding figure). To calculate years remaining, subtract one from the years remaining in Year 2:

C12		fx =C11-1						
	A	B	C	D	E	F	G	H	I
1	MACRS Depreciation								
2									
3		Initial Cost:	$40,000						
4		Service Life:	10	years					
5		F_DB:	200%						
6									
7		Beginning	SL	SL	DDB			End	
8		of Year	Years	Depreciation	Depreciation	Method	Depreciation	of Year	
9	Year	Book Value	Remaining	Factor	Factor	Used	Amount	Book Value	
10	1	$40,000	10	0.050	0.100	DDB	$4,000	$36,000	
11	2	$36,000	9.5	0.105	0.200	DDB	$7,200	$28,800	
12	3	$28,800	8.5						
13	4								

Copy cells D11 through H11 down one row to complete the calculations for Year 3:

	A	B	C	D	E	F	G	H	I
1	MACRS Depreciation								
2									
3		Initial Cost:	$40,000						
4		Service Life:	10	years					
5		F_DB:	200%						
6									
7		Beginning	SL	SL	DDB			End	
8		of Year	Years	Depreciation	Depreciation	Method	Depreciation	of Year	
9	Year	Book Value	Remaining	Factor	Factor	Used	Amount	Book Value	
10	1	$40,000	10	0.050	0.100	DDB	$4,000	$36,000	
11	2	$36,000	9.5	0.105	0.200	DDB	$7,200	$28,800	
12	3	$28,800	8.5	0.118	0.200	DDB	$5,760	$23,040	
13	4								

Years 4 through 10

Copy cells B12 through H12 down to complete the calculations for Years 4 through 10:

	A	B	C	D	E	F	G	H	I
1	MACRS Depreciation								
2									
3		Initial Cost:	$40,000						
4		Service Life:	10	years					
5		F_DB:	200%						
6									
7		Beginning	SL	SL	DDB			End	
8		of Year	Years	Depreciation	Depreciation	Method	Depreciation	of Year	
9	Year	Book Value	Remaining	Factor	Factor	Used	Amount	Book Value	
10	1	$40,000	10	0.050	0.100	DDB	$4,000	$36,000	
11	2	$36,000	9.5	0.105	0.200	DDB	$7,200	$28,800	
12	3	$28,800	8.5	0.118	0.200	DDB	$5,760	$23,040	
13	4	$23,040	7.5	0.133	0.200	DDB	$4,608	$18,432	
14	5	$18,432	6.5	0.154	0.200	DDB	$3,686	$14,746	
15	6	$14,746	5.5	0.182	0.200	DDB	$2,949	$11,796	
16	7	$11,796	4.5	0.222	0.200	SL	$2,621	$9,175	
17	8	$9,175	3.5	0.286	0.200	SL	$2,621	$6,554	
18	9	$6,554	2.5	0.400	0.200	SL	$2,621	$3,932	
19	10	$3,932	1.5	0.667	0.200	SL	$2,621	$1,311	
20	11								
21									

Year 11

Year 11 is the extra half-year created by using the half-year convention on the first year of depreciation. The beginning-of-year book value and the years remaining can be copied from Year 10. The straight-line depreciation factor is 1 (by definition) any time the remaining service life is less than one year, so the value 1 is entered in cell D20:

	D20		f_x	1					
	A	B	C	D	E	F	G	H	I
1	MACRS Depreciation								
2									
3		Initial Cost:	$40,000						
4		Service Life:	10	years					
5		F_DB:	200%						
6								End	
7		Beginning	SL	SL	DDB			of Year	
8		of Year	Years	Depreciation	Depreciation	Method	Depreciation	Book Value	
9	Year	Book Value	Remaining	Factor	Factor	Used	Amount		
10	1	$40,000	10	0.050	0.100	DDB	$4,000	$36,000	
11	2	$36,000	9.5	0.105	0.200	DDB	$7,200	$28,800	
12	3	$28,800	8.5	0.118	0.200	DDB	$5,760	$23,040	
13	4	$23,040	7.5	0.133	0.200	DDB	$4,608	$18,432	
14	5	$18,432	6.5	0.154	0.200	DDB	$3,686	$14,746	
15	6	$14,746	5.5	0.182	0.200	DDB	$2,949	$11,796	
16	7	$11,796	4.5	0.222	0.200	SL	$2,621	$9,175	
17	8	$9,175	3.5	0.286	0.200	SL	$2,621	$6,554	
18	9	$6,554	2.5	0.400	0.200	SL	$2,621	$3,932	
19	10	$3,932	1.5	0.667	0.200	SL	$2,621	$1,311	
20	11	$1,311	0.5	1					
21									

The double-declining-balance depreciation factor would also be divided by two in the final half-year (although it would never be used in the MACRS system):

	E20		f_x	=(C5/C4)/2					
	A	B	C	D	E	F	G	H	I
1	MACRS Depreciation								
2									
3		Initial Cost:	$40,000						
4		Service Life:	10	years					
5		F_DB:	200%						
6								End	
7		Beginning	SL	SL	DDB			of Year	
8		of Year	Years	Depreciation	Depreciation	Method	Depreciation	Book Value	
9	Year	Book Value	Remaining	Factor	Factor	Used	Amount		
10	1	$40,000	10	0.050	0.100	DDB	$4,000	$36,000	
11	2	$36,000	9.5	0.105	0.200	DDB	$7,200	$28,800	
12	3	$28,800	8.5	0.118	0.200	DDB	$5,760	$23,040	
13	4	$23,040	7.5	0.133	0.200	DDB	$4,608	$18,432	
14	5	$18,432	6.5	0.154	0.200	DDB	$3,686	$14,746	
15	6	$14,746	5.5	0.182	0.200	DDB	$2,949	$11,796	
16	7	$11,796	4.5	0.222	0.200	SL	$2,621	$9,175	
17	8	$9,175	3.5	0.286	0.200	SL	$2,621	$6,554	
18	9	$6,554	2.5	0.400	0.200	SL	$2,621	$3,932	
19	10	$3,932	1.5	0.667	0.200	SL	$2,621	$1,311	
20	11	$1,311	0.5	1	0.100				
21									

The rest of the cells in Year 11 can be copied from Year 10. The final result is as follows:

	A	B	C	D	E	F	G	H	I
1	MACRS Depreciation								
2									
3		Initial Cost:	$40,000						
4		Service Life:	10	years					
5		F_DB:	200%						
6								End	
7		Beginning	SL	SL	DDB			of Year	
8		of Year	Years	Depreciation	Depreciation	Method	Depreciation	Book Value	
9	Year	Book Value	Remaining	Factor	Factor	Used	Amount		
10	1	$40,000	10	0.050	0.100	DDB	$4,000	$36,000	
11	2	$36,000	9.5	0.105	0.200	DDB	$7,200	$28,800	
12	3	$28,800	8.5	0.118	0.200	DDB	$5,760	$23,040	
13	4	$23,040	7.5	0.133	0.200	DDB	$4,608	$18,432	
14	5	$18,432	6.5	0.154	0.200	DDB	$3,686	$14,746	
15	6	$14,746	5.5	0.182	0.200	DDB	$2,949	$11,796	
16	7	$11,796	4.5	0.222	0.200	SL	$2,621	$9,175	
17	8	$9,175	3.5	0.286	0.200	SL	$2,621	$6,554	
18	9	$6,554	2.5	0.400	0.200	SL	$2,621	$3,932	
19	10	$3,932	1.5	0.667	0.200	SL	$2,621	$1,311	
20	11	$1,311	0.5	1	0.100	SL	$1,311	$0	
21									

Notice that the MACRS method switched from using the double-declining-balance method to the straight-line method in Year 7, when the depreciation factor for the straight-line method became greater than the double-declining-balance depreciation factor.

Note: The straight-line depreciation amount is constant in Years 7 through 10, but the straight-line depreciation factor is not constant.

Using Excel's *VDB()* Function for MACRS Depreciation

Excel provides the VDB() function to handle MACRS depreciation. The syntax of the VDB() function is

$$=VDB(C_{init}, S, N_{SL}, Per_{start}, Per_{end}, F_{DB}, NoSwitch)$$

where

C_{init}	is the initial cost of the asset (required),
S	is the salvage value (set to zero for MACRS depreciation) (required),
N_{SL}	is the service life of the asset (required),
Per_{start}	is the start of the period over which the depreciation amount will be calculated (required),
Per_{stop}	is the end of the period over which the depreciation amount will be calculated (required), and
F_{DB}	is the declining-balance percentage (optional; Excel uses 200% if omitted).

NoSwitch tells Excel whether to switch to straight-line depreciation when the straight-line depreciation factor is larger than the declining-balance depreciation factor (optional). (Set this to FALSE to switch to straight line, to TRUE not to switch; default is FALSE):

	E6		▼		f_x	=VDB($B$6,0,10,C6,D6,200%,FALSE)	
	A	B	C	D	E	F	G
1	MACRS Depreciation						
2							
3		Beginning	Period	Period		End	
4		of Year	Start	End	Depreciation	of Year	
5	Year	Book Value	Time	Time	Amount	Book Value	
6	1	$40,000	0	0.5	$4,000	$36,000	
7	2	$36,000	0.5	1.5	$7,200	$28,800	
8	3	$28,800	1.5	2.5	$5,760	$23,040	
9	4	$23,040	2.5	3.5	$4,608	$18,432	
10	5	$18,432	3.5	4.5	$3,686	$14,746	
11	6	$14,746	4.5	5.5	$2,949	$11,796	
12	7	$11,796	5.5	6.5	$2,621	$9,175	
13	8	$9,175	6.5	7.5	$2,621	$6,554	
14	9	$6,554	7.5	8.5	$2,621	$3,932	
15	10	$3,932	8.5	9.5	$2,621	$1,311	
16	11	$1,311	9.5	10	$1,311	$0	
17							

The VDB() function in cell E6 tells Excel that the initial cost is $40,000 (always in cell B6—hence the dollar signs), that no salvage value should be used, that the service life is 10 years, that the start period and end period for the depreciation calculation are found in cells C6 and D6, that double-declining-balance depreciation should be used ($R_{DB} = 200\%$), and that NoSwitch is set to FALSE to cause the calculation to switch to straight-line depreciation if that method gives a larger depreciation factor.

Using MACRS Depreciation Factors   Because the MACRS system is fairly complex, the common way of calculating depreciation is through the use of MACRS depreciation factors, as tabulated here:

			MACRS DEPRECIATION FACTORS			
YEAR			**SERVICE LIFE**			
	3	5	7	10	15	20
1	0.3333	0.2000	0.1429	0.1000	0.0500	0.0375
2	0.4444	0.3200	0.2449	0.1800	0.0950	0.0722
3	0.1481	0.1920	0.1749	0.1440	0.0855	0.0668
4	0.0741	0.1152	0.1249	0.1152	0.0770	0.0618
5		0.1152	0.0892	0.0922	0.0693	0.0571
6		0.0576	0.0892	0.0737	0.0623	0.0528
7			0.0892	0.0655	0.0590	0.0489
8			0.0446	0.0655	0.0590	0.0452
9				0.0655	0.0590	0.0446
10				0.0655	0.0590	0.0446
11				0.0328	0.0590	0.0446
12					0.0590	0.0446
13					0.0590	0.0446
14					0.0590	0.0446
15					0.0590	0.0446
16					0.0295	0.0446
17						0.0446
18						0.0446
19						0.0446
20						0.0446
21						0.0223
Rate:	200%	200%	200%	200%	150%	150%

Values in this table were calculated by using the VDB() function with a cost of 1, except for the final value in each column. The final value is for a half-year using straight-line depreciation, so the final factor is always half of the next-to-last year's factor.

The first-year MACRS depreciation factor for a 10-year asset is 0.1000, so the depreciation amount on the $40,000 asset in the first year would be

$$(0.1000)\$40,000 = \$4,000. \tag{11.11}$$

In the second year, the MACRS depreciation factor is 0.1800, and the depreciation amount is calculated as

$$(0.1800)\$40,000 = \$7,200. \tag{11.12}$$

Note that the MACRS depreciation factors always multiply the initial cost, not the current book value.

Using the MACRS depreciation factors, we find that the depreciation amounts each year are easy to calculate:

D9		▼	*fx* =$C$3*C9			
	A	B	C	D	E	F
1	MACRS Depreciation Using Factors					
2						
3		Initial Cost:	$40,000			
4						
5		Beginning			End	
6		of Year	MACRS	Depreciation	of Year	
7	Year	Book Value	Factor	Amount	Book Value	
8	1	$40,000	0.1000	$4,000	$36,000	
9	2	$36,000	0.1800	$7,200	$28,800	
10	3	$28,800	0.1440	$5,760	$23,040	
11	4	$23,040	0.1152	$4,608	$18,432	
12	5	$18,432	0.0922	$3,686	$14,746	
13	6	$14,746	0.0737	$2,949	$11,796	
14	7	$11,796	0.0655	$2,621	$9,175	
15	8	$9,175	0.0655	$2,621	$6,554	
16	9	$6,554	0.0655	$2,621	$3,932	
17	10	$3,932	0.0655	$2,621	$1,311	
18	11	$1,311	0.0328	$1,311	$0	
19						

Cell D9 has been selected to illustrate that the depreciation amount is calculated with the MACRS factor (in C9) and the initial cost (in $C$3).

KEY TERMS

Asset
Book value
Capital expenditures
Declining-balance depreciation
Declining-balance method
Depreciation

Depreciation basis
Half-year convention
Internal rate of return
Modified Accelerated Cost
    Recovery System (MACRS)
Net present value

Payout period
Periodic interest rate
Recovery period
Straight-line depreciation

SUMMARY

Net Present Value

To calculate the net present value, we use

    =NPV(i_P, Incomes1, Incomes2, ...)

where

i_P is a the interest rate per period for the cash flow (required),

Incomes1   is a value, cell address, or range of cells containing one or more
               incomes, with expenses as negative incomes, (required), and

Incomes2, ... is used to include a series of individual incomes (up to 29) if
               the incomes are not stored as a range of cells.

Internal Rate of Return

This is the periodic interest rate that produces the net present value zero for a set of incomes and expenses, so

    =IRR(Incomes, irrGuess)

where

Incomes is a range of cells containing the incomes, with expenses as negative incomes (required), and

irrGuess is a starting value, or guess value, for the internal rate of return. (This is optional, so Excel uses 10% if it is omitted.)

Depreciation of Assets

Straight-Line Depreciation

$$D = \frac{C_{\text{init}} - S}{N_{\text{SL}}}, \tag{11.13}$$

where

$\quad D \quad$ is the depreciation amount,

$\quad C_{\text{init}} \quad$ is the initial cost of the asset,

$\quad S \quad$ is the salvage value at the end of the service life, and

$\quad N_{\text{SL}} \quad$ is the service life of the asset.

Double-Declining-Balance Depreciation

$$D_j = (B_{j-1} - S)F_{\text{DB}},$$

and

$$F_{\text{DB}} = \frac{200\%}{N_{\text{SL}}}, \tag{11.14}$$

where

$\quad D_j \quad$ is the allowable depreciation amount in the current year (year j),

$\quad B_{j-1} \quad$ is the book value at the end of the previous year ($B_0 = C_{\text{init}}$),

$\quad N_{\text{SL}} \quad$ is the service life of the asset,

$\quad F_{\text{DB}} \quad$ is the declining-balance depreciation factor, and

$\quad R_{\text{DB}} \quad$ is the declining-balance percentage. For the double-declining-balance method, $R_{\text{DB}} = 200\%$.

MACRS Depreciation

A rapid depreciation method based on the double-declining-balance and straight-line depreciation methods. The depreciation amount used each year is the larger of the double-declining-balance and straight-line depreciation amounts. MACRS depreciation factors normally are used to determine annual depreciation amounts with this method. Excel's VDB() function can also handle MACRS depreciation:

$$\texttt{=VDB(C}_{\texttt{init}}, \texttt{ S, N}_{\texttt{SL}}, \texttt{ Per}_{\texttt{start}}, \texttt{ Per}_{\texttt{end}}, \texttt{ F}_{\texttt{DB}}, \texttt{ NoSwitch)}$$

Here,

$\quad \texttt{C}_{\texttt{init}} \quad$ is the initial cost of the asset (required),

$\quad \texttt{S} \quad$ is the salvage value, required (set to zero for MACRS depreciation),

$\quad \texttt{N}_{\texttt{SL}} \quad$ is the service life of the asset (required),

$\quad \texttt{Per}_{\texttt{start}} \quad$ is the start of the period over which the depreciation amount will be calculated (required),

$\quad \texttt{Per}_{\texttt{stop}} \quad$ is the end of the period over which the depreciation amount will be calculated (required), and

$\quad \texttt{F}_{\texttt{DB}} \quad$ is the declining-balance percentage (optional; Excel uses 200% if it is omitted).

$\texttt{NoSwitch}$ tells Excel whether to switch to straight-line depreciation when the straight-line depreciation factor is larger than the declining-balance depreciation factor. (This is optional; set it to FALSE to switch to straight line, to TRUE not to switch. The default is FALSE.)

Problems

Internal Rate of Return I

1.   Find out the internal rate of return for the following cash flows:

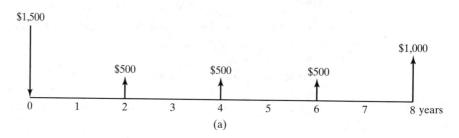

(a)

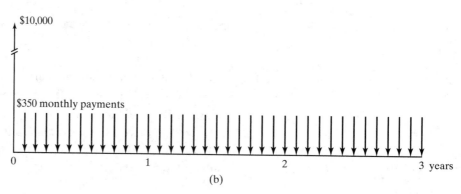

(b)

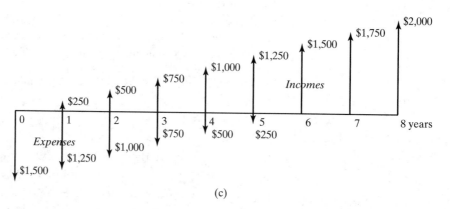

(c)

Internal Rate of Return II

2.   Archway Industries requires projects to provide an internal rate of return of at least 4% to be considered for funding. Sara's proposed project requires an initial investment of $350,000 for new equipment plus $24,000 annually for operating costs (years 1 through 12). The project is expected to generate revenues of $14,000 the first year, $170,000 for years two through six, and $100,000 for years seven through 12. At that time, the equipment will be scrapped at an estimated cost of $400,000.

Will Sara's project be considered for funding?

Internal Rate of Return III

3.   When major expenses come at the end of a project (e.g., nuclear power-plant cleanup costs), something interesting can happen to the internal rate of return:

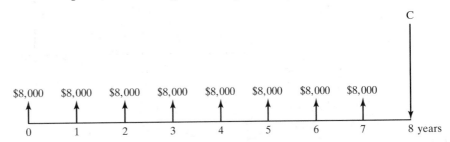

   a.   Calculate the internal rate of return for the preceding cash flow, diagrammed when the end-of-product cost, C, is

   (i) $70,000. (The internal rate of return is near 1% for this case, and the IRR() function might need that value as an initial guess for its iteration.)

   (ii) $80,000.

   (iii) $90,000.

   b.   Calculate the net present value of the cash flow if the interest rate is 7% when C is

   (i) $70,000.

   (ii) $80,000.

   (iii) $90,000.

   c.   Which case—(i), (ii), or (iii)—produces the highest internal rate of return?

   d.   Which case—(i), (ii), or (iii)—produces the highest net present value?

   e.   Which is the best project in which to invest?

Depreciation I

4.   Create a depreciation schedule showing annual depreciation amounts and end-of-year book values for a $26,000 asset with a five-year service life and a $5,000 salvage value, using the straight-line depreciation method.

Depreciation II

5.   Create a depreciation schedule showing annual depreciation amounts and end-of-year book values for a $26,000 asset with a five-year service life, using the MACRS depreciation method. Use MACRS depreciation factors to create the schedule.

Time-Value-of-Money Problems

6.   Find the present value of the following cash flows if the interest rate is 9% APR compounded monthly:

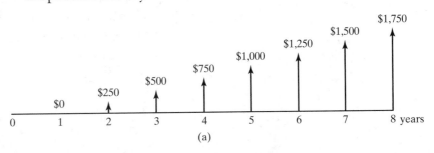

(a)

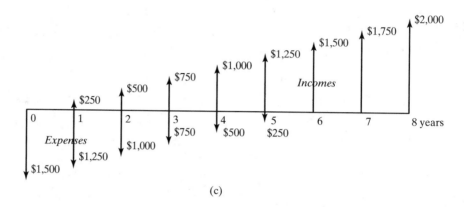

(b)

(c)

Evaluating Options

7. Your company is expanding into a new manufacturing area, but there are two possible ways to do this:

- Option *A*: Computerized manufacturing equipment requiring an initial outlay of $2.3 million for equipment with operating costs (excluding materials, which will be the same for either option) of $20,000 per year for general costs and $80,000 per year in wages for equipment operators. Products from a new computerized line are expected to generate revenues of $650,000 per year over the 20-year life expectancy of the equipment.

- Option *B*: Manual manufacturing equipment requiring an outlay of $0.94 million for equipment with operating costs (excluding materials) of $70,000 per year of general costs, plus $240,000 per year for wages to the skilled technicians required to run the manually operated equipment. Products from a new manually operated line are expected to generate revenues of $530,000 per year over the 20-year life expectancy of the equipment.

a. Which option provides the higher internal rate of return?

b. If the company uses a rate of 6% (eff. ann. rate) for comparing investment options, what is the net present value of each option?

c. What nonfinancial factors should be considered in making the decision on which option to implement?

12

Excel's Statistics Functions

12.1 OVERVIEW

Data analysis is a standard part of an engineer's day-to-day work, aimed at understanding a process better and making better products. Whenever you try to glean information from data sets, you soon find yourself needing some *statistics*.

A statistical analysis is performed on a data set, and the rows and columns of values fit naturally into a spreadsheet. Perhaps it is no surprise, then, that Excel has a variety of built-in functions to help out with statistical calculations. Because of these features, Excel has become a commonly used tool for statistical analysis.

12.2 POPULATIONS AND SAMPLES

If a dairyman has 238 cows that he milks every day, those 238 cows form a *population*—his herd. If he keeps track of the daily milk production for each cow, he has a data set that represents the daily milk production for a population.

If, one day, he's running behind and decides to record the milk production for only 24 cows and then use the values from those 24 cows to estimate the total milk production for the entire herd, then his data set represents a *sample* of the total population.

Whenever a sample is used to predict something about a population, the *sampling method* must be carefully considered. If the hired hand, not understanding what the farmer planned to do with the data, recorded data for the 24 cows that gave the most milk, the data set's not much good for predicting the total milk production for the herd. The farmer needs a *representative sample*. The usual way to try to achieve this is to choose a *random sample*. The key to getting a random sample is to try to make sure that every cow has an equal chance of being a part of the sample.

OBJECTIVES

After reading this chapter, you will know

- How to distinguish samples and populations and how to select a representative sample
- How to use Excel for common statistical calculations, including means, standard deviations, and variances
- How to create histograms with Excel's Data Analysis Package
- How to calculate confidence intervals about Sample Mean Values

12.2.1 Alternative Example

If cows aren't your cup of tea, consider mice—computer mice. If every mouse coming off the assembly line during November was tested (for ease of rolling, button functionality, cursor tracking, etc.), then the data set represents a population: mice produced in November. (It could also represent a sample of mice produced in the fourth quarter.) If the company decided to test only part of the mice coming off the assembly line, the mice that were tested would represent a sample of the total population. If the sample is to be considered a representative sample, some care must be taken in designing the sampling protocol. For example, if mice are collected for testing only during the day shift, defects introduced by sleepy employees on the night shift will never be observed. Again, the usual way to try to get a representative sample is to choose a random sample. The key to getting a random sample is to make sure that every mouse produced has an equal chance of being a part of the sample.

12.2.2 Multiple Instrument Readings

Instruments, such as digital thermometers and scales, are imperfect measuring devices. It is common to weigh something several times to try to get a "better" value than would be obtained from a single measurement. Any time you record multiple readings from an instrument, you are taking a sample from an infinite number of possible readings. The multiple readings should be treated as a sample, not a population. The impact of this will become apparent when the equations for standard deviations and variances are presented.

PRACTICE!

The SO_2 content in the stack gas from a coal-fired boiler is measured halfway up the stack at six different positions across the diameter, and the concentration values are averaged.

- Do these data represent a sample or a population?
- If it's a sample, is it a representative sample? Is it a random sample? Why, or why not?

12.2.3 Arithmetic Mean, or Average

The formulas for the *population mean*, μ (mu), and *sample mean*, $\bar{x}$, are

$$\mu = \frac{\sum_{i=1}^{N_{\text{pop}}} x_i}{N_{\text{pop.}}},$$

$$\bar{x} = \frac{\sum_{i=1}^{N_{\text{sample}}} x_i}{N_{\text{sample}}}. \tag{12.1}$$

Both equations add up all of the values in their respective data set and then divide by the number of values in their data set. The same Excel function, AVERAGE ( ), is used for computing the population mean and the sample mean. Different symbols are used for these means as a reminder that they have different meanings: The population mean is the *arithmetic average* value for the data set; the sample mean should be interpreted as the best estimate of the population mean.

Excel provides the AVERAGE(*range*) function for computing arithmetic means, or averages. The range in the function represents a range of cells in a spreadsheet. The average of the values in a small data set consisting of six random integers is computed in the following spreadsheet:

	B10	▼	f_x =AVERAGE(B3:B8)		
	A	B	C	D	E
1	**Small Data Set**				
2					
3		3			
4		9			
5		1			
6		3			
7		4			
8		6			
9					
10	mean:	4.33			
11					

B10: =AVERAGE(B3:B8)

12.3 STANDARD DEVIATIONS AND VARIANCES

The same function is used to calculate both sample and population means, but separate functions are needed for sample and population *standard deviations* and *variances*. The computational formulas for standard deviations and variances are slightly different for populations and samples, and this is accounted for in the spreadsheet functions.

The variance (and standard deviation, which is just the square root of the variance) provides information about the spread of the values (or dispersion) about the mean. A small variance or standard deviation suggests that all of the data values are clustered closely around the mean value; a large standard deviation tells you that there is a wide range of values in the data set. For example, the two data sets shown in the following spreadsheet have the same mean value, but quite different standard deviations:

	C12	▼	f_x =STDEV(C3:C9)		
	A	B	C	D	E
1		**Set A**	**Set B**		
2					
3		4.02	6.12		
4		3.98	2.53		
5		4.04	5.27		
6		3.93	4.24		
7		3.89	3.50		
8		4.16	5.10		
9		3.95	1.22		
10					
11	mean:	4.00	4.00		
12	stdev:	0.089	1.71		
13	var:	0.008	2.92		
14					

The greater spread around the mean value (4.00) in Set B is easily seen if the data are plotted:

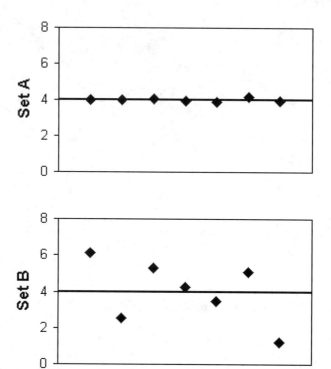

12.3.1 Population Statistics

The Greek symbol σ (sigma) is used for the standard deviation of a population. The population variance is indicated as σ^2:

$$\sigma^2 = \frac{\sum_{i=1}^{N_{\text{pop.}}} (x_i - \mu)^2}{N_{\text{pop.}}},$$

$$\sigma = \sqrt{\sigma^2}. \tag{12.2}$$

Excel provides functions for computing the variance and standard deviation of a population:

```
VARP(range)
STDEVP(range)
```

In the following spreadsheet, these two functions have been used to calculate the standard deviation and variance in the small data set (assumed here to be a population):

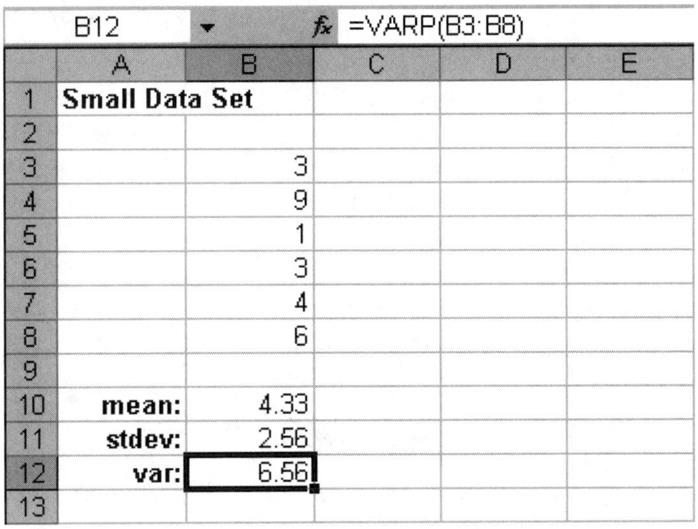

B12	▼	f_x	=VARP(B3:B8)		
	A	B	C	D	E
1	**Small Data Set**				
2					
3		3			
4		9			
5		1			
6		3			
7		4			
8		6			
9					
10	**mean:**	4.33			
11	**stdev:**	2.56			
12	**var:**	6.56			
13					

12.3.2 Samples of a Population

The denominator of the variance equation changes to $N_{sample} - 1$ when your data set represents a sample. The symbol s is used for the standard deviation of a sample, and the sample variance is indicated as s^2:

$$s^2 = \frac{\sum_{i=1}^{N_{sample}} (x_i - \bar{x})^2}{N_{sample} - 1},$$

(12.3)

$$s = \sqrt{s^2}.$$

Excel provides functions for computing the variance and standard deviation of a sample:

VAR(*range*)

STDEV(*range*)

Applying these formulas to the small data set (treated here as a sample) produces the following results:

	B12	▼		f_x	=VAR(B3:B8)	
	A	B	C	D	E	F
1	**Small Data Set**					
2						
3		3				
4		9				
5		1				
6		3				
7		4				
8		6				
9						
10	**mean:**	4.33				
11	**stdev:**	2.80	<< formula changed to STDEV( )			
12	**var:**	7.87	<< formula changed to VAR( )			
13						

PRACTICE!

Two thermocouples are placed in boiling water, and the output voltages are measured five times. Calculate the standard deviations for each thermocouple to see which is giving the most reliable output. Is the voltage data presented here a sample or a population? Why?

TC_A	TC_B
(mV)	(mV)
3.029	2.999
3.179	3.002
3.170	3.007
3.022	3.004
2.928	3.013

12.4 ERRORS, DEVIATIONS, AND DISTRIBUTIONS

A *deviation* is the difference between a measured value in a sample and the sample mean:

$$\text{dev}_i = x_i - \bar{x}. \tag{12.4}$$

A related quantity, *error*, is computed for populations:

$$\text{error}_i = x_i - \mu. \tag{12.5}$$

A *deviation plot* is sometimes used to look at the distribution of deviation (or error) values around the mean. In a quality-control setting, for example, patterns in a deviation plot can offer hints that something is wrong.

Example: Tools made by two machines are supposed to be identical, and tolerances are measured on each tool as it comes off the line. When the deviations were plotted against time (or sample number if the samples were taken at regular intervals), the following graph was obtained:

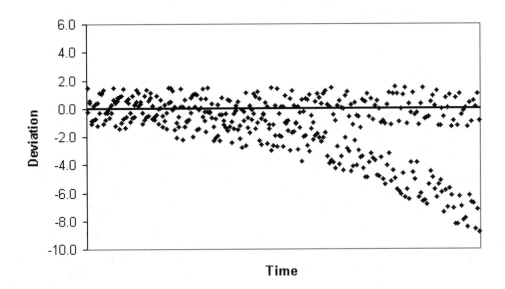

It looks as if something is going wrong with one of the machines. This type of trend could be seen in the tolerance measurements directly, but problems can often be hidden in the scale used to graph the original measurements, as in the following graph:

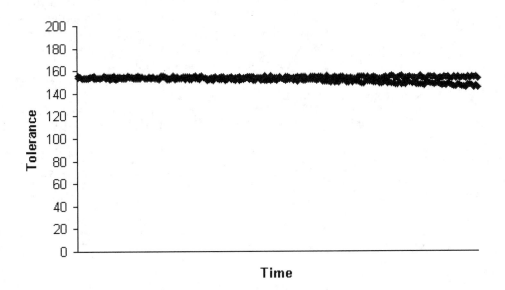

12.4.1 Frequency Distributions

Producing a *frequency distribution*, or *histogram*, can help you understand how the error in your data is distributed. Excel's Data Analysis package will automatically create frequency distributions (and perform several other statistical analyses, such as t-tests and ANOVA).

EXAMPLE 12.1

On a good day, when the tools made by the two machines are nearly identical, the deviation plot looks as follows:

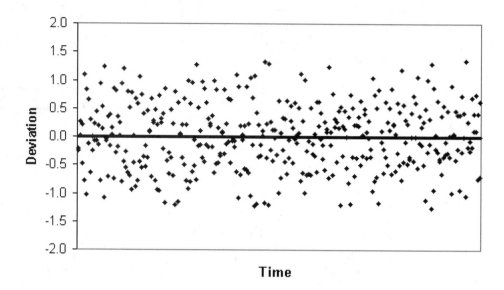

Notice that the magnitudes of the deviations are generally smaller than in the previous deviation plot. The machines are working better.

The basic idea of a frequency distribution is to separate the values into bins and then plot how many values fall into each bin. For deviations, if the bins near zero contain more values than the bins far from zero, then your deviations (sample) or errors (population) might follow a *normal distribution* (the classic bell curve). If all of the bins contain approximately the same number of values, then you might have a *uniform distribution*. Frequency distributions can be created by using deviations, as shown here, or directly from the original data values.

To create a frequency distribution in Excel, use the following steps:

1. Decide how many bins you want to use. The number of bins is arbitrary, but the following list is workable for the deviations just plotted:

BIN 1	FROM −1.4	TO −1.0
2	−1.0	−0.6
3	−0.6	−0.2
4	−0.2	0.2
5	0.2	0.6
6	0.6	1.0
7	1.0	1.4

2.   Create a column of bin limits (cells C2:C9 in this example):

	A	B	C	D	E
1	Value	Deviation	Bins		
2	155.77	0.27	-1.4		
3	155.72	0.22	-1.0		
4	156.60	1.10	-0.6		
5	156.33	0.83	-0.2		
6	156.15	0.65	0.2		
7	155.80	0.30	0.6		
8	155.51	0.01	1.0		
9	155.42	-0.08	1.4		
10	155.94	0.44			
11	155.69	0.19			
12	156.43	0.93			
398	155.60	0.10			
399	155.61	0.11			
400	154.75	-0.75			
401	154.79	-0.71			
402					

Hidden Rows (between rows 12 and 398)

Note the following:

a.   *If you do not create a column of bin limits and do not indicate any bin limits in Step 6, Excel will automatically create them for you.*

b.   *To hide one or more rows, first select the rows to be hidden and then right click on the selected rows and choose Hide from the pop-up menu. When you select a cell range across hidden rows, the hidden rows will be included in the selected cell range by default.*

3.   Activate Excel's histogram analysis procedure by using Tools / Data Analysis . . . .

Note: The data-analysis package is installed, but not as an activated part of Excel, by default. If the Data Analysis menu option does not appear under the Tools menu, it has not been activated. Use Tools/Add-Ins . . . , and activate the Analysis option. This has to be done only once.

4.   Excel offers a wide variety of analysis options. Choose "Histogram" from the selection list:

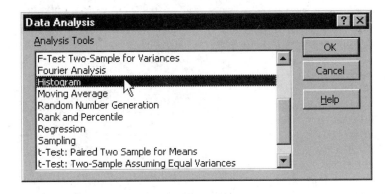

5. The Histogram dialog box will be displayed. You need to first indicate the range containing the deviations. You can either type in the range or click on the spreadsheet icon (as shown next) to go to the spreadsheet and indicate the range by using the mouse:

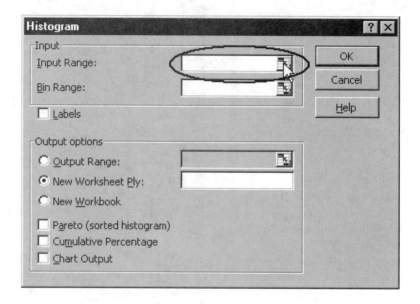

Back at the spreadsheet, indicate the range of cells that contains the deviation values (part of the range has been hidden to save space, but all of the deviation values were used in the analysis):

	A	B	C	D	E	F
1	Value	Deviation	Bins			
2	155.77	0.27	-1.4			
3	155.72	0.22	-1.0			
4	156.60	1.10	-0.6			
5	156.33	0.83	-0.2			
6	156.15	0.65	0.2			
7	155.80	0.30	0.6			
8	155.51	0.01	1.0			
9	155.42	-0.08	1.4			
10	155.94	0.44				
11	155.69	0.19				
12	156.43	0.93				
398	155.60	0.10				
399	155.61	0.11				
400	154.75	-0.75				
401	154.79	-0.71				
402						

Histogram ? ✕
$B$2:$B$401

Once you have indicated the range containing the deviations, either click on the dialog box icon (right side of the data entry field in the Histogram window) or just press [Enter] to return to the Histogram dialog box.

6.   In the same manner, indicate the cell range containing the bin limit values:

	A	B	C	D	E	F
1	**Value**	**Deviation**	**Bins**			
2	155.77	0.27	-1.4			
3	155.72	0.22	-1.0			
4	156.60	1.10	-0.6			
5	156.33	0.83	-0.2			
6	156.15	0.65	0.2			
7	155.80	0.30	0.6			
8	155.51	0.01	1.0			
9	155.42	-0.08	1.4			
10	155.94	0.44				
11	155.69	0.19				
12	156.43	0.93				
398	155.60	0.10				
399	155.61	0.11				
400	154.75	-0.75				
401	154.79	-0.71				
402						

Histogram `[?][X]`
`$C$2:$C$9`

7.   Now tell Excel how to handle the output data. Typically, you will want to put the results on a "New Worksheet Ply," and you will want "Chart Output." Both of these options have been indicated on the dialog box shown here:

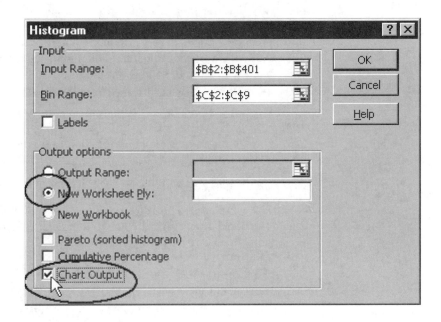

8. Click **OK** to create the histogram plot. The new histogram looks like this:

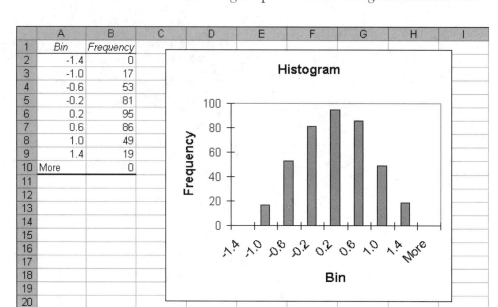

This data set shows the "bell curve" shape suggesting a normal distribution.

12.5 CONFIDENCE INTERVALS

The public-opinion polls used by the news media usually are attempts to find out how the general population is thinking about issues by asking a small sample—typically only a few hundred people are actually polled. With such a small sample, there is always a possibility that the sample results do not accurately reflect the opinions of the entire population, so the poll results usually indicate an uncertainty in the result (plus or minus four percentage points is very common). There is also the question of how the sample was obtained. If you choose random telephone listings, your sample will exclude people whose names aren't in telephone directories. If you use voting records, you exclude non-voters. Coming up with a representative sample is sometimes difficult.

The public opinion polls point out a couple of things you should keep in mind:

- We often use sample information to predict something about the population.
- Any time you use a sample, you introduce some uncertainty into the result.
- If you use a sample, you need to take care how you choose the sample.

A pollster who wants to be 100% confident (zero uncertainty) would have to poll everyone in the population. If you are not going to poll everyone, you have to accept a level of uncertainty in your result—you choose a *confidence level*.

12.5.1 Confidence Levels and the Level of Significance

When you cannot be 100% confident, you must decide how much uncertainty you are willing to accept, or—put another way—how often you are willing to be wrong. It is very common to use a confidence level of 95%, implying that you are willing to be wrong 5% of the time, or one time out of every 20. There are other times when 95% confidence is totally unrealistic. No one would design a pitched curve on a highway, for example, such that 19 out of every 20 cars would successfully negotiate the curve, with one out of 20 going over the edge.

A confidence level of 95% will be used for the remainder of this chapter, but remember that the choice of confidence level is up to you and should be chosen to suit the accuracy required in your result.

The *level of significance*, α, is computed from the confidence level. The level of significance, α, corresponding to a confidence level of 95% is

$$\alpha = 1 - 0.95 = 0.05. \tag{12.6}$$

12.5.2 Confidence Intervals: Bounding the Extent of Uncertainty

The small data set (3, 9, 1, 3, 4, 6) was collected as a sample and has sample mean $\bar{x} = 4.33$; standard deviation, $s = 2.80$; and variance, $s^2 = 7.87$.

What is the population mean, μ?

Who knows? If the sample was well chosen, the population mean is probably close to 4.33. Sometimes "probably close to" is not good enough, and we need to say that the population mean will be between two limits, a *confidence interval*. In setting the limits, there are a couple of things to keep in mind:

- The most likely value of the population mean is the mean we calculated from the sample data, $\bar{x}$, so the range of possible population means will be centered on $\bar{x}$.
- The size of the range will depend on how willing you are to be wrong. If you want to avoid ever being wrong, then set the limits to $\pm \infty$. It's not a very useful result, but you can be sure the population mean will always fall in that range.
- The more uncertainty you are willing to accept (i.e., the lower your confidence level), the narrower the range of possible population mean values becomes. So lower confidence levels produce narrower confidence intervals.

The confidence interval uses sample information ($\bar{x}$, s) to say something about the population mean, μ. Specifically, the confidence interval indicates the range of likely population mean values.

With only six values in the small data set and a wide range of values in the set (ranging from 1 to 9), you should expect a pretty wide confidence interval (unless you are willing to be wrong a high percentage of the time). We'll calculate the 95% confidence interval ($\alpha = 0.05$) for this data set. The procedure is as follows:

1. Compute the mean and standard deviation of the sample (done).
2. Count the number of values in the sample. With the small data set used here, the number of values is obviously six. For larger data sets, it is convenient to have the spreadsheet count the number of values by using

   ```
   =COUNT(range)
   ```

3. Decide on a confidence level. A confidence level of 95%, or 0.95, will be used here.
4. Compute the level of significance from the confidence level:

   $$\alpha = 1 - 0.95 = 0.05. \tag{12.7}$$

5. Compute the *t-value*, using the number of values in the sample and the level of significance.

Note: The t-value is usually obtained from tables, but the tables in various texts are not all set up the same. Some are based on confidence level (e.g., 95%), some are based on level of significance (e.g., α = 0.05). Some tables are one-tailed, some are two-tailed (this will be explained later). You need to be careful to match the equation for a confidence interval and the t-distribution table in the same text. Excel will compute t-values, and the equations shown in this text work with Excel. They will also work with the tables in some, but not all, texts.

The *t*-value is computed in Excel by using the `TINV()` function, which returns the *t*-value for a two-tailed distribution. This function takes two parameters:

`=TINV(α, DOF)`

α   level of significance

DOF   *degrees of freedom* $= N_{\text{sample}} - 1$

One-Tailed and Two-Tailed *t* Tables   A confidence level of 95% implies a 5% chance of being wrong. The t *distribution* is used as a probability map when calculating a confidence interval about a sample mean. Ninety-five percent of the area under the *t* distribution will be centered on the sample mean, with the remaining 5% divided in the two tails of the *t* distribution. When you are placing two *bounds* (upper and lower) on a mean, the likelihood of error (5% in this example) is divided between the two tails:

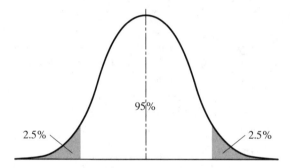

Excel's `TINV()` function returns *t*-values for a two-tailed *t* distribution—that is, the probability of error is divided between the tails on each side of the *t* distribution. For confidence intervals, that's exactly what we want.

Aside   If the probability of error is all on one side of the distribution (illustrated in the accompanying figure), you want *t*-values from a one-tailed *t* distribution. One-tailed distributions are used (for example) when you want to see whether the mean value is less than a specified value.

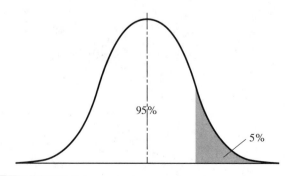

For this example, we need a t-value corresponding to a level of significance of 0.05 (spread between two tails) with five degrees of freedom. We can let Excel find the t-value by using =TINV(0.05,5). The function returns the value $t = 2.571$, as is seen in the following spreadsheet:

B3	▼	f_x =TINV(0.05,5)	
	A	B	C	D
1	't' Statistic			
2				
3		2.571		
4				
5				

6. The upper and lower bounds on the confidence interval are found by using the formulas

$$B_L = \bar{x} - t \frac{s}{\sqrt{N_{\text{sample}}}},$$

and (12.8)

$$B_U = \bar{x} + t \frac{s}{\sqrt{N_{\text{sample}}}}.$$

The quantity $\dfrac{s}{\sqrt{N_{\text{sample}}}}$ has a name: It is called the *standard error of the sample*, or *SES*. So the t-value multiplied by the standard error of the sample, $t \cdot \text{SES}$ is the quantity that is added to and subtracted from the sample mean to calculate the upper and lower bounds of the confidence interval. In this example, the t-value is 2.571, and SES is

$$\text{SES} = \frac{s}{\sqrt{N_{\text{sample}}}} = \frac{2.80}{\sqrt{6}} = 1.143.$$

Then $t \cdot \text{SES} = 2.571 \cdot 1.143 = 2.939$. This is the value that is added to and subtracted from the sample mean to determine the confidence interval.

7. The confidence interval can be written as

$$(\bar{x} - t \cdot \text{SES}) \leq \mu \leq (\bar{x} + t \cdot \text{SES}).$$

and

$$B_L \leq \mu \leq B_U. \tag{12.9}$$

which, for the small data set, results in the 95% confidence interval

$$1.39 \leq \mu \leq 7.28. \tag{12.10}$$

This says that, with 95% confidence, the mean value for the complete population (from which the small sample was taken) is between 1.39 and 7.28. This is a pretty large band, but that's not surprising because of the small number of values and the large range of values in the data set.

Notice that the confidence interval uses sample information ($\bar{x}$ and s) to infer something about the population. The size of the confidence interval can be reduced by

(1) taking more samples or (2) using a lower confidence level (i.e., if you have better data or are willing to be wrong more frequently).

The following spreadsheet illustrates how the upper and lower bounds are computed for the test data set:

	A	B	C	D	E	F	G	H
1	**Small Data Set: Confidence Interval**							
2								
3		3						
4		9		**Mean:**	4.33		=AVERAGE(B3:B8)	
5		1		**St. Dev.:**	2.80		=STDEV(B3:B8)	
6		3		**N$_{sample}$:**	6		=COUNT(B3:B8)	
7		4						
8		6		**α:**	0.05		0.05	
9				**DOF:**	5		=E6-1	
10				**t:**	2.57		=TINV(E8,E9)	
11								
12				**B$_L$:**	1.39		=E4-E10*(E5/SQRT(E6))	
13				**B$_U$:**	7.28		=E4+E10*(E5/SQRT(E6))	
14								

The formulas used in column E are shown in column G.

Using Excel's Descriptive Statistics Package

One of Excel's data-analysis packages is designed to calculate descriptive statistics for a data set. To use the Descriptive Statistics package, select Data Analysis... from the Tools menu, and then select Descriptive Statistics from the list of analysis tools:

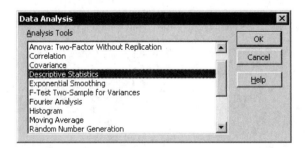

The Descriptive Statistics dialog will open. The required input is simply the range containing the data set (B3:B8), an indication of where you want the results placed and a selection of the type of results that should be calculated. Here, summary statistics (mean, standard deviation, variance, standard error of the sample, etc.) have been requested along with a confidence level of the mean, which has been set to 95% confidence:

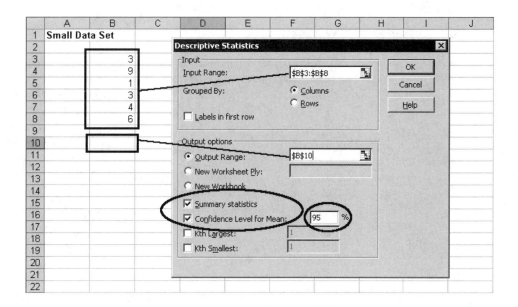

When the OK button is clicked, the results are calculated and presented on the spreadsheet:

	A	B	C	D
1	**Small Data Set**			
2				
3			3	
4			9	
5			1	
6			3	
7			4	
8			6	
9				
10		*Column1*		
11				
12		Mean	4.333333	
13		Standard Error	1.145038	
14		Median	3.5	
15		Mode	3	
16		Standard Deviation	2.804758	
17		Sample Variance	7.866667	
18		Kurtosis	0.670066	
19		Skewness	0.876235	
20		Range	8	
21		Minimum	1	
22		Maximum	9	
23		Sum	26	
24		Count	6	
25		Confidence Level(95.0%)	2.943413	
26				

The descriptive statistics are the same as those calculated earlier in the chapter by using Excel's functions, so the Descriptive Statistics package is simply an alternative method for calculating these quantities. But notice the value 2.94 in cell C25. The label is "Confidence

Level (95%)", but that value is actually $t \cdot \text{SES}$. This is a very handy way of calculating a confidence interval, because the confidence interval is simply

$$(\bar{x} - t \cdot \text{SES}) \leq \mu \leq (\bar{x} + t \cdot \text{SES}).$$

PRACTICE!

Two thermocouples are placed in boiling water and the output voltages are measured five times. Calculate the sample mean and standard deviation for each thermocouple. Then compute the confidence interval for the population mean for each thermocouple, assuming a 95% confidence level. The following table shows the measured values:

TC_A	TC_B
(mV)	(mV)
3.029	2.999
3.179	3.002
3.170	3.007
3.022	3.004
2.928	3.013

A More Realistic Example To generate a more realistic data set, I searched the lab for the oldest, most beat-up weight set I could find. Then I chose the worst looking weight in the set (the 2-gram weight) and weighed it 51 times on a very sensitive balance, but I left the scale doors open with air moving through the room to generate some noise. The results of the repeated sampling of the mass were a sample mean $\bar{x}$ of 1.9997 grams and a standard deviation s of 0.0044 grams, or

$$\bar{x} = 1.9997 \pm 0.0044 \text{ grams.} \tag{12.11}$$

A 95% confidence interval can be used to put bounds on the expected value of the true mass. The confidence interval for this example is computed as follows:

	A	B	C	D	E	F	G
1	**Two-Gram Weight Data: Confidence Interval**						
2							
3							
4		**Mean:**	1.9998	grams			
5		**St. Dev.:**	0.0044	grams			
6		**N$_{sample}$:**	51				
7							
8		**α:**	0.05				
9		**DOF:**	50			=C6-1	
10		**t:**	2.0086			=TINV(C8,C9)	
11							
12		**B$_L$:**	1.9986	grams		=C4-C10*(C5/SQRT(C6))	
13		**B$_U$:**	2.0010	grams		=C4+C10*(C5/SQRT(C6))	
14							

The formulas used in column C are shown in column F.

The 95% confidence interval for this example can be written as

$$1.9986\,g \leq \mu \leq 2.0010\,g. \tag{12.12}$$

For an old, beat-up weight, it's still pretty close to 2 grams (with 95% certainty).

KEY TERMS

Arithmetic average	Level of significance	Sampling method
Confidence interval	Normal distribution	Standard deviation (sample, population)
Confidence level	Population	Standard error of the sample (SES)
Degrees of freedom	Population mean	Statistics
Deviation	Random sample	t distribution (one-tailed, two-tailed)
Deviation plot	Range	Uniform distribution
Error	Representative sample	Variance (sample, population)
Frequency distribution	Sample	
Histogram	Sample mean	

SUMMARY

Population	When a data set includes information from each member of a group, or all possible measurements, it is said to represent the population.
Sample	When only a subset of the entire population is measured or only some of all possible measurements are taken, the data set is called a sample. Whenever a sample is used, care should be taken to ensure that the sample is representative of the population. This generally means that a random sample should be taken.
Random Sampling	Each member of a population should have an equal chance of being part of the sample.
Error	The difference between a value and the true value—typically, the population mean.
Deviation	The difference between a value and another value—typically, the sample mean.
Arithmetic Mean	The average value in the data set. Describes the central value in the data: AVERAGE(range)
Standard Deviation	A value that gives a sense of the range of scatter of the values in the data set around the mean: STDEV(range) sample STDEVP(range) population
Variance	The square of the standard deviation: VAR(range) sample VARP(range) population
Confidence Level	Statistical tests are based on probabilities. For each test you choose a confidence level—the percentage of the time you want to be right. A confidence level of 95%, or 0.95, is very common.
Level of Significance	One minus the (fractional) confidence level.
Confidence Interval	If you know the mean and standard deviation of your sample and choose a confidence level, you can calculate the range of values (centered on the sample mean) in which the population mean will probably lie. The range of possible values is called the confidence interval for the population mean.

Problems

Exam Scores

1. Calculate the average and standard deviation for the set of exam scores shown in the accompanying table. Using a 90% = A, 80% = B, and so on, grading scale, what is the average grade on the exam?

SCORES
92
81
72
67
93
89
82
98
75
84
66
90
55
90
91

Thermocouple Reliability I

2. Two thermocouples are supposed to be identical, but they don't seem to be giving the same results when they are both placed in a beaker of water at room temperature. Calculate the mean and standard deviation for each thermocouple. Which one would you use in an experiment?

TC_1	TC_2
(mv)	(mV)
0.3002	0.2345
0.2998	0.1991
0.3001	0.2905
0.2992	0.3006
0.2989	0.1559
0.2996	0.1605
0.2993	0.5106
0.3006	0.3637
0.2980	0.4526
0.3014	0.4458
0.2992	0.3666
0.3001	0.3663
0.2992	0.2648
0.2976	0.2202
0.2998	0.2889

Thermocouple Reliability II

3. The two thermocouples mentioned in the previous problem are supposed to be identical, but they don't seem to be giving the same results when they are both placed in a beaker of water at room temperature. What is the confidence interval for each set of readings ($\alpha = 0.05$)? Which thermocouple would you use in an experiment?

Sample Size and Confidence Interval

4. Some PRACTICE! Problems in this chapter have used five readings from a thermocouple. Those are actually the first five readings from a larger data set, which is reproduced here and available on the text's website, *http://www.coe. montana.edu/che/Excel*:

TC$_A$ (mV)					
3.029	2.838	3.039	2.842	3.083	3.013
3.179	2.889	2.934	2.962	3.076	2.921
3.170	3.086	2.919	2.893	3.009	2.813
3.022	2.792	2.972	3.163	2.997	2.955
2.928	2.937	2.946	3.127	3.004	2.932

a. Calculate the confidence interval for the population mean (95% confidence), using the 5 data points in the left column, the 10 points in the left two columns, and all 30 data points.

b. How significant is the number of data points for the size of the confidence interval?

Thermocouple Calibration Curve

5. A thermocouple was calibrated by using the system illustrated. The thermocouple and a thermometer were dipped in a beaker of water on a hot plate. The power was set at a preset level (known only as 1, 2, 3 . . . , on the dial) and the thermocouple readings were monitored on the computer screen. When steady state had been reached, the thermometer was read, and 10 thermocouple readings were recorded. Then the power level was increased and the process repeated.

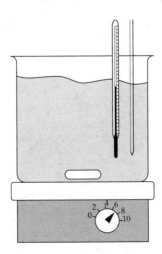

The accumulated calibration data (steady-state data only) is tabulated as follows and is available electronically at *http://www.coe.montana.edu/che/Excel*.

POWER SETTING	THERMO-METER	THERMOCOUPLE READINGS									
	(°C)	(mV)									
0	24.6	1.253	1.175	1.460	1.306	1.117	1.243	1.272	1.371	1.271	1.173
1	38.2	1.969	1.904	2.041	1.804	2.010	1.841	1.657	1.711	1.805	1.670
2	50.1	2.601	2.506	2.684	2.872	3.013	2.832	2.644	2.399	2.276	2.355
3	60.2	3.141	2.920	2.931	2.795	2.858	2.640	2.870	2.753	2.911	3.180
4	69.7	3.651	3.767	3.596	3.386	3.624	3.511	3.243	3.027	3.181	3.084
5	79.1	4.157	4.322	4.424	4.361	4.115	4.065	4.169	4.376	4.538	4.809
6	86.3	4.546	4.384	4.376	4.548	4.654	4.808	4.786	4.535	4.270	4.156
7	96.3	5.087	5.197	5.390	5.624	5.634	5.335	5.525	5.264	4.993	5.267
8	99.8	5.277	5.177	4.991	5.190	5.228	5.274	4.990	5.010	4.916	4.784

Note: This is very noisy thermocouple data, so that the error bars will be visible when graphed. Actual thermocouples are much more precise than the values reported here.

a. Calculate the average and standard deviation of the thermocouple readings at each power setting.

b. Plot the thermocouple calibration curve with temperature on the *x*-axis and average thermocouple reading on the *y*-axis.

c. Add a linear trendline to the graph, and have Excel display the equation for the trendline and the R^2 value.

d. Use the standard deviation values to add error bars (±1 std. dev.) to the graph.

Bottling Plant Recalibration

6. A bottling plant tries to put just a bit over 2 liters in each bottle. If the filling equipment is delivering less than 2.00 liters per bottle, it must be shut down and recalibrated.

The last 10 sample volumes are listed here:

VOLUME (LITERS)
1.92
2.07
2.03
2.04
2.11
1.94
2.01
2.03
2.13
1.99

   a.   Design a random-sampling protocol for a bottling plant filling 20,000 bottles per day. How many samples should you take? How often? How would you choose which bottles to sample?

   b.   Calculate the mean and standard deviation of the last 10 sample volumes.

   c.   Should the plant be shut down for recalibration?

Extracting Vanilla

7.   When you want to extract vanilla from vanilla beans, you usually grind the bean to speed the extraction process, but you want very uniformly sized particles so that the vanilla is extracted from all of the particles at the same rate. If you have a lot of different-sized particles, the small particles extract faster (removing the vanilla *and* some poorer flavored components), while the larger particles could leave the process incompletely extracted.

A sample was collected from the vanilla-bean grinder and split into 50 parts. The average particle size of each part was determined. The following data were collected, and they are available on the text's website:

AVERAGE PARTICLE SIZE (μm)				
91	147	150	165	130
157	117	114	139	131
105	64	137	115	97
116	94	99	163	144
135	112	116	138	129
138	115	106	139	76
147	116	115	101	107
129	122	120	131	104
88	98	157	110	138
152	113	128	130	125

   a.   Calculate the standard deviation of the particle sizes in the sample. Is the standard deviation less than 15 μm? (If not, the grinder should be repaired.)

   b.   How would you determine the average particle size of the ground vanilla beans?

13 Numerical Differentiation Using Excel

13.1 INTRODUCTION

One of the basic data manipulations is differentiation—it can also be one of the more problematic if there is noise in the data. This chapter presents two methods for numerical differentiation: using finite differences on the data values themselves, and using *curve fitting* to obtain a best-fit equation through the data. It is then possible to take the derivative of the equation. Curve fitting is one way to handle differentiation of noisy data; another is *filtering*. Data-filtering methods are also presented in this chapter.

13.2 FINITE DIFFERENCES

Finite differences are simply algebraic approximations of derivatives and are used to calculate approximate derivative values from data sets. These may be used directly or as a first step in integrating a differential equation. However, some care should be taken when calculating derivatives from data, because noise in the data can have a significant impact on the derivative values, even to the extent of making them meaningless.

13.2.1 Derivatives as Slopes

In the following graph, a tangent line has been drawn to the curve at $x = 0.8$:

OBJECTIVES

After reading this chapter, you will know

- How to calculate derivative values from a data set by using finite differences
- How to use Excel's Data Analysis package to filter noisy data
- How to calculate derivative values from a data set by using curve fitting

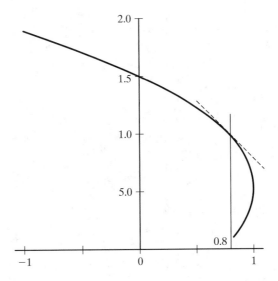

The *slope* of the tangent line is the value of the *derivative* of the function (a parabola) evaluated at $x = 0.8$. When you have a continuous curve, the tangent line is fairly easy to draw, and the slope can be calculated.

Because the plotted function is a parabola, described by the equation

$$y = 0.5 + \sqrt{(1 - x)}, \tag{13.1}$$

it is possible to take the derivative of the function,

$$\frac{dy}{dx} = -\frac{1}{2}(1 - x)^{-1/2}, \tag{13.2}$$

and evaluate the derivative at $x = 0.8$:

$$\left.\frac{dy}{dx}\right|_{x=0.8} = -\frac{1}{2}(1 - 0.8)^{-1/2} = -1.118. \tag{13.3}$$

But when the function is represented by a series of data points instead of a continuous curve, how do you calculate the slope at $x = 0.8$?

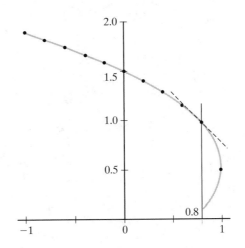

The usual method for estimating the slope at $x = 0.8$ is to use the x-and y-values at adjacent points and calculate the slope, using

$$\text{slope} \approx \frac{\Delta y}{\Delta x}. \tag{13.4}$$

The "approximately equal to" symbol is a reminder that the deltas in the equation make this a *finite-difference* approximation of the slope and will give the true value of the slope only in the limit as Δx goes to zero. There are several options for using adjacent points to estimate the slope at $x = 0.8$. You could use the points at $x = 0.6$ and $x = 0.8$. This is called a *backwards difference* (technically, a *backwards finite difference approximation of a first derivative*):

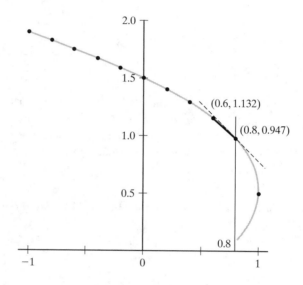

$$\text{slope}_B \approx \frac{\Delta y}{\Delta x} = \frac{0.947 - 1.132}{0.8 - 0.6} = -0.925. \tag{13.5}$$

Or you could use the points at $x = 0.8$ and $x = 1.0$ to estimate the slope at $x = 0.8$. This is called a *forward-difference* approximation:

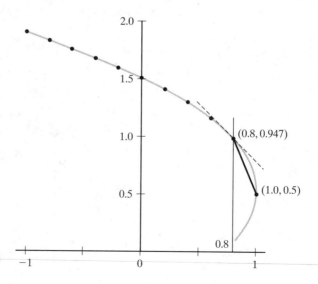

$$\text{slope}_F \approx \frac{\Delta y}{\Delta x} = \frac{0.5 - 0.947}{1.0 - 0.8} = -2.235. \qquad (13.6)$$

Or, you could use the points around $x = 0.8$: $x = 0.6$ and $x = 1.0$. This is called a *central-difference* approximation:

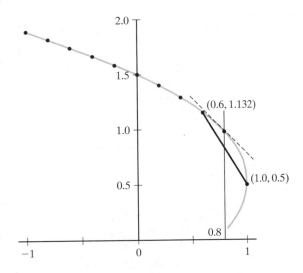

$$\text{slope}_C \approx \frac{\Delta y}{\Delta x} = \frac{0.5 - 1.132}{1.0 - 0.6} = -1.58. \qquad (13.7)$$

Which is the correct way to calculate the slope at $x = 0.8$? Well, all three finite-differences are approximations, so none is actually "correct." But all three methods can be and are used. Central-difference approximations are the most commonly used.

In this case, none of the methods gives a particularly good approximation, because the points are widely spaced and the curvature (slope) changes dramatically between the points. These finite-difference approximations will give better results if the points are closer together (and approximate the true slope as Δx goes to zero).

13.2.2 Caution on Using Finite Differences with Noisy Data

Finite-difference calculations can give very poor results with noisy data. For example, if a thermocouple is left in a pot of boiling water, you would expect the temperature to remain constant over time. If the thermocouple measures precisely, the temperature versus time plot would be a horizontal line, and the slope calculated using any finite-difference approximation would be zero:

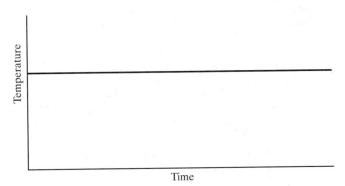

But if the thermocouple is returning a noisy signal because of corrosion, or of loose connections, or of the thermocouple bouncing in and out of the water, or of whatever, the data might look different:

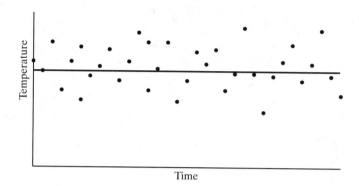

If you try to use finite-difference approximations on adjacent points to estimate the derivative with data like these, you will get all sorts of values, but none is likely to be correct. If you have noisy data, your options are as follows:

- Try to clean up the data as they are being taken (fix the equipment or filter the signal electronically);
- Try to clean up the data after they have been taken (numerical filtering of the data); or
- Fit a curve to the data and then differentiate the equation of the curve.

Some of these options will be discussed here.

13.2.3 Common Finite-Difference Forms

We've already seen three equations for common finite-difference forms for first derivatives. Now let's write them in standard mathematical form. If you want to evaluate the derivative at point i, the adjacent points can be identified as points $i - 1$ (to the left) and $i + 1$ (to the right):

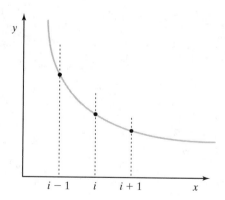

The finite-difference equations can then be written in terms of these generic points and applied wherever needed.

First-Derivative Finite-Difference Approximations

Backward difference

$$\frac{dy}{dx}\bigg|_i \approx \frac{y_i - y_{i-1}}{x_i - x_{i-1}}$$

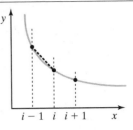

Central difference

$$\frac{dy}{dx}\bigg|_i \approx \frac{y_{i+1} - y_{i-1}}{x_{i+1} - x_{i-1}}$$

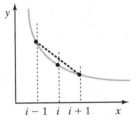

Forward difference

$$\frac{dy}{dx}\bigg|_i \approx \frac{y_{i+1} - y_i}{x_{i+1} - x_i}$$

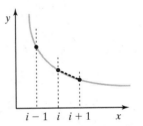

Finite-difference approximations involving more points are sometimes used to obtain better approximations of derivatives. These multipoint approximations are called *higher order approximations*. A few higher order approximations for first derivatives are tabulated as follows:

Higher Order Finite-Difference Approximations for First Derivatives

Asymmetric
$$\frac{dy}{dx}\bigg|_i \approx \frac{-y_{i-2} + 4y_{i-1} - 3y_i}{2\,\Delta x}$$

Asymmetric
$$\frac{dy}{dx}\bigg|_i \approx \frac{-3y_i + 4y_{i+1} - y_{i+2}}{2\,\Delta x}$$

Central
$$\frac{dy}{dx}\bigg|_i \approx \frac{-y_{i-2} + 8y_{i-1} - 8y_{i+1} + y_{i+2}}{12\,\Delta x}$$

Second-Derivative Finite-Difference Approximations   The commonly used finite-difference approximation for a second derivative is the central-difference approximation:

$$\frac{d^2y}{dx^2}\bigg|_i \approx \frac{y_{i+1} - 2y_i + y_{i+1}}{(\Delta x)^2}. \tag{13.8}$$

This equation is derived by assuming uniform spacing between x-values ($\Delta x =$ constant) and so is written with $(\Delta x)^2$ in the denominator. The parentheses in the denominator are a reminder that it is Δx that is squared, not the difference of squared x-values.

This equation can be derived fairly easily, and the process will be helpful if other finite difference approximations are required (e.g., higher order derivatives or nonuniformly spaced x-values).

We begin the derivation by adding two temporary points, call them A and B, halfway between the existing points:

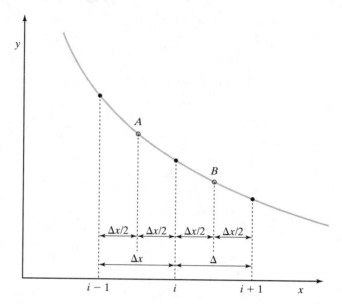

A central-difference first-derivative approximation at point A uses points at i and $i - 1$:

$$\left.\frac{dy}{dx}\right|_A \approx \frac{y_i - y_{i-1}}{2\frac{\Delta x}{2}}. \tag{13.9}$$

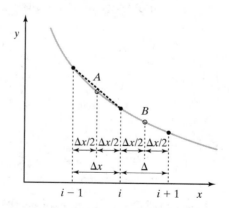

Similarly, a central-difference first-derivative approximation at point B uses points at $i + 1$ and i:

$$\left.\frac{dy}{dx}\right|_B \approx \frac{y_{i+1} - y_i}{2\frac{\Delta x}{2}}. \tag{13.10}$$

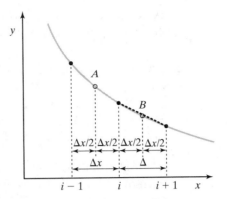

But a second derivative is the derivative of first a derivative, so a second-derivative central finite-difference approximation can be created from the first-derivative equations:

$$\frac{d^2y}{dx^2}\bigg|_i = \frac{d}{dx}\frac{dy}{dx}\bigg|_i \approx \frac{\dfrac{dy}{dx}\bigg|_B - \dfrac{dy}{dx}\bigg|_A}{2\frac{\Delta x}{2}} = \frac{\dfrac{y_{i+1} - y_i}{2\frac{\Delta x}{2}} - \dfrac{y_i - y_{i-1}}{2\frac{\Delta x}{2}}}{2\frac{\Delta x}{2}} = \frac{y_{i+1} - 2y_i + y_{i-1}}{(\Delta x)^2}.$$

(13.11)

Central-difference approximations for second derivatives are the most commonly used, but forward and backward approximations are also available:

Second-Derivative Finite-Difference Approximations

Backward difference	$\dfrac{d^2y}{dx^2}\bigg|_i \approx \dfrac{y_i - 2y_{i-1} + y_{i-2}}{(\Delta x)^2}$
Central difference	$\dfrac{d^2y}{dx^2}\bigg|_i \approx \dfrac{y_{i+1} - 2y_i + y_{i-1}}{(\Delta x)^2}$
Forward difference	$\dfrac{d^2y}{dx^2}\bigg|_i \approx \dfrac{y_{i+2} - 2y_{i+1} + y_i}{(\Delta x)^2}$

Higher order approximations involving additional points are available for second derivatives also. A few of these are tabulated here:

Higher Order Finite-Difference Approximations for Second Derivatives

Asymmetric	$\dfrac{d^2y}{dx^2}\bigg|_i \approx \dfrac{-y_{i-3} + 4y_{i-2} - 5y_{i-1} + 2y_i}{(\Delta x)^2}$
Asymmetric	$\dfrac{d^2y}{dx^2}\bigg|_i \approx \dfrac{2y_i - 5y_{i+1} + 4y_{i+2} - y_{i+3}}{(\Delta x)^2}$
Central	$\dfrac{d^2y}{dx^2}\bigg|_i \approx \dfrac{-y_{i-2} + 16y_{i-1} - 30y_i + 16y_{i+1} - y_{i+2}}{12(\Delta x)^2}$

Many other finite-difference approximations of various orders of derivatives are available. Only a few common forms have been shown here.

13.2.4 Accuracy and Error Management

Finite differences are approximations of derivatives, so some error associated with the use of these equations is to be expected. The finite differences turn into derivatives in the limit as Δx goes to zero. So the first rule of error management is to *keep Δx small*. If you are taking the data that will have to be differentiated, take lots of data points.

Central-difference approximations are more accurate than forward and backward differences, so rule two is to *use central-difference approximations for general-purpose calculations*. There are some situations for which forward or backward approximations are preferable. For example, if point i is at a right-side boundary, then point $i + 1$ doesn't exist. Also, backward differences are often used for wave-propagation problems, to keep from performing calculations using information from regions the wave has not yet reached.

Finally, *you need clean data* if you are going to use finite-difference approximations to calculate derivatives. Filtering noisy data is an option, but filtering *can* change the calculated derivative values. (Technically, it *does* change the calculated derivative values, but a little bit of filtering is possible in many cases. Heavy-handed filtering will always change the calculated derivative values.)

13.3 FILTERING DATA

Excel actually provides a very simple filter called a *moving average*, available as a trendline and as part of the Data Analysis package (Analysis ToolPak). Exponential smoothing is also available through the Data Analysis package.

13.3.1 Moving Average

Because the moving average is available as a trendline on a graph, you can observe the effect of the moving average on your data, but you can't use the calculated point values, because they are not recorded in the spreadsheet. Fortunately, Excel's Data Analysis package also provides a moving-average filter, and it does output the results to the spreadsheet.

Excel's moving-average trendline displays the averaged value at the x-value at the end of the N-point region, where N is the number of values used in the moving average. Because of this, there is an N-point gap at the left side of the graph when a moving average is used, and the trendline appears to be N points to the right of the data points. If N is small, this offset might not be visible.

To show how a moving average can be used to clean up a noisy data set and what happens when N is large, consider the following cosine curve ($0 \leq x \leq 2\pi$) with some random noise added on:

Cosine Data

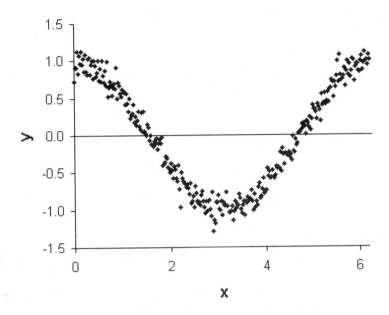

First, a moving-average trendline is added to the graph by right clicking on any data point and selecting Add Trendline from the pop-up menu. On the Add Trendline dialog box, Moving Average is selected, and the period (the number of averaged values) is set to $N = 5$, a reasonable value for a moving-average filter:

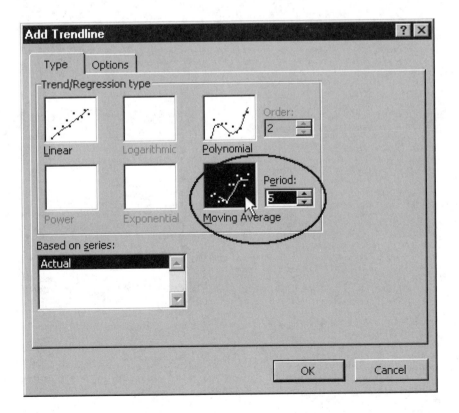

In the following graph, the original data values are presented in gray so that the moving-average results can be more easily observed:

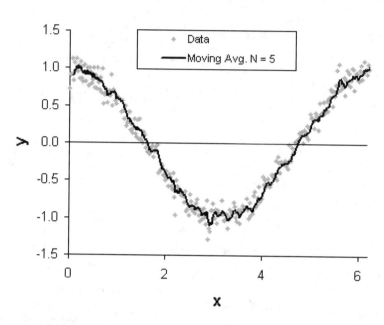

Now, the number of values used in the moving average is changed to a very large value, $N = 40$, by right clicking on the trendline and selecting Format Trendline from the pop-up menu.

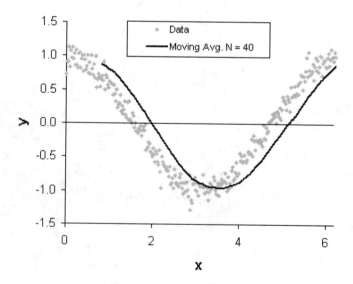

The trendline is moving to the right of the data. If you used the moving average to calculate derivatives, you would find that the derivative would not be zero at $x = \pi$ but closer to $x = 3.6$. This is an artifact of the moving-average filter.

Moving-Average Filtering with the Data Analysis Package   Excel's Data Analysis Package also provides a moving-average filter. It is available by selecting Tools/Data Analysis and then selecting Moving Average from the Data Analysis list of options.

Note: By default, the data analysis package is installed, but is not an activated part in Excel. If the Data Analysis menu option does not appear under the Tools menu, it has not been activated. Select TOOLS/ADD-INS ..., and activate the Analysis option. This need be done only once.

The moving-average dialog asks for the column of values to be averaged and the number of values to include in the average. Here, the y-values in column B are to be averaged, first with a fairly small value of N ($N = 5$):

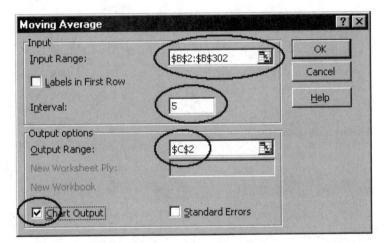

You can automatically create a graph of the original data and the filtered data. Unfortunately, the graph is created as a Line Plot rather than as an XY Scatter plot. In the following spreadsheet, the graph type has been changed to XY Scatter:

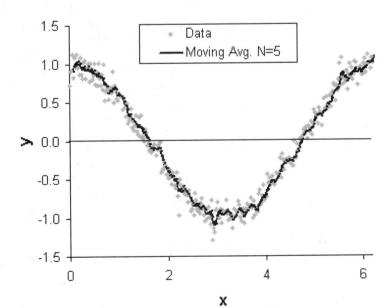

When the value of N is increased to 40, it becomes apparent that the moving filter used by the Data Analysis Package has the same tendency to walk to the right, as does the moving-average trendline:

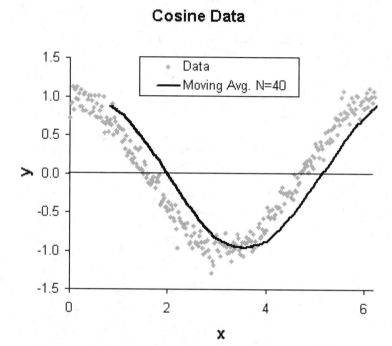

But when you use the Data Analysis package moving-average filter, the filtered y-values are stored in the spreadsheet—you can edit the spreadsheet to move the filtered y-values by $N/2$ cells to realign the original and filtered data:

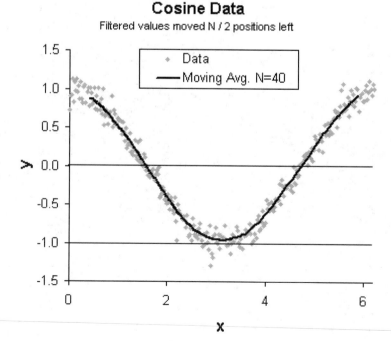

Notice that the filtered curve has been realigned with the original data, but the minimum value is not -1; heavy filtering flattens curves.

The moving-average filter weights each value in the average equally. An alternative is to weight nearby points more heavily than more distant points. One implementation of this approach is *exponential smoothing*, which is also available as part of the Data Analysis package.

13.3.2 Exponential Smoothing

Excel's version of the exponential smoothing equation is

$$y_{F_i} = (1 - \beta)y_i + \beta y_{F_{i-1}}, \tag{13.12}$$

where

- y is an original data value,
- y_F is a filtered data value,
- β is the *damping factor* ($0 < \beta < 1$; small β means little smoothing)

The exponential-smoothing equation is often written in terms of a *smoothing constant*, α, where

$$\alpha = 1 - \beta. \tag{13.13}$$

Exponential smoothing is also available by selecting Tools/Data Analysis and then choosing Exponential Smoothing from the Data Analysis list of options. The dialog asks for the location of the column of values to be smoothed (B2:B302 has been used here), the desired damping factor ($\beta = 0.75$ is used here, quite heavy filtering), and the location to be used for the output (the filtered values, starting at cell C2 and going down the column as far as needed). A graph (line plot) of the original and filtered values is optional:

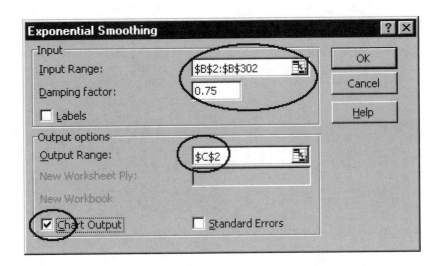

Again, the graph type has been changed to XY Scatter type:

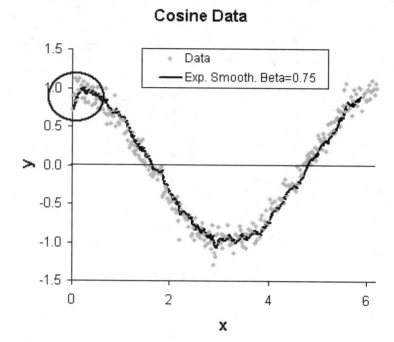

Notice the unusual curvature near $x = 0$. In this data set, the value at $x = 0$ is fairly low (0.714). This value is used directly by Excel and has a strong impact on nearby calculations. Because the value at $x = 0$ was quite low compared with its neighbors, it had a dramatic impact on the curve. This is an artifact of the way Excel implements exponential smoothing for the first data points in the smoothed column.

Filtering is an option for noisy data, but it needs to be done carefully. Heavy-handed filtering will skew the data.

13.4 CURVE FITTING AND DIFFERENTIATION

You can have Excel perform a linear regression on a data set either by adding a trendline to graphed data or by using the Data Analysis package (Regression option). The trendline approach is simpler and will be used here.

For the equation of a regression line via a trendline, the data are first plotted (the cosine data will be used again), and the Add Trendline dialog is called up by right clicking on a data point and selecting Add Trendline ... from the pop-up menu:

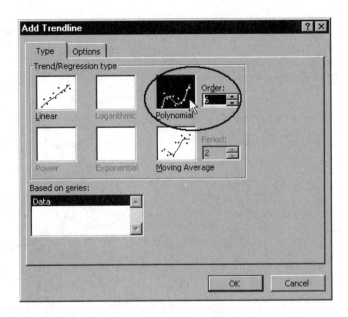

The logarithmic, power, and exponential fits are not available for the cosine data, because they require operations that are invalid for these data. The moving-average trendline is available, but it does not produce an equation. So linear and polynomial fits are the only potentially useful options (for this data set) available from the Add Trendline dialogbox. The linear fit will put a straight line through the data set, so that's out. That leaves Polynomial as the only viable option for this data set.

With the polynomial fit, you need to try various orders of polynomial to see which fits the data best. Excel's polynomial trendline allows orders between 2 and 6. Order 6 will be used here, but the fit is only minimally better than a fourth-order fit:

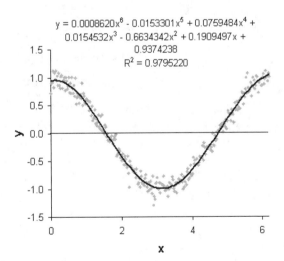

The R^2 value and the equation of the polynomial regression line have been displayed on the graph. This was selected on the Options panel of the Add Trendline (or Format Trendline) dialog box.

The fit doesn't look too bad, although the polynomial misses $y = 1$ at both ends of the graph. Predicted values can be calculated by using the regression equation so that a residual plot can be prepared:

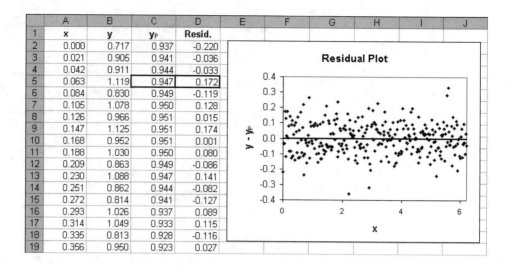

Typical calculations for predicted values (column C) and residuals (column D) are

```
C5: = 0.000862*A5^6 − 0.0153301*A5^5 + 0.0759484*A5^4 +
         0.0154532*A5^3 − 0.6634342*A5^2 + 0.1909497*A5 +
         0.9374238
D5: =B5−C5
```

The lack of apparent patterns in the residual plot suggests that this polynomial is fitting the data about as well as possible.

The polynomial regression equation (shown on the graph) can be differentiated to obtain an equation for calculating derivative values at any x:

$$y = 0.000862x^6 - 0.0153301x^5 + 0.0759484x^4 + 0.0154532x^3$$
$$- 0.6634342x^2 + 0.1909497x + 0.9374238$$
$$\frac{dy}{dx} = 6(0.000862)x^5 - 5(0.0153301)x^4 + 4(0.0759484)x^3 + 3(0.0154532)x^2$$
$$- 2(0.6634342)x + 0.1909497$$
$$= 0.0051720x^5 - 0.0766505x^4 + 0.3037936x^3 + 0.0463596x^2$$
$$- 1.3268684x + 0.1909497. \tag{13.14}$$

This equation has been used to calculate "predicted" derivatives for each x-value in column E, and the theoretical value of the derivative has been calculated in column F. Typical formulas are

```
E5: = 0.005172*A5^5 − 0.07665*A5^4 + 0.303794*A5^3 +
         0.04636*A5^2 − 1.32687*A5 + 0.19095
F5: = −SIN(A5)
```

In the next graph, this actual and predicted derivative values are compared in the graph shown next. Note that the predicted derivative values are off by nearly 0.20 at $x = 0$ and $x = 2\pi$, although the results are better for the rest of the range:

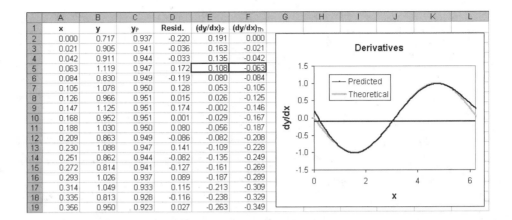

	A	B	C	D	E	F
1	x	y	y$_P$	Resid.	(dy/dx)$_P$	(dy/dx)$_{Th}$
2	0.000	0.717	0.937	-0.220	0.191	0.000
3	0.021	0.905	0.941	-0.036	0.163	-0.021
4	0.042	0.911	0.944	-0.033	0.135	-0.042
5	0.063	1.119	0.947	0.172	0.108	-0.063
6	0.084	0.830	0.949	-0.119	0.080	-0.084
7	0.105	1.078	0.950	0.128	0.053	-0.105
8	0.126	0.966	0.951	0.015	0.026	-0.125
9	0.147	1.125	0.951	0.174	-0.002	-0.146
10	0.168	0.952	0.951	0.001	-0.029	-0.167
11	0.188	1.030	0.950	0.080	-0.056	-0.187
12	0.209	0.863	0.949	-0.086	-0.082	-0.208
13	0.230	1.088	0.947	0.141	-0.109	-0.228
14	0.251	0.862	0.944	-0.082	-0.135	-0.249
15	0.272	0.814	0.941	-0.127	-0.161	-0.269
16	0.293	1.026	0.937	0.089	-0.187	-0.289
17	0.314	1.049	0.933	0.115	-0.213	-0.309
18	0.335	0.813	0.928	-0.116	-0.238	-0.329
19	0.356	0.950	0.923	0.027	-0.263	-0.349

APPLICATIONS: TRANSPORT PHENOMENA

Conduction Heat Transfer

The next figure depicts a device that could be used to measure thermal conductivity of a material. A rod of the material to be tested is placed between a resistance heater (on the right) and a block containing a cooling coil (on the left). Five thermocouples are inserted into the rod at evenly spaced intervals. The entire apparatus is placed under a bell jar, and the space around the rod is evacuated to reduce energy losses.

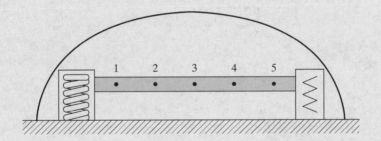

To run the experiment, a known amount of power is sent to the heater, and the system is allowed to reach steady state. Once the temperatures are steady, the power level and the temperatures are recorded. A data sheet from an experiment with the device is reproduced here:

Thermal Conductivity Experiment	
Rod Diameter:	2 CM
Thermocouple Spacing:	5 CM
Power:	100 Watts

Thermocouple	Temperature (K)
1	348
2	387
3	425
4	464
5	503

The thermal conductivity can be determined via Fourier's law,

$$\frac{q}{A} = -k\frac{dT}{dx},$$

(13.15)

where

q is the power applied to the heater and
A is the cross-sectional area of the rod.

The q/A is the *energy flux* and is a vector quantity, having both magnitude and direction. With the energy source on the right, the energy moves to the left (in the negative x-direction), so the flux in this problem is negative.

Calculate the Cross-Sectional Area of the Rod and the Energy Flux

C7		fx	=-C5/C6			
	A	B	C	D	E	F	G
1	**Conduction Heat Transfer**						
2							
3		Rod Diameter:	2	cm			
4		Thermocouple Spacing:	5	cm			
5		Power:	100	watts			
6		Cross-Section:	3.142	cm²			
7		Flux:	-31.8	watts/cm²	(flux in -x direction)		
8							

C6: =PI()/4*C3^2

Estimate *dT/dx*

Central-difference approximations have been used on the interior points:

E12		fx	=(D13-D11)/(C13-C11)			
	A	B	C	D	E	F	G
1	**Conduction Heat Transfer**						
2							
3		Rod Diameter:	2	cm			
4		Thermocouple Spacing:	5	cm			
5		Power:	100	watts			
6		Cross-Section:	3.142	cm²			
7		Flux:	-31.8	watts/cm²	(flux in -x direction)		
8							
9		**Thermocouple**	**Position**	**Temp.**	**dT/dx**		
10			(cm)	(K)	(K/cm)		
11		1	0	348			
12		2	5	387	7.7		
13		3	10	425	7.7		
14		4	15	464	7.8		
15		5	20	503			
16							

Calculate the Thermal Conductivity of the Material

| F12 | ▼ | *fx* =-($C$7/E12)*100 |

	A	B	C	D	E	F	G
1	Conduction Heat Transfer						
2							
3		Rod Diameter:	2	cm			
4		Thermocouple Spacing:	5	cm			
5		Power:	100	watts			
6		Cross-Section:	3.142	cm²			
7		Flux:	-31.8	watts/cm²	(flux in -x direction)		
8							
9		Thermocouple	Position	Temp.	dT/dx	k	
10			(cm)	(K)	(K/cm)	(watt/m K)	
11		1	0	348			
12		2	5	387	7.7	413.39	
13		3	10	425	7.7	413.39	
14		4	15	464	7.8	408.09	
15		5	20	503			
16							

The "100" in the formula in cell F12 is used to convert centimeters to meters to allow the thermal conductivity to be reported in watts/m K:

```
F12: =($6 $7/E12)*100
```

KEY TERMS

Backwards difference
Central difference
Curve fitting
Damping factor
Derivative

Energy flux
Exponential smoothing (filter)
Filtering
Finite difference
Forward difference

Higher order approximations
Moving average (filter)
Slope
Smoothing constant

SUMMARY

Common Finite-Difference Approximations

First-Derivative Finite-Difference Approximations

Backward difference
$$\left.\frac{dy}{dx}\right|_i \approx \frac{y_i - y_{i-1}}{x_i - x_{i-1}}$$

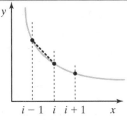

Central difference
$$\left.\frac{dy}{dx}\right|_i \approx \frac{y_{i+1} - y_{i-1}}{x_{i+1} - x_{i-1}}$$

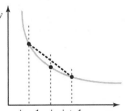

Forward difference $\quad\dfrac{dy}{dx}\bigg|_i \approx \dfrac{y_{i+1} - y_i}{x_{i+1} - x_i}$

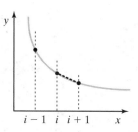

Second-Derivative Finite-Difference Approximations

Backward difference $\qquad\dfrac{d^2y}{dx^2}\bigg|_i \approx \dfrac{y_i - 2y_{i-1} + y_{i-2}}{(\Delta x)^2}$

Central difference $\qquad\dfrac{d^2y}{dx^2}\bigg|_i \approx \dfrac{y_{i+1} - 2y_i + y_{i-1}}{(\Delta x)^2}$

Forward difference $\qquad\dfrac{d^2y}{dx^2}\bigg|_i \approx \dfrac{y_{i+2} - 2y_{i+1} + y_i}{(\Delta x)^2}$

Data Smoothing

- Excel provides moving-average filters as trendlines and as part of the Data Analysis package.
- Exponential smoothing is available in the Data Analysis package.

Curve Fitting and Differentiation

- Use a trendline or regression analysis to fit a curve to the data.
- Particularly useful (necessary) for noisy data.

Problems

Finite Differences or Regression and Differentiation

1. Given the following data sets, which are candidates for finite-difference calculations and which should be fit with a regression function to obtain derivative values?

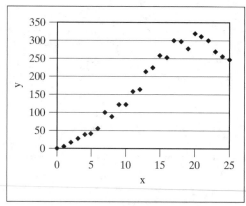

(a)

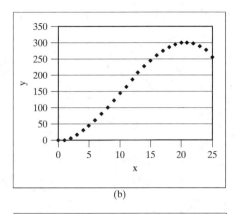

(b)

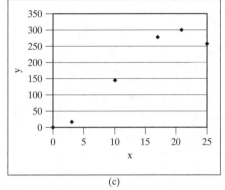

(c)

Conduction Heat Transfer II

2.   The application example in this chapter described a device that could be used to measure thermal conductivity of a material:

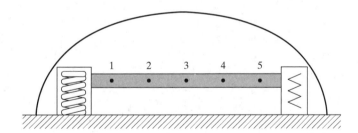

A data sheet from an experiment with the device is reproduced here:

Thermal Conductivity Experiment	
Rod Diameter:	**2 cm**
Thermocouple Spacing:	**5 cm**
Power:	**100 Watts**
THERMOCOUPLE	**TEMPERATURE (K)**
1	287
2	334
3	380
4	427
5	474

a. Calculate the cross-sectional area of the rod and the energy flux.

b. Use central finite-difference approximations to estimate the dT/dx at points 2 through 4.

c. Use Fourier's law and the results of part (b) to calculate the thermal conductivity of the material at points 2 through 4.

Calculating Heat Capacity I

3. The heat capacity at constant pressure is defined as

$$C_P = \left(\frac{\partial \hat{H}}{\partial T}\right)_P, \qquad (13.16)$$

where

$\hat{H}$ is specific enthalpy.

T is absolute temperature.

Data for carbon dioxide at a pressure of 1 atmosphere are tabulated as follows:[1]

TEMP.	$\hat{H}$
(°C)	(KJ/MOL)
100	2.90
200	7.08
300	11.58
400	16.35
500	21.34
600	26.53
700	31.88
800	37.36
900	42.94
1000	48.60
1100	54.33
1200	60.14
1300	65.98
1400	71.89
1500	77.84

a. Use finite-difference approximations to calculate the molar heat capacity of CO_2 at each temperature.

b. Plot the specific enthalpy on the y-axis against absolute temperature on the x-axis, and add a linear trendline with the data. Compare the slope of the trendline with the heat capacity values you calculated in part (a). Are they equal? Should they be equal?

c. Does it appear that the heat capacity of CO_2 is constant over this temperature range? Why, or why not?

---

[1]This data is from a steam table in *Elementary Principles of Chemical Engineering*, 3d ed., R. M. Felder and R. W. Rousseau, New York: Wiley, 2000.

Calculating Heat Capacity II

4. The heat capacity at constant pressure is defined as

$$C_P = \left(\frac{\partial \hat{H}}{\partial T}\right)_P, \tag{13.17}$$

where

$\hat{H}$ is specific enthalpy and

T is absolute temperature.

If enthalpy data are available as a function of temperature at constant pressure, the heat capacity can be computed. For common combustion gases, the data are available. Enthalpy data for carbon dioxide at a pressure of 1 atmosphere are tabulated in the previous problem.

a. Use finite-difference approximations to calculate the molar heat capacity of CO_2 at each temperature.

b. Plot the specific enthalpy on the y-axis against absolute temperature on the x-axis, and add a polynomial trendline to the data. Ask Excel to display the equation of the line and the R^2 value.

c. Record the equations for second- and third-order polynomial fits to the data.

   Note: You might need to have Excel display a lot of decimal places to see anything other than zero for the higher order coefficients. Simply select the text box displaying the regression equation and then right click to set the format of the data labels. Use the Number panel to set the number of displayed decimal places.

d. Differentiate the equations from part (c) to obtain two equations for the heat capacity as a function of temperature.

e. Calculate heat-capacity values with each equation from part (d) at each temperature.

f. The heat capacity of carbon dioxide at temperatures between 0 and 1,500°C can be calculated from the equation[2]

$$C_P = 36.11 + 0.04233\,T - 2.887 \times 10^{-5}\,T^2 + 7.464 \times 10^{-9}\,T^3. \tag{13.18}$$

   Calculate heat-capacity values at each temperature, using this equation. Be aware that the units on the heat capacity from this equation are J/kg K rather than kJ/kg K.

g. Do your calculated heat-capacity values from parts (a), (e), and (f) agree? If not, which of the numerical methods gives the best agreement with the result from part (f)?

---

[2]Equation from *Basic Principles and Calculations in Chemical Engineering*, 6th ed., D. M. Himmelblau, Upper Saddle River, NJ: Prentice Hall PTR, 1996.

14

Numerical Integration Using Excel

14.1 INTRODUCTION

There are two types of integration that engineers routinely perform: integration for the area under a curve and integration of a differential equation. The two are related mathematically, but procedurally they are handled quite differently. This chapter presents two numerical methods to find the area beneath a curve: using small regions (e.g., trapezoids) to compute the total area beneath the curve, and using regression to fit the curve. The regression equation can then be integrated to determine the area beneath the curve.

Integrating differential equations is the subject of the next chapter.

14.2 INTEGRATING FOR AREA UNDER A CURVE

The function $y = 3 + 1.5x - 0.25x^2$ is plotted as the heavy curve in the following graph:

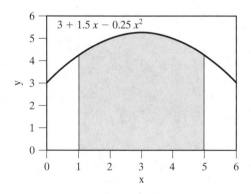

The integral

$$\int_1^5 (3 + 1.5x - 0.25x^2)\, dx \qquad (14.1)$$

OBJECTIVES

After reading this chapter, you will know

- How to use geometric regions to approximate the area between a curve and the *x*-axis
- How to fit an equation to a curve via regression and then integrate that equation to find out the area
- How integrating a function is equivalent to calculating the area between a plot of the function and the *x*-axis

is physically represented by the area between the x-axis and the curve representing the function—the shaded area on the graph. Any method we can find that calculates that area can be used to integrate the function. The geometric integration methods we will use in this chapter are all methods for estimating the area under the function.

14.3   INTEGRATING FOR AREA BETWEEN TWO CURVES

There are times when the area you need to calculate does not go all the way to the x-axis. In the next example, the desired area is the area between two functions: $y = 3 + 1.5x - 0.25x^2$ and $y = 0 + 1.5x - 0.25x^2$. (The zero is obviously not needed in the last equation; it is there as a reminder that the two functions have the same form, just a different offset from the x-axis.)

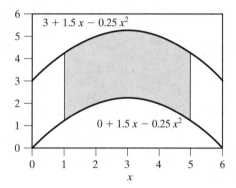

We can calculate the hatched area in this latter plot by first calculating the area between $y = 3 + 1.5x - 0.25x^2$ and the x-axis, then subtracting out the area between $y = 0 + 1.5x - 0.25x^2$ and the x-axis:

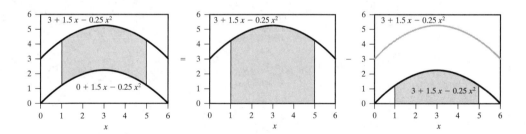

Mathematically, this can be written as

$$\text{Area} = \int_1^5 (3 + 1.5x - 0.25x^2)\, dx - \int_1^5 (0 + 1.5x - 0.25x^2)\, dx. \qquad (14.2)$$

In this case, most of the terms cancel out, leaving a very simple result:

$$\text{Area} = \int_1^5 3\, dx \qquad (14.3)$$

$$= 12.$$

In general, however, the two functions would be integrated separately, then subtracted to find the area between the curves.

14.4 NUMERICAL INTEGRATION METHODS

Implementation of the following numerical integration techniques, or "rules," will be discussed here:

- Approximating areas by using rectangles
- Approximating areas by using trapezoids
- Simpson's rule

Each of these methods will be illustrated on cosine data, to allow comparison of the numerical integration results with the analytical result of integrating

$$y = \cos(x) \tag{14.4}$$

between 0 and $\pi/2$:

$$\int_0^{\pi/2} y\,dx = \int_0^{\pi/2} \cos(x)\,dx$$

$$= \sin(\pi/2) - \sin(0) \tag{14.5}$$

$$= 1.$$

Approximating the area under a curve with a series of rectangular or trapezoidal shapes defined by a series of data pairs is very straightforward in a spreadsheet. The process requires the following steps:

1. Enter or import the data.
2. Enter the formula for the area of one rectangle or trapezoid.
3. Copy the formula over all *intervals* (*not* all data points).
4. Sum the areas.

Implementation of Simpson's rule is slightly more involved because two data intervals are used for each integration step; thus, a distinction must be made between an *interval* (between data points) and an *integration step* (two intervals). Simpson's rule is discussed in detail later.

14.4.1 Integration by Using Rectangles

Step 1. Enter or Import the Raw Data
Eleven data pairs of cosine data have been entered in the following spreadsheet:

	A	B	C	D	E
1	**Rectangles**				
2	Height set by left side y value				
3					
4	**x**	**y**			
5	0.0000	1.0000			
6	0.1571	0.9877			
7	0.3142	0.9511			
8	0.4712	0.8910			
9	0.6283	0.8090			
10	0.7854	0.7071			
11	0.9425	0.5878			
12	1.0996	0.4540			
13	1.2566	0.3090			
14	1.4137	0.1564			
15	1.5708	0.0000			
16					

Step 2. Enter the Formula for One Rectangle

The area below the cosine data can be approximated with rectangles as follows:

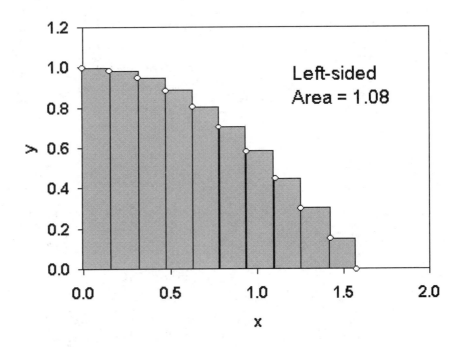

The height of the first rectangle is 1.0 (using the y-value on the left side of the first interval, which is in cell B5). The integration step is the width of the rectangle $x_2 - x_1$ (cell A6 minus cell A5). The formula for the area of the first rectangle is =B5*(A4-A5). This formula is entered into cell C5:

C5			f_x =B5*(A6-A5)		
	A	B	C	D	E
1	**Rectangles**				
2	Height set by left side y value				
3					
4	**x**	**y**	**Area**		
5	0.0000	1.0000	0.1571		
6	0.1571	0.9877			
7	0.3142	0.9511			
8	0.4712	0.8910			
9	0.6283	0.8090			
10	0.7854	0.7071			
11	0.9425	0.5878			
12	1.0996	0.4540			
13	1.2566	0.3090			
14	1.4137	0.1564			
15	1.5708	0.0000			
16					

Step 3. Copy the Area Formula to All Intervals

Remember that the number of intervals is one less than the number of data pairs. There are 11 data pairs, so there are 10 intervals. Copy the formula in cell C5 over the range C5:C14 (areas for a total of 10 rectangles should be computed):

C14			f_x =B14*(A15-A14)		
	A	B	C	D	E
1	**Rectangles**				
2	Height set by left side y value				
3					
4	**x**	**y**	**Area**		
5	0.0000	1.0000	0.1571		
6	0.1571	0.9877	0.1551		
7	0.3142	0.9511	0.1494		
8	0.4712	0.8910	0.1400		
9	0.6283	0.8090	0.1271		
10	0.7854	0.7071	0.1111		
11	0.9425	0.5878	0.0923		
12	1.0996	0.4540	0.0713		
13	1.2566	0.3090	0.0485		
14	1.4137	0.1564	0.0246		
15	1.5708	0.0000			
16					

Step 4. Sum the Individual Areas

Summing the rectangle areas in cells C5 through C14 yields the estimated area under the entire curve. For this data set, using rectangles with the left side aligned to the data points overestimates the true area (obtained from analytical integration) by 7.65%, giving the area 1.0765 instead of 1:

	C17	▼	f_x	=SUM(C5:C14)	
	A	B	C	D	E
1	**Rectangles**				
2	Height set by left side y value				
3					
4	**x**	**y**	**Area**		
5	0.0000	1.0000	0.1571		
6	0.1571	0.9877	0.1551		
7	0.3142	0.9511	0.1494		
8	0.4712	0.8910	0.1400		
9	0.6283	0.8090	0.1271		
10	0.7854	0.7071	0.1111		
11	0.9425	0.5878	0.0923		
12	1.0996	0.4540	0.0713		
13	1.2566	0.3090	0.0485		
14	1.4137	0.1564	0.0246		
15	1.5708	0.0000			
16					
17		**Total Area:**	1.0765		
18					

Aligning the Right Side of the Rectangles with the Data   We will now use the y-value on the *right* side of the interval to determine the height of the rectangle. As evident in the graph, for this cosine data, using y-values on the right causes the rectangles to underestimate the area under the cosine curve:

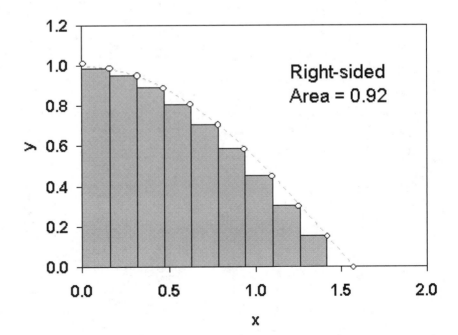

The *y*-value on the right side of the first rectangle is in cell B6. The width of the rectangle is still A6 − A5. The formula for the area of the first rectangle is then =B6*(A6-A5), which is placed in cell C6:

	C6	▼	*fx* =B6*(A6-A5)		
	A	B	C	D	E
1	**Rectangles**				
2	Height set by right side y value				
3					
4	**x**	**y**	**Area**		
5	0.0000	1.0000			
6	0.1571	0.9877	0.1551		
7	0.3142	0.9511			
8	0.4712	0.8910			
9	0.6283	0.8090			
10	0.7854	0.7071			
11	0.9425	0.5878			
12	1.0996	0.4540			
13	1.2566	0.3090			
14	1.4137	0.1564			
15	1.5708	0.0000			
16					

The equation in cell C6 should be copied over the range C6:C15 to find the area of each of the 10 rectangles. The estimate of the total area under the curve is obtained by summing the areas for each interval. In this example, the formula =SUM(C6:C15) was placed in cell C17 to perform this task. The result, Area = 0.9194, is lower than the true value by about 8%:

	C17	▼	*fx* =SUM(C5:C14)		
	A	B	C	D	E
1	**Rectangles**				
2	Height set by right side y value				
3					
4	**x**	**y**	**Area**		
5	0.0000	1.0000			
6	0.1571	0.9877	0.1551		
7	0.3142	0.9511	0.1494		
8	0.4712	0.8910	0.1400		
9	0.6283	0.8090	0.1271		
10	0.7854	0.7071	0.1111		
11	0.9425	0.5878	0.0923		
12	1.0996	0.4540	0.0713		
13	1.2566	0.3090	0.0485		
14	1.4137	0.1564	0.0246		
15	1.5708	0.0000	0.0000		
16					
17		**Total Area:**	0.9194		
18					

At this point, many people want to try to improve the results by averaging the left-aligned and right-aligned results. It is certainly possible to do that—it is also equivalent to using trapezoids to approximate the area below the curve.

14.4.2 Integration by Using Trapezoids

Step 1. Enter or Import the Data
No difference.

Step 2. Enter the Formula for the Area of One Trapezoid
The area of a *trapezoid* is governed by the data pairs on the left $(x, y)_L$ and on the right $(x, y)_R$. The equation for the area of a trapezoid is

$$A_{trap} = \frac{y_L + y_R}{2}(x_R - x_L).\qquad(14.6)$$

The formula entered into cell C5 is =0.5*(B5+B6)*(A6-A5). We then get

C5		f_x =0.5*(B5+B6)*(A6-A5)		
A	**B**	**C**	**D**	**E**
1 **Trapezoids**				
2 Height set by averaging left and right y values				
3				
4 **x**	**y**	**Area**		
5 0.0000	1.0000	0.1561		
6 0.1571	0.9877			
7 0.3142	0.9511			
8 0.4712	0.8910			
9 0.6283	0.8090			
10 0.7854	0.7071			
11 0.9425	0.5878			
12 1.0996	0.4540			
13 1.2566	0.3090			
14 1.4137	0.1564			
15 1.5708	0.0000			
16				

Step 3. Copy the Formula Over All the Intervals

The formula in cell C5 is copied over all 10 intervals in the range C5:C14:

	C14	▼		f_x =0.5*(B14+B15)*(A15-A14)	
	A	B	C	D	E
1	**Trapezoids**				
2	Height set by averaging left and right y values				
3					
4	**x**	**y**	**Area**		
5	0.0000	1.0000	0.1561		
6	0.1571	0.9877	0.1523		
7	0.3142	0.9511	0.1447		
8	0.4712	0.8910	0.1335		
9	0.6283	0.8090	0.1191		
10	0.7854	0.7071	0.1017		
11	0.9425	0.5878	0.0818		
12	1.0996	0.4540	0.0599		
13	1.2566	0.3090	0.0366		
14	1.4137	0.1564	0.0123		
15	1.5708	0.0000			
16					

Step 4. Sum the Areas of All Trapezoids

The formula =SUM(C5:C14) is entered into cell C17 to compute the estimate of the area under the cosine curve. The result, Area = 0.9979, is lower than the true value by 0.2%, which is a considerable improvement over either of the results using rectangles:

	C17	▼		f_x =SUM(C5:C14)	
	A	B	C	D	E
1	**Trapezoids**				
2	Height set by averaging left and right y values				
3					
4	**x**	**y**	**Area**		
5	0.0000	1.0000	0.1561		
6	0.1571	0.9877	0.1523		
7	0.3142	0.9511	0.1447		
8	0.4712	0.8910	0.1335		
9	0.6283	0.8090	0.1191		
10	0.7854	0.7071	0.1017		
11	0.9425	0.5878	0.0818		
12	1.0996	0.4540	0.0599		
13	1.2566	0.3090	0.0366		
14	1.4137	0.1564	0.0123		
15	1.5708	0.0000			
16					
17		**Total Area:**	0.9979		
18					

APPLICATIONS: CALCULATING REQUIRED VOLUME

Concrete Retaining Wall

Forms for a nine-inch-thick retaining wall with a complex shape have been constructed, and the contractor is ready to order the concrete. Given the dimensions shown on the following drawing (units are in feet), how much concrete should be ordered?

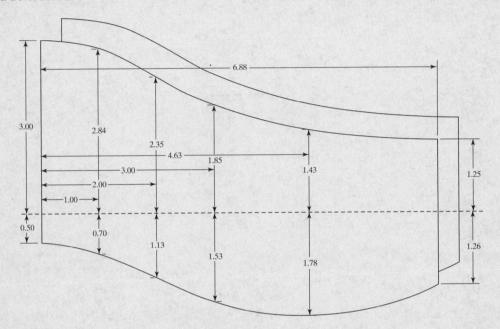

The solution to this problem will involve a few steps:

1. Enter the dimension data into a spreadsheet.
2. Calculate the surface area of the upper portion of the form.
3. Calculate the surface area of the lower portion of the form.
4. Calculate the total surface area of the form.
5. Multiply by the depth (nine inches) to find the required volume.

Step 1. The dimensions are entered into a spreadsheet as x-values, y_U-(upper) values and y_L-(lower) values:

	A	B	C	D	E	F
1	**Concrete Retaining Wall**					
2						
3	x	y_U	y_L			
4	(ft)	(ft)	(ft)			
5						
6	0.00	3.00	0.50			
7	1.00	2.84	0.70			
8	2.00	2.35	1.13			
9	3.00	1.85	1.53			
10	4.63	1.40	1.78			
11	6.88	1.25	1.26			
12						

Step 2. Calculate the surface area of the upper portion of the form, using trapezoids:

	D6	▼	f_x	=0.5*(B6+B7)*(A7-A6)			
	A	B	C	D	E	F	G
1	**Concrete Retaining Wall**						
2							
3	x	y_U	y_L	A_{U_Trap}			
4	(ft)	(ft)	(ft)	(ft²)			
5							
6	0.00	3.00	0.50	2.92			
7	1.00	2.84	0.70	2.60			
8	2.00	2.35	1.13	2.10			
9	3.00	1.85	1.53	2.65			
10	4.63	1.40	1.78	2.98			
11	6.88	1.25	1.26				
12				A_{UPPER}			
13			**Total:**	13.25			
14							

```
D13:  =SUM(D6:D10)
```

Step 3. Calculate the surface area of the lower portion of the form, using trapezoids:

	E6	▼	f_x	=0.5*(C6+C7)*(A7-A6)			
	A	B	C	D	E	F	G
1	**Concrete Retaining Wall**						
2							
3	x	y_U	y_L	A_{U_Trap}	A_{L_Trap}		
4	(ft)	(ft)	(ft)	(ft²)	(ft²)		
5							
6	0.00	3.00	0.50	2.92	0.60		
7	1.00	2.84	0.70	2.60	0.92		
8	2.00	2.35	1.13	2.10	1.33		
9	3.00	1.85	1.53	2.65	2.70		
10	4.63	1.40	1.78	2.98	3.42		
11	6.88	1.25	1.26				
12				A_{UPPER}	A_{LOWER}		
13			**Total:**	13.25	8.96		
14							

Step 4. Calculate the total surface area of the form:

	F13	▼	f_x	=D13+E13		
	A	B	C	D	E	F	G

	A	B	C	D	E	F	G
1	**Concrete Retaining Wall**						
2							
3	x	y_U	y_L	A_{U_Trap}	A_{L_Trap}		
4	(ft)	(ft)	(ft)	(ft^2)	(ft^2)		
5							
6	0.00	3.00	0.50	2.92	0.60		
7	1.00	2.84	0.70	2.60	0.92		
8	2.00	2.35	1.13	2.10	1.33		
9	3.00	1.85	1.53	2.65	2.70		
10	4.63	1.40	1.78	2.98	3.42		
11	6.88	1.25	1.26				
12				A_{UPPER}	A_{LOWER}	A_{FORM}	
13			Total:	13.25	8.96	22.21	ft^2
14							

Step 5. Multiply by the depth (nine inches) to find the required volume:

	F15	▼	f_x	=F13*(9/12)		
	A	B	C	D	E	F	G

	A	B	C	D	E	F	G
1	**Concrete Retaining Wall**						
2							
3	x	y_U	y_L	A_{U_Trap}	A_{L_Trap}		
4	(ft)	(ft)	(ft)	(ft^2)	(ft^2)		
5							
6	0.00	3.00	0.50	2.92	0.60		
7	1.00	2.84	0.70	2.60	0.92		
8	2.00	2.35	1.13	2.10	1.33		
9	3.00	1.85	1.53	2.65	2.70		
10	4.63	1.40	1.78	2.98	3.42		
11	6.88	1.25	1.26				
12				A_{UPPER}	A_{LOWER}	A_{FORM}	
13			Total:	13.25	8.96	22.21	ft^2
14							
15					Volume:	16.7	ft^3
16							

F15: =F13*(9/12)

The "12" converts 9 inches to feet.

14.4.3 Integration by Using Smooth Curves (Simpson's Rule)

Note: Simpson's rule, presented here, requires (1) an odd number of data points, and (2) evenly spaced data (i.e., uniform Δx).

The procedure for Simpson's rule is quite similar to that used in the preceding cases, but you must be careful to distinguish between the *data interval* between adjacent x-values and the integration step which, for Simpson's rule, incorporates two data intervals. It is because of this that Simpson's rule is restricted to an even number of intervals (an odd number of data points).

14.4.4 Procedure For Simpson's Rule

1. Enter or import the data.
2. Compute h, the distance between any two adjacent x-values (because the x-values are uniformly spaced).
3. Enter the area formula for one *integration step* (*not* interval).
4. Copy the formula over all integration regions.
5. Sum the areas of all integration regions.

Step 1. Enter or Import the Data

No change from previous cases.

Step 2. Compute h, the Distance between Adjacent x Values

The value of h can be computed by using any two adjacent x-values, because Simpson's rule requires uniformly spaced x-values. The first two x-values in cells A5 and A6 have been used here, and h is computed with the formula =A6−A5 and stored in cell D2.

Note: h is the width of the interval between any two data points, not the entire integration step:

$$h = \Delta x. \tag{14.7}$$

C2		f_x =A6-A5		
A	**B**	**C**	**D**	**E**
1 **Simpson's Rule**				
2	h:	0.1571		
3				
4 **x**	**y**	**Step**		
5 0.0000	1.0000	1		
6 0.1571	0.9877	1		
7 0.3142	0.9511	2		
8 0.4712	0.8910	2		
9 0.6283	0.8090	3		
10 0.7854	0.7071	3		
11 0.9425	0.5878	4		
12 1.0996	0.4540	4		
13 1.2566	0.3090	5		
14 1.4137	0.1564	5		
15 1.5708	0.0000			
16				

Step 3. Enter the Formula for the Area of One Integration Step

The formula for Simpson's rule is

$$A_s = \frac{h}{3}(y_L + 4y_M + y_R), \tag{14.8}$$

where A_s is the area of an integration step for Simpson's rule, or two data intervals. The subscripts L, M, and R stand for left, middle, and right sides of the integration step, respectively. (The integration steps are numbered on the spreadsheet as a reminder.) The first area formula is entered into cell D6 as =($D$2/3)*(B5+4*B6+B7).

Note: The dollar signs on the reference to cell $D$2 indicate that the reference to h *should not be changed during any subsequent copy:*

	D5	▼		*fx*	=($C$2/3)*(B5+4*B6+B7)	
	A	B	C	D	E	
1	**Simpson's Rule**					
2		**h:**	0.1571			
3						
4	**x**	**y**	**Step**	**Area**		
5	0.0000	1.0000	1	0.3090		
6	0.1571	0.9877	1			
7	0.3142	0.9511	2			
8	0.4712	0.8910	2			
9	0.6283	0.8090	3			
10	0.7854	0.7071	3			
11	0.9425	0.5878	4			
12	1.0996	0.4540	4			
13	1.2566	0.3090	5			
14	1.4137	0.1564	5			
15	1.5708	0.0000				
16						

Step 4. Copy the Area Formula to Each Integration Step

The formula in cell D6 must be copied to each integration step (i.e., every other row). A quick way to do this is to copy the formula to the Windows clipboard; then, select the destination cells by first clicking on cell D8 and holding down the [Ctrl] key while clicking on cells D10, D12, and D14. Excel uses the [Ctrl] key to select multiple, noncontiguous regions. Once all the destination cells have been selected, paste the formula into the cells from the clipboard:

D13			f_x =($C$2/3)*(B13+4*B14+B15)		
	A	B	C	D	E
1	**Simpson's Rule**				
2		**h:**	0.1571		
3					
4	**x**	**y**	**Step**	**Area**	
5	0.0000	1.0000	1	0.3090	
6	0.1571	0.9877	1		
7	0.3142	0.9511	2	0.2788	
8	0.4712	0.8910	2		
9	0.6283	0.8090	3	0.2212	
10	0.7854	0.7071	3		
11	0.9425	0.5878	4	0.1420	
12	1.0996	0.4540	4		
13	1.2566	0.3090	5	0.0489	
14	1.4137	0.1564	5		
15	1.5708	0.0000			
16					

Step 5. Sum the Areas for Each Region

You can use =SUM(D6:D14) to compute this sum, because empty cells are ignored during the summing process. The resulting estimate for the area under the curve is 1.0 (actually 1.000003), which is virtually equivalent to the true value, 1:

D17			f_x =SUM(D5:D13)		
	A	B	C	D	E
1	**Simpson's Rule**				
2		**h:**	0.1571		
3					
4	**x**	**y**	**Step**	**Area**	
5	0.0000	1.0000	1	0.3090	
6	0.1571	0.9877	1		
7	0.3142	0.9511	2	0.2788	
8	0.4712	0.8910	2		
9	0.6283	0.8090	3	0.2212	
10	0.7854	0.7071	3		
11	0.9425	0.5878	4	0.1420	
12	1.0996	0.4540	4		
13	1.2566	0.3090	5	0.0489	
14	1.4137	0.1564	5		
15	1.5708	0.0000			
16					
17			**Total Area:**	1.0000	
18					

Simpson's Rule: Summary   *Simpson's rule* is pretty straightforward to apply and does a good job of fitting most functions, but there are two important limitations on Simpson's rule. You must have the following:

1.   an odd number of data points (even number of intervals); and
2.   uniformly spaced x-values.

The trapezoidal rule is slightly less accurate, but it is very simple to use and does not have the restrictions just listed. Sometimes Simpson's method is used with an even number of data points (with uniformly spaced x-values) by using the Simpson's rule formula for all intervals except the last one. A trapezoid is used to compute the area of the final interval (half-Simpson integration step).

14.5   USING REGRESSION EQUATIONS FOR INTEGRATION

Another general approach to integrating a data set is to use *regression* to fit a curve to the data, then analytically integrate the regression equation. Excel provides a couple of alternatives for linear regression: *trendlines* with limited regression forms, and regression by using the Data Analysis package.

For the cosine data used here ($y = \cos(x), 0 \leq x \leq \pi/2$), the obvious fitting function is a cosine. To give the regression analysis something to work with, significant random noise has been added to the data values:

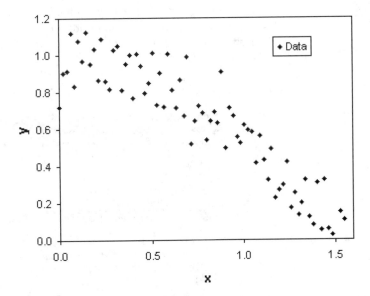

A cosine fit is not an option with Excel's trend lines, so we will use the regression option in the Data Analysis package. The *regression model* is

$$y_p = b_0 + b_1 \cos(x). \tag{14.9}$$

We are fitting cosine data, so the intercept b_0 should be unnecessary and will be set equal to zero. To prepare for the regression, we first take the cosine of each x-value in the data set:

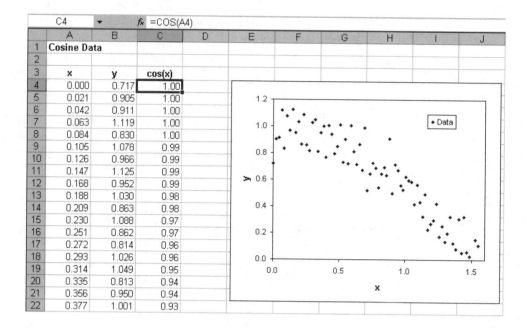

Then, open the list of options available in the Data Analysis package by selecting Tools/Data Analysis ...

Note: By default, the analysis package is installed in Excel, but not activated. If the Data Analysis menu option does not appear under the Tools menu, it has not been activated. Choose TOOLS/ADD-INS ..., and activate the Analysis option. This need be done only once.

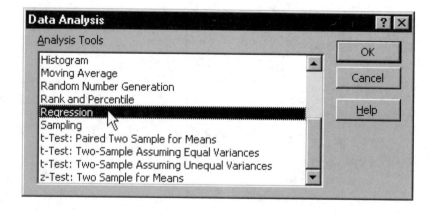

Select the Regression tool. This opens the Regression dialog box:

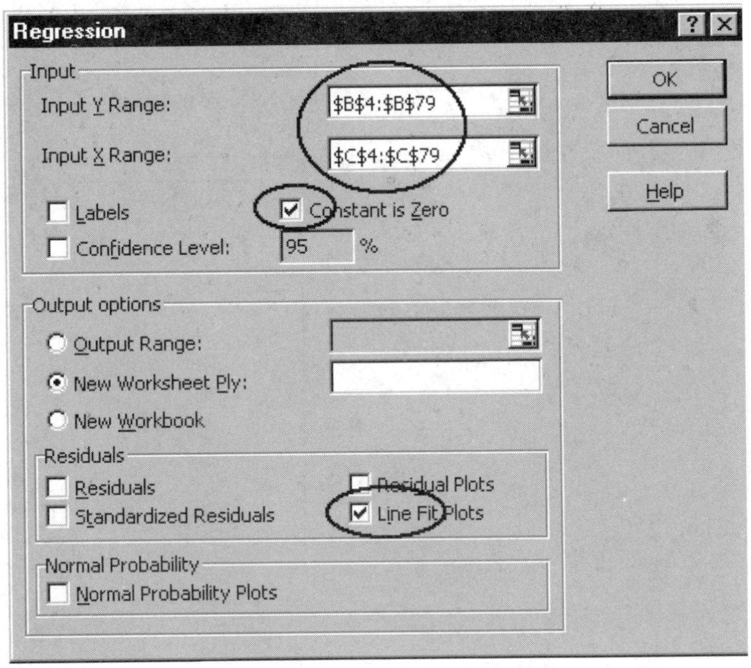

The cell ranges containing the y-values (B4:B79) and the x-values for the regression model (i.e., the $\cos(x)$ values C4:C79) are indicated. The Constant is Zero checkbox is checked to force the intercept (b_0) to zero, and a line-fit plot is requested.

When the OK button is clicked, the regression is performed, and the output is sent to a new spreadsheet:

	A	B	C
1	SUMMARY OUTPUT		
2			
3	*Regression Statistics*		
4	Multiple R	0.9403	
5	R Square	0.8842	
6	Adjusted R Square	0.8709	
7	Standard Error	0.1112	
8	Observations	76	
9			
10	ANOVA		
11		*df*	*SS*
12	Regression	1	7.0780
13	Residual	75	0.9270
14	Total	76	8.0050
15			
16		*Coefficients*	*Standard Error*
17	Intercept	0	#N/A
18	X Variable 1	0.9806	0.01803
19			
20			

The intercept is zero (as requested), and $b_1 = 0.9806$. The R^2 value, 0.8842, reflects the scatter in the data. The requested line-fit plot is as follows:

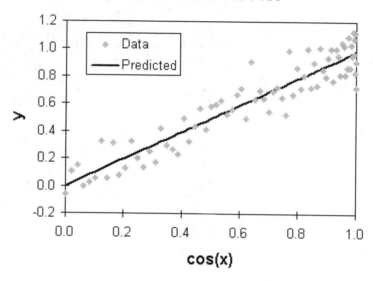

The label on the x-axis has been changed to $\cos(x)$ to indicate that the regression analysis used $\cos(x)$ as the independent variables (the x-values in the regression analysis). The regressed slope ($b_1 = 0.9806$) can be used to calculate predicted y-values. The predicted values have been superimposed on the data in the following graph. The regression equation has been shown in cell D4 as `=0+0.9806*COS(A4)`. The zero intercept included in the formula serves no purpose except to remind us that the intercept was forced to zero as part of the regression analysis:

Once the regression equation is obtained, the integration for the area under the curve is straightforward:

$$A = \int_0^{\pi/2} [0.9806 \cos(x)]\, dx$$

$$= 0.9806 \int_0^{\pi/2} \cos(x)\, dx$$

$$= 0.9806 \left[\sin\left(\tfrac{\pi}{2}\right) - \sin(0)\right]$$

$$= 0.9806\, [1 - 0]$$

$$= 0.9806. \tag{14.10}$$

The result is low by about 2%, which is not too bad considering the noise in the data.

KEY TERMS

Data interval	Regression	Tensile test
Extension	Regression model	Trapezoid
Integration step	Simpson's rule	Trendline
Interval	Strain	
Pitot tube	Stress	

SUMMARY

Numerical Integration by Using Geometric Regions

Rectangles

$$\text{Area} = \sum_{i=1}^{N-1} [y_i(x_{i+1} - x_i)], \text{ or}$$

$$\text{Area} = \sum_{i=1}^{N-1} [y_{i+1}(x_{i+1} - x_i)]. \tag{14.11}$$

Trapezoids

$$\text{Area} = \sum_{i=1}^{N-1} \left[\left(\frac{y_i + y_{i+1}}{2}\right)(x_{i+1} - x_i)\right]. \tag{14.12}$$

Simpson's Rule

Requirements

1. An odd number of data points (even number of intervals)
2. Uniformly spaced x-values.

Area Equation for One Integration Region ($2\, \Delta x$)

$$A_s = \frac{h}{3}(y_L + 4y_M + y_R), \tag{14.13}$$

where $h = \Delta x$ (must be a constant).

Total Area Equation

$$\text{Area} = \sum_{j=1}^{M} [A_{S_j}], \tag{14.14}$$

where

M is the number of integration regions $M = \frac{N-1}{2}$ and
N is the number of data points.

Integration by Using Regression Equations

1. Perform a regression analysis on the data to fit a curve to the data. Options:
   - Add a trendline to an XY Scatter graph of your data and request the equation of the trendline.
   - Use the Regression tool in the Data Analysis package.
2. Integrate the regression equation analytically.

Problems

Area Beneath a Curve

1. Approximate the area under curve 1, using trapezoids:

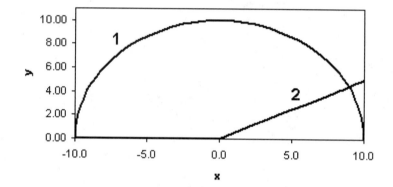

x	y₁	y₂
−10.0	0.00	0.0
−9.8	1.99	0.0
−9.4	3.41	0.0
−9.0	4.36	0.0
−8.0	6.00	0.0
−7.0	7.14	0.0
−6.0	8.00	0.0
−5.0	8.66	0.0
−4.0	9.17	0.0
−3.0	9.54	0.0
−2.0	9.80	0.0
−1.0	9.95	0.0
0.0	10.00	0.0
1.0	9.95	0.5
2.0	9.80	1.0
3.0	9.54	1.5
4.0	9.17	2.0
5.0	8.66	2.5
6.0	8.00	3.0
7.0	7.14	3.5

x	y_1	y_2
8.0	6.00	4.0
9.0	4.36	4.5
9.4	3.41	4.7
9.8	1.99	4.9
10.0	0.00	5.0

Area Between Curves

2.　Using the data in the previous problem, approximate the area between curves 1 and 2, using numerical integration.

Work Required to Stretch a Spring

3.　The device shown in the accompanying figure can be used to estimate the work required to extend a spring. This device consists of a spring, a spring balance, and a ruler. Before the stretching of the spring, its length is measured and found to be 1.3 cm. The spring is then stretched 0.4 cm at a time, and the force indicated on the spring balance is recorded. The resulting data set is shown in the following table:

SPRING DATA			
MEASUREMENT	UNEXTENDED LENGTH	EXTENSION	FORCE
(cm)	(cm)	(cm)	(N)
1.3	1.3	0.0	0.00
1.7	1.3	0.4	0.88
2.1	1.3	0.8	1.76
2.5	1.3	1.2	2.64
2.9	1.3	1.6	3.52
3.3	1.3	2.0	4.40
3.7	1.3	2.4	5.28
4.1	1.3	2.8	6.16
4.5	1.3	3.2	7.04
4.9	1.3	3.6	7.92

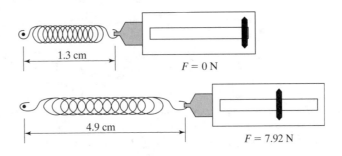

The work required to extend the spring can be computed as

$$W = \int F \, dx, \tag{14.15}$$

where x is the *extension* (distance stretched) of the spring.

Calculate the work required to extend the spring from 0 to 3.6 cm of extension.

Enthalpy Required to Warm a Gas

4. The enthalpy required to warm n moles of a gas from T_1 to T_2 can be calculated from the heat capacity of the gas:

$$\Delta H = n \int_{T_i}^{T_2} C_P(T) \, dT. \tag{14.16}$$

But the heat capacity is a function of temperature. An equation commonly used to describe the change in heat capacity with temperature is a simple third-order polynomial in T:[1]

$$C_P = a + bT + cT^2 + dT^3. \tag{14.17}$$

If the coefficients a through d are known for a particular gas, you can calculate the heat capacity of the gas at any T (within an allowable range). The coefficients for a few common gases are as follows:

HEAT-CAPACITY COEFFICIENTS

Gas	a	b	c	d	UNITS ON T	VALID RANGE
Air	28.94×10^{-3}	0.4147×10^{-5}	0.3191×10^{-8}	-1.965×10^{-12}	°C	0–1500°C
CO_2	36.11×10^{-3}	4.233×10^{-5}	-2.887×10^{-8}	7.464×10^{-12}	°C	0–1500°C
CH_4	34.31×10^{-3}	5.469×10^{-5}	0.3661×10^{-8}	-11.00×10^{-12}	°C	0–1200°C
H_2O	33.46×10^{-3}	0.6880×10^{-5}	0.7604×10^{-8}	-3.593×10^{-12}	°C	0–1500°C

The units on the heat-capacity values computed from Equation (14.17) are kJ/gmol °C.

a. Calculate the heat capacity of methane (CH_4) between 200 and 800°C, using 20° intervals.

b. Integrate the heat-capacity data, using a numerical integration technique, to find the energy (i.e., enthalpy change) required to warm 100 gmol of methane from 200 to 800°C.

c. Substitute the polynomial for C_P into the enthalpy integral and integrate analytically, to check your numerical integration result.

Stress–Strain Curve

5. Strength testing of materials often involves a *tensile test* in which a sample of the material is held between two mandrels and increasing force (actually,

---

[1]From *Elementary Principles of Chemical Processes*, 3d ed., R. M. Felder and R. W. Rousseau, New York: Wiley, 2000.

stress = force per unit area) is applied. A stress-vs.-strain curve for a typical ductile material is shown in the following figure; the sample first stretches reversibly (*A* to *B*), then irreversibly (*B* to *D*), before it finally breaks (point *D*):

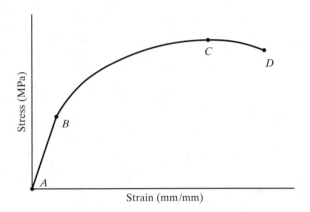

The *strain* is the amount of elongation of the sample (mm) divided by the original sample length (mm). A little reworking of the data transforms the stress–strain curve to a force–displacement curve, so

$$F = \text{stress} \times A_{\text{cross section,}}$$

and

$$D = \text{strain} \times L_{\text{original,}} \qquad (14.18)$$

where

F	is the force applied to the sample,
$A_{\text{cross section}}$	is the cross-sectional area of the sample (100 mm^2),
D	is the displacement, and
L_{original}	is the original sample length (40 mm).

For a sample 10 mm by 10 mm by 40 mm, the force and displacement data can be calculated and a force-vs.-displacement graph can be produced:

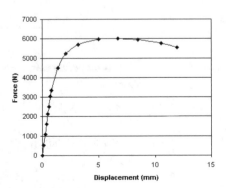

The area under this curve represents the work done on the sample by the testing equipment. Use a numerical integration technique to estimate the work done on the sample. Use the following data:

STRAIN	STRESS	FORCE	DISP.
(MM/MM)	(MPa)	(N)	(mm)
0.0000	0.00	0	0.00
0.0030	5.38	538	0.12
0.0060	10.76	1076	0.24
0.0090	16.14	1614	0.36
0.0120	21.52	2152	0.48
0.0140	25.11	2511	0.56
0.0170	30.49	3049	0.68
0.0200	33.34	3334	0.80
0.0350	44.79	4479	1.40
0.0520	52.29	5229	2.08
0.0790	57.08	5708	3.16
0.1240	59.79	5979	4.96
0.1670	60.10	6010	6.68
0.2120	59.58	5958	8.48
0.2640	57.50	5750	10.56
0.3000	55.42	5542	12.00

Fluid Velocity from Pitot Tube Data

6.   A pitot tube is a device that measures local velocity. The theory behind the operation of the pitot tube comes from Bernoulli's equation (without the potential energy terms) and relates the change in kinetic energy to the change in fluid pressure:

$$\frac{p_a}{\rho} + \frac{u_a^2}{2} = \frac{p_b}{\rho} + \frac{u_b^2}{2}. \tag{14.19}$$

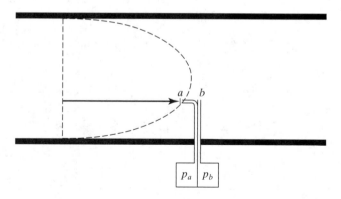

The flow that strikes the tube at point a hits a dead end, and its velocity goes to zero (stagnation). At point b, the flow slides past the pitot tube and does not slow down at all (free-stream velocity). The pressure at point b is the free-stream pressure. At point a, the energy carried by the fluid has to be conserved, so the kinetic energy of the moving fluid is transformed to pressure energy as the velocity goes to zero. The pitot tube measures a higher pressure at point a than at point b, and the pressure difference can be used to calculate the free-stream velocity at point b.

When the velocity at point a has been set to zero and Bernoulli's equation is rearranged to solve for the local velocity at point b, we get

$$u_b = \sqrt{\frac{2}{\rho}(p_a - p_b)}. \tag{14.20}$$

By moving the pitot tube across the diameter of a pipe, we find that a pressure-difference profile can be measured across the pipe. From the pressure differences, local velocities can be calculated. In this way, a pitot tube can be used to measure the velocity profile across a pipe. If the local velocities are integrated, the average velocity of the fluid in the pipe can be approximated:

$$\overline{V} = \frac{\displaystyle\int_{r=0}^{R}\int_{\theta=0}^{2\pi} ur\,dr\,d\theta}{\displaystyle\int_{r=0}^{R}\int_{\theta=0}^{2\pi} r\,dr\,d\theta}$$

$$= \frac{2\pi\displaystyle\int_{r=0}^{R} ur\,dr}{\pi R^2}. \tag{14.21}$$

Given the pitot-tube pressure differences shown in the accompanying table,

a.   calculate and graph the local velocities at each point if the fluid has a density of 810 kg/m^3; and

b.   use trapezoids to numerically integrate the local velocities to estimate the average velocity for the flow.

PITOT-TUBE DATA	
r (mm)	$(p_a - p_b)$ (kPa)
125	0.00
100	1.18
75	3.77
50	6.51
25	8.49
0	9.22
−25	8.50
−50	6.49
−75	3.76
−100	1.19
−125	0.01

15

Numerical Integration Techniques for Differential Equations Using Excel

15.1 INTRODUCTION

There are two types of integration that engineers routinely perform: integration for the area under a curve and integration of a differential equation. The two are related mathematically, but procedurally they are handled quite differently. This chapter presents numerical methods for integrating differential equations. Three methods are presented: Euler's method, Runge–Kutta methods, and an implicit method. The process of transforming a second-order differential equation into two first-order equations is also presented, because this technique makes the first-order solution methodologies applicable to second- (and higher) order differential equations.

15.2 EULER'S METHOD

Mathematical models of physical processes are often described by differential equations. Although complex models are better handled outside of Excel, many simple differential equations can be integrated in a spreadsheet. The simplest numerical integration method is *Euler's method*.

15.2.1 Choosing an Integration Step Size

Numerical integration methods typically work forward step by step through either space or time (the *independent variable*). To demonstrate Euler's method, we will consider a specific example (Example 15.1), shown next.

OBJECTIVES

After reading this chapter, you will know

- How to integrate a single, first-order ordinary differential equation by using Euler's method
- How to use a fourth-order Runge–Kutta method to integrate a single, first-order ordinary differential equation
- How to use a fourth-order Runge–Kutta method to integrate simultaneous ordinary differential equations
- How to use Excel's matrix math functions to integrate a partial differential equation, using an implicit method

EXAMPLE 15.1

Single Ordinary Differential Equation

This first differential equation describes the washout of an inert material, called component A, from a stirred tank:

$$\frac{dC_A}{dt} = \frac{1}{\tau}[C_{A_{\text{inf}}} - C_A]. \tag{15.1}$$

Here,

C_A	is the concentration of A in the tank and in the tank effluent (mg/mL),
$C_{A_{\text{inf}}}$	is the concentration of A in the influent to the tank (mg/mL),
τ	is the residence time of the system $(\tau = V/\dot{V})$,
V	is the tank volume (constant) (mL), and
$\dot{V}$	is the influent and effluent volumetric flow rate (mL/s).

To integrate the equation, we must know the initial concentration in the tank and some parameter values, such as the concentration of A in the influent, the tank volume, and the flow rate:

Required Information	
Concentration of A in tank initially:	100 mg/mL
Concentration of A in influent:	0 mg/mL
Tank volume:	10 L
Flow rate:	100 mL/sec

The differential equation can be integrated analytically to yield

$$C_A = C_{A_{\text{inf}}} + (C_{A_{\text{inf}}} - C_{A_{\text{inf}}})e^{-t/\tau}, \tag{15.2}$$

where

$C_{A_{\text{init}}}$	is the concentration of A initially in the tank (mg/mL).

The analytical solution will be used to check the accuracy of the numerical integration results.

In the preceeding example, we have a time derivative of concentration, so we will integrate forward in time. Euler's method will be used to work forward from a known concentration at a point in time to compute new concentrations (*dependent variable*) at new times; this is the solution to our problem. We must choose an *integration step size*, or *time step*, Δt. The choice needs to be made with some care. Euler's method uses current information and an algebraic approximation of the differential equation to predict what will happen in the future. The method will be much more successful at predicting into the near future than into the distant future.

Problems generally have some kind of information that gives you a clue about the *time scale* of the problem. In this example, it's in the vessel *residence time*, τ. The residence time can be viewed as the length of time required to fill the tank (10 L, or 10,000 mL) at the stated flow rate (100 mL/sec). Euler's method will succeed at predicting the changing concentration in the tank only if the integration time step is small compared with τ. The residence time for Example 15.1 is $\tau = 100$ seconds. Our choice of Δt should be small compared with this.

For this problem, a reasonable time step size might be one second; that is only 1% of τ. Instead, we will use a 20-second time step to demonstrate the impact of a relatively large time step on the accuracy of the solution. The time step will be built into a cell of the spreadsheet, so that it can easily be changed later to obtain a more accurate solution, which will be demonstrated later in this chapter.

Solution Procedure The following steps are required to use Euler's method to integrate the differential equation:

1. Enter problem parameters (tank volume, flow rate, influent concentration) into the spreadsheet.
2. Enter the integration parameter (time step, Δt).
3. Compute a working variable (residence time, τ)
4. Enter the initial condition (concentration in the tank at time zero).
5. Compute the new time and concentration in the tank (and tank effluent) after one time step.
6. Copy the formulas used in Step 5 to compute the effluent concentrations at later times.

The numerical solution is complete at this point; however, the following steps will be performed to check the accuracy of the computed solution:

7. Compute the concentration at each time, using the analytical solution.
8. Calculate the percentage error at each time.

Step 1. Entering Problem Parameters into the Spreadsheet

It is common to place *parameter* values (values common to the entire problem) near the top of the spreadsheet. This is simply a matter of style, not a requirement, but it does keep the parameters easy to reach when entering formulas.

The parameters for the problem include the 10 L (10,000 mL) volume of the tank, the influent and effluent flow rates (they are equal in this problem and so the liquid volume stays constant)—100 mL/s, and the concentration of component A in the influent stream, which is zero in this problem. These are entered into the spreadsheet as follows:

	A	B	C	D	E	F
1	Single ODE: Euler's Method					
2						
3		Parameter	Value	Units		
4		V	10000	mL		
5		V_{dot}	100	mL/sec		
6		C_{A_inf}	0	mg/mL		
7						

Step 2. Enter the Integration Parameter, Δt

The size of the integration step (a time step in this problem) was chosen to be 20 seconds for this example, which, as mentioned before, is a fairly large time step relative to the residence time. This large step size was chosen primarily to demonstrate a noticeable error in the numerical solution results when compared with the analytical results. The integration step size can easily be reduced later to produce a more accurate solution. The chosen time step is entered in cell C7:

	A	B	C	D	E	F
1	**Single ODE: Euler's Method**					
2						
3		**Parameter**	**Value**	**Units**		
4		V	10000	mL		
5		V_{dot}	100	mL/sec		
6		C_{A_inf}	0	mg/mL		
7		Δt	20	sec		
8						

Step 3. Compute Working Variables

The residence time τ appears in the differential equation and will be used for each calculation, so it is convenient to calculate it from the problem parameter values:

C8		f_x	=C4/C5			
	A	B	C	D	E	F
1	**Single ODE: Euler's Method**					
2						
3		**Parameter**	**Value**	**Units**		
4		V	10000	mL		
5		V_{dot}	100	mL/sec		
6		C_{A_inf}	0	mg/mL		
7		Δt	20	sec		
8		τ	100	sec		
9						

Step 4. Enter the Initial Conditions

The *initial conditions* begin the actual presentation of the solution. To integrate the differential equation, a starting value of the dependent variable (concentration in this example) must be available. The problem statement specifies that the concentration of A in the tank initially is 100 mg/mL. This value is entered in cell C13 as the first concentration value in the solution, and a zero is entered as the initial time in cell B13:

	A	B	C	D	E	F
9						
10	**Solution**					
11		**Time**	C_A			
12		(sec)	(mg/mL)			
13		0	100.00			
14						

Step 5. Compute Time and Effluent Concentrations After One Integration Step

The time at the end of the integration step is simply the previous time plus the time step

$$t_{new} = t_{previous} + \Delta t, \tag{15.3}$$

which can be entered into cell B14 as

    B14: =B13+$C$7

The dollar signs in this formula will allow it to be copied down the spreadsheet to compute future times; the copied formulas will always reference the time step in cell C7.

   To compute the new effluent concentration at the end of the integration step, we use Euler's method. Euler's method simply involves replacing the differential equation by an algebraic approximation. Using a forward-finite-difference approximation to the time derivative, we see that our differential equation becomes

$$\frac{C_{A_{new}} - C_{A_{old}}}{\Delta t} = \frac{1}{\tau}(C_{A_{inf}} - C_{A_{old}}). \tag{15.4}$$

Either the new or the old concentration value could be used on the right side of this equation. When the old (i.e., currently known) concentration value is used, the integration technique is called an *explicit technique*. When the new (currently unknown) concentration is used, the technique is called *implicit*. For now, we will use an explicit technique, with the known value of concentration, $C_{A_{old}}$, on the right side of the finite-difference equation.

   Solving the finite-difference equation for the new concentration at the end of the time step gives

$$C_{A_{new}} = C_{A_{old}} + \frac{\Delta t}{\tau}(C_{A_{inf}} - C_{A_{old}}). \tag{15.5}$$

This equation is built into cell C14 as the formula

    C14:=C13+($C$7/$C$8)*($C$6−C13)

When this formula is copied down the spreadsheet to new cells, it will always reference the cell just above it as the "old" C_A, but the absolute referencing (dollar signs) on $\Delta t/\tau$ and $C_{A_{inf}}$ ensures that the parameter values will always be used in the copied formulas:

C14		▼	f_x =C13+($C$7/$C$8)*($C$6-C13)		
	A	B	C	D	E	F
3		**Parameter**	**Value**	**Units**		
4		V	10000	mL		
5		V$_{dot}$	100	mL/sec		
6		C$_{A\_inf}$	0	mg/mL		
7		Δt	20	sec		
8		τ	100	sec		
9						
10	**Solution**					
11		**Time**	**C$_A$**			
12		(sec)	(mg/mL)			
13		0	100.00			
14		20	80.00			
15						

Step 6. Copy the Formulas Down the Spreadsheet

At this point, the rest of the integration can be accomplished simply by copying the formulas entered in Step 5 down to rows 15 and below. Copying the formulas in cells B14 and C14 through row 38 results in computing the results for 25 integration steps (500 sec). The first few time steps of the complete solution are shown in the following spreadsheet, and the entire solution is shown graphically:

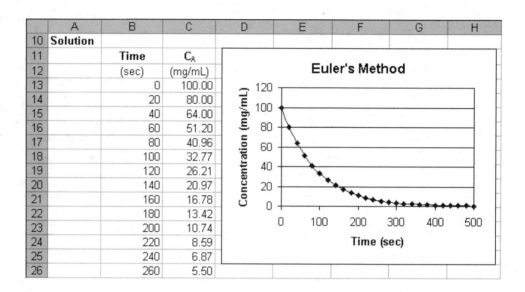

	A	B	C	D	E	F	G	H
10	Solution							
11		Time	C$_A$					
12		(sec)	(mg/mL)					
13		0	100.00					
14		20	80.00					
15		40	64.00					
16		60	51.20					
17		80	40.96					
18		100	32.77					
19		120	26.21					
20		140	20.97					
21		160	16.78					
22		180	13.42					
23		200	10.74					
24		220	8.59					
25		240	6.87					
26		260	5.50					

Five hundred seconds was chosen as the span of the integration, because it represents five residence times (5τ) for this system. For a wash-out from a perfectly mixed tank, the concentration should fall off by 95% after three residence times. (That is easily seen from the analytical solution). Therefore, most of the "action" would be expected to occur within the first three residence times, and the effluent concentration should be nearly zero by 5τ. This appears to be the case in the figure.

15.2.2 Checking the Accuracy of the Result

It is not common to have an analytical solution available. (If an analytical solution is available, there is little point in integrating the differential equation numerically.) It is common, however, to develop solution procedures using differential equations for which analytical solutions are available to provide a true value against which to judge the accuracy of the solution method. This is the case here. The analytical solution, shown here, can be incorporated into the spreadsheet:

$$C_A = C_{A_{\inf}} + (C_{A_{\inf}} - C_{A_{\inf}})e^{-t/\tau}. \tag{15.6}$$

This equation was entered into cell D14 as the formula

```
D14:  =$C$6+($D$13-$C$6)*EXP(-B14/$C$8)
```

and then copied to cells D14 through D38:

D14	▼		f_x =$C$6+($D$13-$C$6)*EXP(-B14/$C$8)		
	A	B	C	D	E	F
1	**Single ODE: Euler's Method**					
2						
3		**Parameter**	**Value**	**Units**		
4		V	10000	mL		
5		V_{dot}	100	mL/sec		
6		C_{A_inf}	0	mg/mL		
7		Δt	20	sec		
8		τ	100	sec		
9						
10	**Solution**		**Euler**	**Analytical**		
11		**Time**	C_A	C_A		
12		(sec)	(mg/mL)	(mg/mL)		
13		0	100.00	100.00		
14		20	80.00	81.87		
15		40	64.00	67.03		
16		60	51.20	54.88		
17		80	40.96	44.93		
18		100	32.77	36.79		

15.2.3 Error

Fractional error can be computed as

$$\text{Fractional Error} = \frac{C_{A_{\text{true}}} - C_{A_{\text{Euler}}}}{C_{A_{\text{true}}}}. \qquad (15.7)$$

A typical cell entry is

```
E13:  =(D13-C13)/D13
```

The results (partial listing) are shown next. Note that the fractional error has been displayed in percent formatting; effectively, the fractional error has been increased by a factor of 100 and shown with a percent symbol:

E13	▼		f_x =(D13-C13)/D13		
	A	B	C	D	E	F
9						
10	**Solution**		**Euler**	**Analytical**		
11		**Time**	C_A	C_A	**Error**	
12		(sec)	(mg/mL)	(mg/mL)	(%)	
13		0	100.00	100.00	0.0%	
14		20	80.00	81.87	2.3%	
15		40	64.00	67.03	4.5%	
16		60	51.20	54.88	6.7%	
17		80	40.96	44.93	8.8%	
18		100	32.77	36.79	10.9%	

The errors are clearly significant, and getting worse (on a percentage basis) as the integration proceeds:

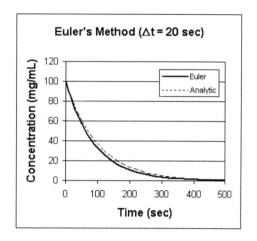

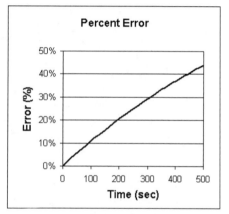

To improve the accuracy of the result, the integration step size in cell C7 should be reduced. This will require more integration steps to cover the same span (i.e., 500 sec), but adding the steps is accomplished simply by copying the last row of formulas down the spreadsheet as far as necessary. The next few graphs show the results when the step size is reduced to $\Delta t = 2$ seconds. There is still some error, but it has been greatly reduced by using the smaller integration step:

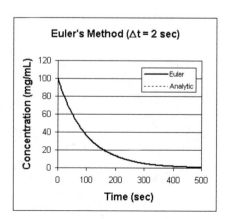

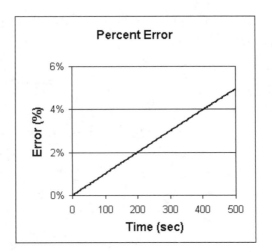

15.3 FOURTH-ORDER RUNGE–KUTTA METHOD

Runge–Kutta methods also use what is currently known about the variables (e.g., time and concentration) and the differential equation to predict the new values of the variable(s) at the end of the integration step. The difference is in how they compute the new values. The heart of the fourth-order Runge–Kutta method is a weighted-average derivative estimate, called dD_{avg} here (this nomenclature comes about by using D to stand for some arbitrary dependent variable value and dD to represent the derivative of that dependent variable):

$$dD_{\mathrm{avg}} = \tfrac{1}{6}[dD_1 + 2\,dD_2 + 2\,dD_3 + dD_4]. \tag{15.8}$$

Those four derivative values, dD_1 through dD_4, must be calculated from the differential equation and the currently known values of concentration (the dependent variable in our example). They could be computed directly in a spreadsheet, but the solution process gets messy. Instead, we will write a function to carry out one integration step by Runge–Kutta's fourth-order technique, and we will call it `rk4()`. But first we need another function that can be used to calculate the derivative value at any point. We'll call this function `dDdI()`, because it will be used to calculate the derivative of a dependent variable with respect to an independent variable. In the tank-washout problem, (Example 15.1) the dependent variable is concentration, and the independent variable is time.

15.3.1 Evaluating the Derivative: Function `dDdI()`

Our differential equation is

$$\frac{dC_A}{dt} = \frac{1}{\tau}(C_{A_{\mathrm{inf}}} - C_A)\cdot \tag{15.9}$$

The left side is the derivative we need to evaluate, and it is equal to the algebra on the right side of the equation. We will use the algebra on the right side to evaluate the values of the derivative at any point. The `dDdI()` function, written in VBA® (Visual Basic for Applications) from Excel, looks as follows:

```
Public Function dDdI(Ival As Double, Dval As Double) As Double
    Dim Tau As Double
    Dim Cinf As Double
    Dim C As Double
    Tau  = 100      'seconds
    Cinf = 0        'mg/ml
    C    = Dval
    dDdI = (1 / Tau) * (Cinf - C)
End Function
```

The independent and dependent variable values are passed into the function as parameters Ival and Dval, and the function returns a double-precision result through variable dDdI. The value passed into the function as Dval has been transferred to variable C just to make the last equation more recognizable as coming from the right side of our differential equation.

The independent variable passed into the function as Ival has not been used in this function. The independent variable is being passed to function dDdI() for compatibility with function rk4(). It is common practice to write Runge–Kutta functions that pass both the independent and dependent values to the derivative function in case the derivative needs both values.

15.3.2 How the Fourth-Order Runge–Kutta Method Works

The fourth-order Runge–Kutta method uses four estimates of the derivative to compute a weighted-average derivative that is used with the current value of concentration (in our example) to predict the next concentration value at the end of the time step. The four derivative estimates include the following:

First-Derivative Estimate The dDdI() function is used with the concentration value at the beginning of the integration step, $C_{A_{old}}$, to find the first-derivative estimate, called dD_1:

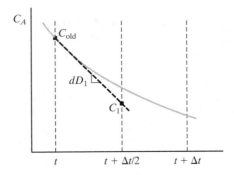

This derivative estimate, shown as a slope in the previous figure, is used to project a concentration, C_1, halfway across the integration step:

$$C_1 = C_{old} + dD_1 \cdot \frac{\Delta t}{2}. \tag{15.10}$$

Second-Derivative Estimate The dDdI() function is used with the concentration value C_1 to calculate the second-derivative estimate, dD_2:

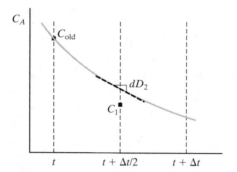

Then slope dD_2 is used back at time t to project another concentration estimate C_2 halfway across the integration step:

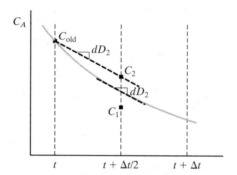

$$C_2 = C_{\text{old}} + dD_2 \cdot \frac{\Delta t}{2}. \tag{15.11}$$

Third-Derivative Estimate   The dDdI() function is used with the concentration value C_2 to calculate the third-derivative estimate, dD_3:

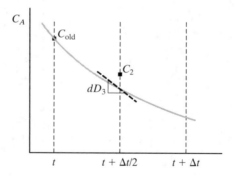

Shouldn't the second- and third-derivative values be the same? They are calculated for the same point in time, but they are based on different *estimates* of the concentration at $t + \Delta t/2$. If the estimates are equal, the derivatives calculated by using the estimates will be equal, but this is not the usual situation.

The third-derivative estimate, dD_3, is used with the original concentration to project the concentration at the end of the time step. This concentration estimate is called C_3:

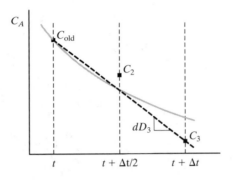

$$C_3 = C_{old} + dD_3 \cdot \Delta t. \tag{15.12}$$

Fourth-Derivative Estimate   The dDdI() function is used with the concentration value C_3 to calculate the fourth-derivative estimate, dD_4:

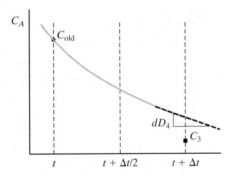

Once the four derivative estimates are available, a weighted-average derivative, dD_{avg}, is calculated. The derivative estimates at the center of the time step are weighted more heavily than those at the beginning and end of the step:

$$dD_{avg} = \tfrac{1}{6}[dD_1 + 2\,dD_2 + 2\,dD_3 + dD_4]. \tag{15.13}$$

The fourth-order Runge–Kutta method uses this weighted average to predict the new concentration at the end of the integration step:

$$C_{new} = C_{old} + dD_{avg} \cdot \Delta t. \tag{15.14}$$

15.3.3 Automating a Fourth-Order Runge–Kutta: Function rk4()

The fourth-order Runge–Kutta method requires four calls of function dDdI(), to calculate derivative values dD_1 through dD_4, and three dependent-value (concentration) calculations before calculating the weighted-average derivative value and the final prediction of the dependent variable value at the end of the integration step. These calculations can be done in the spreadsheet, but a function can be written to simplify the process:

```
Public Function rk4(Dt As Double, Ival As Double, Dval As
Double) As Double

    Dim D(3) As Double
    Dim dD(4) As Double
    Dim dDavg As Double

    dD(1) = dDdI(Ival, Dval)
    D(1) = Dval + dD(1) * Dt / 2

    dD(2) = dDdI(Ival, D(1))
    D(2) = Dval + dD(2) * Dt / 2

    dD(3) = dDdI(Ival, D(2))
    D(3) = Dval + dD(3) * Dt

    dD(4) = dDdI(Ival, D(3))

    dDavg = (1 / 6) * (dD(1) + 2 * dD(2) + 2 * dD(3) + dD(4))
    rk4 = Dval + dDavg * Dt

End Function
```

If this function is written in VBA from Excel, it will be stored as part of the workbook and can be called from any spreadsheet in the workbook.

Note: The program lines in the `dDdI()` function depend on the equation that is being solved; the function must be modified for each differential equation you solve. The `rk4()` function is independent of the equation being solved. This `rk4()` function handles only one ordinary differential equation (ODE), not simultaneous ODEs.

15.3.4 Implementing Fourth-Order Runge–Kutta in a Spreadsheet

The process of creating the spreadsheet begins, much as does the Euler method, by stating the parameters and initial conditions, as is illustrated in the following spreadsheet:

	A	B	C	D	E	F
1	**Single ODE: Fourth-Order Runge-Kutta Method**					
2						
3		**Parameter**	**Value**	**Units**		
4		V	10000	mL		
5		V_{dot}	100	mL/s		
6		C_{A_inf}	0	mg/mL		
7		Δt	20	sec		
8		τ	100	sec		
9						
10	**Solution**		**RK4**	**Analytical**		
11		**Time**	C_A	C_A	**Error**	
12		(sec)	(mg/mL)	(mg/mL)	(%)	
13		0	100.00	100.00	0.0%	
14						

The time at the end of the first integration step is computed by adding Δt to the previous time value, the zero stored in cell B13:

    B14: =B13 + $C$7

The concentration at the end of the first integration step is computed by using the rk4() function in cell C14:

    C14: =rk4($C$7, B13, C13)

	C14	▼	f_x =rk4($C$7,B13,C13)		
	A	B	C	D	E	F
1	Single ODE: Fourth-Order Runge-Kutta Method					
2						
3		Parameter	Value	Units		
4		V	10000	mL		
5		V_{dot}	100	mL/s		
6		C_{A_inf}	0	mg/mL		
7		Δt	20	sec		
8		τ	100	sec		
9						
10	Solution		RK4	Analytical		
11		Time	C_A	C_A	Error	
12		(sec)	(mg/mL)	(mg/mL)	(%)	
13		0	100.00	100.00	0.0%	
14		20	81.87			
15						

The analytical solution and percent-error calculation are unchanged from the Euler solution presented earlier.

Once the formulas in cells B14 through E14 have been entered, simply copy those cells down the page to complete the solution. The result is shown (partially) here, and the complete solution (500 sec) has been graphed:

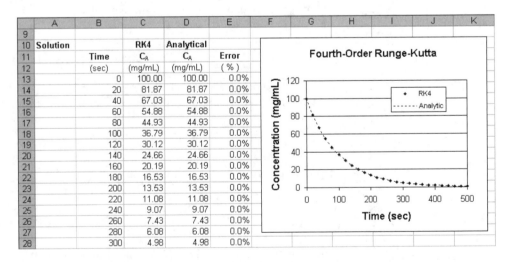

	A	B	C	D	E	F	G	H	I	J	K
9											
10	Solution		RK4	Analytical							
11		Time	C_A	C_A	Error						
12		(sec)	(mg/mL)	(mg/mL)	(%)						
13		0	100.00	100.00	0.0%						
14		20	81.87	81.87	0.0%						
15		40	67.03	67.03	0.0%						
16		60	54.88	54.88	0.0%						
17		80	44.93	44.93	0.0%						
18		100	36.79	36.79	0.0%						
19		120	30.12	30.12	0.0%						
20		140	24.66	24.66	0.0%						
21		160	20.19	20.19	0.0%						
22		180	16.53	16.53	0.0%						
23		200	13.53	13.53	0.0%						
24		220	11.08	11.08	0.0%						
25		240	9.07	9.07	0.0%						
26		260	7.43	7.43	0.0%						
27		280	6.08	6.08	0.0%						
28		300	4.98	4.98	0.0%						

Once the rk4() function has been created, using the fourth-order Runge–Kutta technique is no more effort than using Euler's method, but the results are much better. In this example, the percent error, even with a time step of 20 seconds, is zero (to two decimal places) for the entire solution.

15.4 INTEGRATING TWO SIMULTANEOUS ODES BY USING THE RUNGE–KUTTA METHOD

Solving two *simultaneous ODEs* by using the fourth-order Runge–Kutta technique will be demonstrated here with the conduction–convection problem specified in Example 15.2.

EXAMPLE 15.2

Conduction with Convection

A copper rod is placed between two containers: one contains boiling water and the other ice water. The rod is one meter in length and is of diameter 4 cm (or 0.04 m). Ambient air at 25°C flows over the surface of the rod, causing the rod to exchange energy with the air stream. The heat-transfer coefficient for this convective heat transfer between the rod and the air is $h = 50$ W/m^2 K. Calculate the temperature profile along the rod at steady state.

This problem can be described mathematically as

$$\frac{d^2T}{dx^2} = \frac{4\,h}{D\,k}(T - T_{amb}) \quad \text{(second-order ODE)},$$

with *boundary conditions*

$T = 100$ C at $x = 0$,

$T = 0$ C at $x = 1$ m.

The thermal conductivity of copper is $k = 390$ W/m K.

The differential equation described in Example 15.2 is a second-order ODE. To use Runge–Kutta methods, we need to rewrite it as two first-order ODEs.

Rewriting One Second-Order ODE as Two First-Order ODEs Because good solution methods (such as the fourth-order Runge–Kutta method) exist for first-order ODEs, it is common to rewrite higher-order differential equations as a series of first-order differential equations. This is easily done if you remember that a second-order derivative can be written as the derivative of a derivative:

$$\frac{d^2T}{dx^2} = \frac{d}{dx}\left[\frac{dT}{dx}\right]. \tag{15.15}$$

If the portion of the equation in brackets is simply given a new name, say, F for first derivative, the original differential equation can be rewritten as

$$\frac{dF}{dx} = \frac{4\,h}{D\,k}(T - T_{amb}), \tag{15.16}$$

and the definition of F becomes the second differential equation:

$$\frac{dT}{dx} = F. \tag{15.17}$$

The original second-order differential equation has become two simultaneous first-order differential equations.

Boundary-Value Problems and the Shooting Method The two first-order differential equations must be solved simultaneously. With the Runge–Kutta method used here, starting values of each dependent variable (T and F) must be known in order to carry out the integration. Boundary-value problems such as this one (Example 15.2) specify the value of a dependent variable at each end of the system (each boundary). We know the temperature at $x = 0$ and $x = 1$ m. We don't know anything about F at $x = 0$. For problems such as this, the *shooting method* is required.

The shooting method gets its name from artillery practice. The gunners aim the cannon, shoot, see where the shell lands, and adjust the angle on the cannon and try again. Eventually, they hit their target. The equivalent of "aiming the cannon" is setting the value of F. We will choose a value for F at $x = 0$, integrate the differential equation from $x = 0$ to $x = 1$ m, and see whether we get the right value of T (100°C) at $x = 1$ m. If we don't get the right boundary value, we choose a new F value at $x = 0$ and integrate again.

Modifying the `rk4()` Function for Two ODEs: Function `rk4two()`   The changes required to function `rk4()` are the following:

- To receive an integer code to indicate which equation is being integrated. (eqN = 1 for the dT/dx equation; eqN = 2 for the dF/dx equation).
- To receive two dependent variables (T and F in this example) that will be given generic names D1val and D2val.
- To call two different derivative functions, which we will call `dD1dI()` and `dD2dI()`, to calculate the derivative of each of the two differential equations.

Note: No programmer will consider this an elegant solution. It is a simple modification of function `rk4()` that allows two ODEs to be integrated. There is a lot of room for improvement in this function, such as making it possible to handle more than two ODEs by passing the dependent variable values as a matrix. Many advanced math packages provide built-in Runge-Kutta integration routines.

The modified `rk4()` function will be called `rk4two()` and is listed here:

```
Public Function rk4two(eqN As Integer, Dt As Double, Ival As
    Double, D1val As Double, D2val As Double) As Double

    Dim D(3) As Double
    Dim dD(4) As Double
    Dim dDavg As Double

    Select Case eqN
        Case 1  ' first differential equation

            dD(1) = dD1dI(Ival, D1val, D2val)
            D(1) = Dval + dD(1) * Dt / 2

            dD(2) = dD1dI(Ival, D(1), D2val)
            D(2) = Dval + dD(2) * Dt / 2

            dD(3) = dD1dI(Ival, D(2), D2val)
            D(3) = Dval + dD(3) * Dt

            dD(4) = dD1dI(Ival, D(3), D2val)

            dDavg = (1 / 6) * (dD(1) + 2 * dD(2) + 2 * dD(3) + dD(4))
            rk4two = D1val  + dDavg  * Dt

        Case 2  ' second differential equation

            dD(1) = dD2dI(Ival, D1val, D2val)
            D(1) = Dval + dD(1) * Dt / 2

            dD(2) = dD2dI(Ival, D1val, D(1))
            D(2) = Dval + dD(2) * Dt / 2
```

```
        dD(3) = dD2dI(Ival, D1val, D(2))
        D(3) = Dval + dD(3) * Dt

        dD(4) = dD2dI(Ival, D1val, D(3))

        dDavg = (1 / 6) * (dD(1) + 2 * dD(2) + 2 * dD(3) + dD(4))
        rk4two = D2val + dDavg * Dt
    End Select
End Function
```

New Derivative Functions: `dD1dI()` and `dD2dI()`

The derivative functions are specific to the problem being solved. The functions required for this problem are listed here:

```
Public Function dD1dI(Ival As Double, D1val As Double, D2val
As Double) As Double

' used with equation dT/Dx = F
'
' D1val is T
' D2val is F

    dD1dI = D2val

End Function

Public Function dD2dI(Ival As Double, D1val As Double, D2val
As Double) As Double

' used with equation dF/dx = (4 h)/(D k) * (T - Tamb)
'
' D1val is T
' D2val is F

    Dim h As Double
    Dim D As Double
    Dim k As Double
    Dim T As Double
    Dim Tamb As Double

    h = 50        'W/m2 K
    D = 0.04      'm
    k = 386       'W/m K
    Tamb = 25     '°C
    T = D1val

    dD2dI = (4 * h) / (D * k) * (T - Tamb)

End Function
```

Solving the ODEs

To allow the integration step size to be varied, it is specified as a parameter near the top of the spreadsheet. Then the initial position ($x = 0$), temperature ($100°C$), and guess for F are entered. If the final temperature profile was a straight line from $100°C$ at $x = 0$ to $0°C$ at $x = 1$ m, the F value would be $-100°C/m$. The temperature profile will not be a straight line, because of the energy loss to the moving air, but $100°C/m$ is a reasonable first guess for F:

	A	B	C	D	E
1	**Simultaneous ODEs with Runge-Kutta**				
2					
3		Δx:	0.1	m	
4					
5		**Position**	**Temp.**	**F**	
6		(m)	(°C)	(°C/m)	
7		0.00	100.00	-100.00	
8					

The position at the end of the first integration step is calculated as

    B8: =B7+$C$3

B8		▼	f_x =B7+$C$3		
	A	B	C	D	E
1	**Simultaneous ODEs with Runge-Kutta**				
2					
3		Δx:	0.1	m	
4					
5		**Position**	**Temp.**	**F**	
6		(m)	(°C)	(°C/m)	
7		0.00	100.00	-100.00	
8		0.10			
9					

The temperature at the end of the first integration step is found, using `rk4two()`, to be

    C8: =rk4two(1,$C$3,B7,C7,D7),

where the 1 tells the function to integrate the first differential equation ($dT/dx = F$), the $C$3 passes the integration step size ($\Delta x = 0.1$ m) to the function, and the remaining three arguments pass the current values of position, temperature, and F to the function:

C8		▼	f_x =rk4two(1,$C$3,B7,C7,D7)		
	A	B	C	D	E
1	**Simultaneous ODEs with Runge-Kutta**				
2					
3		Δx:	0.1	m	
4					
5		**Position**	**Temp.**	**F**	
6		(m)	(°C)	(°C/m)	
7		0.00	100.00	-100.00	
8		0.10	90.00		
9					

The value of F at the end of the first integration step is calculated in similar fashion:

> D8: =rk4two(2,$C$3,B7,C7,D7)

Here the 2 tells `rk4two()` to integrate the second differential equation (dF/dx):

	D8	▼		f_x =rk4two(2,$C$3,B7,C7,D7)	
	A	B	C	D	E
1	Simultaneous ODEs with Runge-Kutta				
2					
3		Δx:	0.1	m	
4					
5		Position	Temp.	F	
6		(m)	(°C)	(°C/m)	
7		0.00	100.00	-100.00	
8		0.10	90.00	-2.85	
9					

To complete the integration and solve for temperatures across the rod, simply copy the formulas in row 8 down the sheet:

	A	B	C	D	E
1	Simultaneous ODEs with Runge-Kutta				
2					
3		Δx:	0.1	m	
4					
5		Position	Temp.	F	
6		(m)	(°C)	(°C/m)	
7		0.00	100.00	-100.00	
8		0.10	90.00	-2.85	
9		0.20	89.72	81.35	
10		0.30	97.85	165.17	
11		0.40	114.37	259.54	
12		0.50	140.32	375.30	
13		0.60	177.85	524.68	
14		0.70	230.32	722.67	
15		0.80	302.59	988.63	
16		0.90	401.45	1348.20	
17		1.00	536.27	1835.83	
18					

The temperature at $x = 1$ m came out to be 536°C, not quite the 0°C we were shooting for. A little trial and error with the starting F value in cell D7 yields this result:

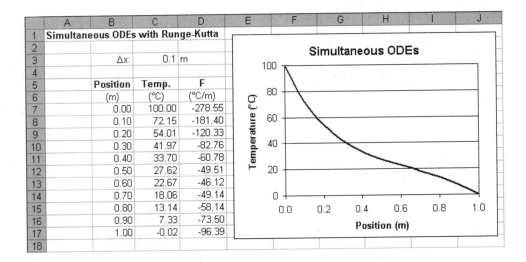

15.4.1 Checking for Accuracy

Numerical integration of differential equations always produces approximate results. The approximation gets better as the size of the integration step goes to zero. It never hurts to reduce the size of the integration step to see whether the calculated results change significantly. If they do, the size of the step is affecting the results, and your integration step size might need to be made even smaller.

Here, the step size has been reduced by the factor 2. Twice as many integration steps were required, so the formulas were copied to more cells:

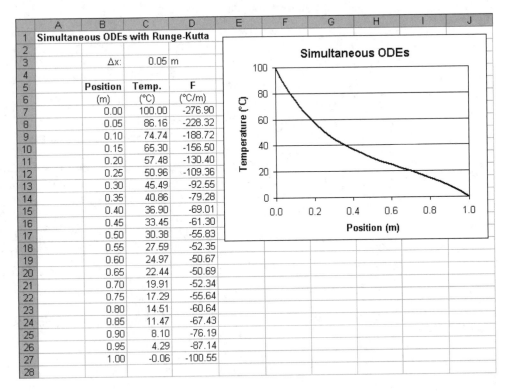

A slightly different value of F was required to get the temperature to 0°C at $x = 1$ m, so the solution definitely was affected by the size of the integration step. It would be a good idea to continue to make the step size smaller until the changes in the solution are insignificant.

15.5 IMPLICIT METHODS

15.5.1 Using the Method of Lines to Get the General Finite-Difference Equation

The *method of lines* is used to convert a *partial differential equation (PDE)* into a series of ordinary differential equations (ODEs), one per *grid point*, where the number of grid points is chosen as part of the solution process.

EXAMPLE 15.3

Partial Differential Equation: Unsteady Conduction

This example considers unsteady conduction along a metal rod. Mathematically, the problem can be summarized as

$$\frac{\partial T}{\partial t} = \alpha \frac{\partial^2 T}{\partial x^2} \qquad \text{partial differential equation (PDE)} \qquad (15.18)$$

subject to the initial condition

$$T = 35°C \qquad \text{throughout the rod at time zero} \qquad (15.19)$$

and the two boundary conditions

$$T = 100°C \qquad \text{at } x = 0,$$

and

$$T = 100°C \qquad \text{at } x = 1 \text{ m}, \qquad (15.20)$$

where α is the thermal diffusivity of the metal and has a prespecified value of 0.01 m²/min. The goal is to find out how the temperature changes with time for the first 10 minutes after the end temperatures are raised to 100°C.

The example we have just considered describes the conduction of energy down a rod. Eleven (arbitrary) grid points will be used here, uniformly distributed over x between $x = 0$ and $x = 1$ m:

When the differential equation is written in terms of finite differences, it looks as follows:

$$\frac{T_{i_{new}} - T_{i_{old}}}{\Delta t} = \alpha \frac{T_{i-1} - 2T_i + T_{i+1}}{(\Delta x)^2} \qquad (i \text{ ranges from 1 to 11}). \qquad (15.21)$$

Here, Δx is the distance between grid points, and Δt is the size of the integration step, chosen as part of the solution process.

The temperatures on the right can be the "old" temperatures (known at the beginning of each integration step) or the "new" temperatures (to be computed during each integration step). If the old temperature values are used on the right side, each of the 11 equations can be solved directly for the new temperatures, an *explicit solution* technique. If the new temperatures are used on the right side, the 11 equations must be solved simultaneously for the 11 new temperature values, an implicit technique. We will use the implicit solution technique here:

$$\frac{T_{i_{new}} - T_{i_{old}}}{\Delta t} = \alpha \frac{T_{i-1_{new}} - 2T_{i_{new}} + T_{i+1_{new}}}{(\Delta x)^2} \quad \text{general finite-difference equation.}$$

(15.22)

Eliminating Fictitious Points by Using Boundary Conditions

Whenever you use the method of lines to turn a PDE into a series of ODEs, you generate a problem on each end of the spatial domain. At $x = 0$ ($i = 1$), the equation becomes

$$\frac{T_{1_{new}} - T_{1_{old}}}{\Delta t} = \alpha \frac{\{T_{0_{new}}\} - 2\,T_{1_{new}} + T_{2_{new}}}{(\Delta x)^2}, \text{ when } i = 1,$$

(15.23)

and the temperature at $i = 0$ is needed, but $i = 0$ doesn't exist; it is outside of the spatial domain of the problem. This is called a *fictitious point*. Boundary conditions are used to eliminate fictitious points before integration. In this case, the temperature at $x = 0$ ($i = 1$) is constant at 100°C throughout the integration. If the temperature at $i = 1$ never changes, then

$$T_{1_{new}} = T_{1_{old}} \quad \text{at } i = 1 \ (x = 0).$$

(15.24)

This simpler equation can be used instead of the general finite-difference equation at $i = 1$.

Similarly, when $x = 1$ m ($i = 11$), the general finite-difference equation becomes

$$\frac{T_{11_{new}} - T_{11_{old}}}{\Delta t} = \alpha \frac{T_{10_{new}} - 2\,T_{11_{new}} + \{T_{12_{new}}\}}{(\Delta x)^2}, \text{ when } i = 11,$$

(15.25)

and there is a fictitious point (T_{12}). Again, in this problem, the temperature at $x = 1$ m ($i = 11$) is constant at 100°C, so we can replace the general finite-difference equation evaluated at $i = 11$ with

$$T_{11_{new}} = T_{11_{old}} \quad \text{at } i = 11 (x = 1 \text{ m}).$$

(15.26)

Summary of Equations to Be Solved

The 11 equations to be solved simultaneously are summarized in the following table:

LOCATION	GRID POINT	EQUATION
$x = 0$ m	$i = 1$	$T_{1_{new}} = T_{1_{old}}$
$x = 0.1$ m to 0.9 m	$i = 2$ to 10	$\dfrac{T_{i_{new}} - T_{i_{old}}}{\Delta t} = \alpha \dfrac{T_{i-1_{new}} - 2T_{i_{new}} + T_{i+1_{new}}}{(\Delta x)^2}$
$x = 1$ m	$i = 11$	$T_{11_{new}} = T_{11_{old}}$

In these equations, the T_{new} values will be the unknowns. Moving all of the unknown temperatures to the left side and the known (old) temperatures to the right side prepares the way for solving the equations via matrix methods:

LOCATION	GRID POINT	EQUATION
$x=0$ m	$i=1$	$1T_{1_{new}} = T_{1_{old}}$
$x=0.1$ m to 0.9 m	$i=2$ to 10	$-\Phi T_{i-1_{new}} + (1 + 2\Phi)T_{i_{new}} - \Phi T_{i+1_{new}} = T_{i_{old}}$
$x=1$ m	$i=11$	$1T_{11_{new}} = T_{11_{old}}$

Here

$$\Phi = \frac{\alpha \, \Delta t}{(\Delta x)^2}.\tag{15.27}$$

These equations can be written in matrix form as a coefficient matrix C multiplying an unknown vector T_{new} set equal to a right-hand-side vector r. So

$$[C][T_{new}] = [r],\tag{15.28}$$

where

$$C = \begin{bmatrix} 1 & 0 & 0 & 0 & 0 & 0 & 0 & 0 & 0 & 0 & 0 \\ -\Phi & 1+2\Phi & -\Phi & 0 & 0 & 0 & 0 & 0 & 0 & 0 & 0 \\ 0 & -\Phi & 1+2\Phi & -\Phi & 0 & 0 & 0 & 0 & 0 & 0 & 0 \\ 0 & 0 & -\Phi & 1+2\Phi & -\Phi & 0 & 0 & 0 & 0 & 0 & 0 \\ 0 & 0 & 0 & -\Phi & 1+2\Phi & -\Phi & 0 & 0 & 0 & 0 & 0 \\ 0 & 0 & 0 & 0 & -\Phi & 1+2\Phi & -\Phi & 0 & 0 & 0 & 0 \\ 0 & 0 & 0 & 0 & 0 & -\Phi & 1+2\Phi & -\Phi & 0 & 0 & 0 \\ 0 & 0 & 0 & 0 & 0 & 0 & -\Phi & 1+2\Phi & -\Phi & 0 & 0 \\ 0 & 0 & 0 & 0 & 0 & 0 & 0 & -\Phi & 1+2\Phi & -\Phi & 0 \\ 0 & 0 & 0 & 0 & 0 & 0 & 0 & 0 & -\Phi & 1+2\Phi & -\Phi \\ 0 & 0 & 0 & 0 & 0 & 0 & 0 & 0 & 0 & 0 & 1 \end{bmatrix},$$

$$T_{new} = \begin{bmatrix} T_{1_{new}} \\ T_{2_{new}} \\ T_{3_{new}} \\ T_{4_{new}} \\ T_{5_{new}} \\ T_{6_{new}} \\ T_{7_{new}} \\ T_{8_{new}} \\ T_{9_{new}} \\ T_{10_{new}} \\ T_{11_{new}} \end{bmatrix}, \quad r = \begin{bmatrix} T_{1_{old}} \\ T_{2_{old}} \\ T_{3_{old}} \\ T_{4_{old}} \\ T_{5_{old}} \\ T_{6_{old}} \\ T_{7_{old}} \\ T_{8_{old}} \\ T_{9_{old}} \\ T_{10_{old}} \\ T_{11_{old}} \end{bmatrix}\tag{15.29}$$

The initial right-hand-side vector contains the "old" temperatures that are known from the problem statement's initial condition. The problem doesn't actually start until the instant the end temperatures change to 100°C, so the end-grid-point initial temperatures are set at 100°C, while all interior temperatures are initially set at 35°C.

Solving the Matrix Problem Using Excel   As usual, the solution process begins
with the entering of problem parameters into the spreadsheet:

	C8	▼		f_x =(C3*C4)/C7^2								
	A	B	C	D	E	F	G	H	I	J	K	L	M
1	**Partial Differential Equation, Implicit Solution: Unsteady Conduction**											
2												
3		α:	0.01	m²/min								
4		Δt:	1	min		chosen						
5		L:	1	m		rod length						
6		N:	11	points		chosen						
7		Δx:	0.1									
8		Φ:	1									
9												

The formulas entered in column C are

$$C7: \ =C5/(C6-1) \qquad \text{calculates } \Delta x$$

$$C8: \ =(C3*C4)/C7\wedge2 \qquad \text{calculates } \Phi \qquad\qquad (15.30)$$

The coefficient matrix is mostly zeroes (a *sparse matrix*), with nonzero elements located
on the diagonal and one element off the diagonal. This type of matrix is called a
tridiagonal matrix. An 11×11 tridiagonal matrix can quickly be created in a spreadsheet by creating a column of 11 zeroes followed by an 11×11 array of zeroes. Finally,
the values and formulas on and near the diagonal are entered:

	C11	▼		f_x =1+2*$C$8								
	A	B	C	D	E	F	G	H	I	J	K	L	M
1	**Partial Differential Equation, Implicit Solution: Unsteady Conduction**											
2												
3		α:	0.01	m²/min								
4		Δt:	1	min		chosen						
5		L:	1	m		rod length						
6		N:	11	points		chosen						
7		Δx:	0.1									
8		Φ:	1									
9												
10	[C]	1	0	0	0	0	0	0	0	0	0	0
11		-1	3	-1	0	0	0	0	0	0	0	0
12		0	-1	3	-1	0	0	0	0	0	0	0
13		0	0	-1	3	-1	0	0	0	0	0	0
14		0	0	0	-1	3	-1	0	0	0	0	0
15		0	0	0	0	-1	3	-1	0	0	0	0
16		0	0	0	0	0	-1	3	-1	0	0	0
17		0	0	0	0	0	0	-1	3	-1	0	0
18		0	0	0	0	0	0	0	-1	3	-1	0
19		0	0	0	0	0	0	0	0	-1	3	-1
20		0	0	0	0	0	0	0	0	0	0	1
21												

The coefficients on the diagonal (except the ones at the top left and bottom right) all contain the same formula:

C11: =1+2*$C$8   diagonal elements

The adjacent coefficients (off-diagonal) contain the value of $-\Phi$, pulled from the parameter list with this formula:

B11: =-$C$8

The ones in the top left and bottom right corner represent the coefficients on the left side of the first and last equations:

$$i = 1 \qquad 1T_{1_{new}} = T_{1_{old}},$$
$$i = 11 \qquad 1T_{11_{new}} = T_{11_{old}}.$$

In solving the simultaneous equations by using matrix methods, the coefficient matrix must be inverted. This is done by using Excel's MINVERSE() function.

This is an array function, so there are a couple of things to remember when entering the function:

- Select the cells that will receive the inverted matrix to set the size of the result matrix *before* entering the array function.

- After typing the formula into the top left cell of the result matrix, press [Ctrl-Shift-Enter] to enter the formula into each of the cells in the result matrix.

While having to determine the size of the result is something of a nuisance, Excel's array functions do have advantages. The principal advantages are the ability to recalculate your matrix calculations automatically if any of the input data change, and the ability to copy matrix formulas by using relative addressing to repeat calculations. This will be extremely useful as we finish this example.

The coefficient matrix was inverted by using

{B22:L32} =MINVERSE(B10:L20) [Ctrl-Shift-Enter] to finish

to give

	B22		f_x {=MINVERSE(B10:L20)}									
	A	B	C	D	E	F	G	H	I	J	K	L	M
10	[C]	1	0	0	0	0	0	0	0	0	0	0	
11		-1	3	-1	0	0	0	0	0	0	0	0	
12		0	-1	3	-1	0	0	0	0	0	0	0	
13		0	0	-1	3	-1	0	0	0	0	0	0	
14		0	0	0	-1	3	-1	0	0	0	0	0	
15		0	0	0	0	-1	3	-1	0	0	0	0	
16		0	0	0	0	0	-1	3	-1	0	0	0	
17		0	0	0	0	0	0	-1	3	-1	0	0	
18		0	0	0	0	0	0	0	-1	3	-1	0	
19		0	0	0	0	0	0	0	0	-1	3	-1	
20		0	0	0	0	0	0	0	0	0	0	1	
21													
22	[C$_{inv}$]	1	0	0	0	0	0	0	0	0	0	0	
23		0.38	0.38	0.15	0.06	0.02	0.01	0	0	0	0	0	
24		0.15	0.15	0.44	0.17	0.06	0.02	0.01	0	0	0	0	
25		0.06	0.06	0.17	0.45	0.17	0.07	0.02	0.01	0	0	0	
26		0.02	0.02	0.06	0.17	0.45	0.17	0.07	0.02	0.01	0	0	
27		0.01	0.01	0.02	0.07	0.17	0.45	0.17	0.07	0.02	0.01	0.01	
28		0	0	0.01	0.02	0.07	0.17	0.45	0.17	0.06	0.02	0.02	
29		0	0	0	0.01	0.02	0.07	0.17	0.45	0.17	0.06	0.06	
30		0	0	0	0	0.01	0.02	0.06	0.17	0.44	0.15	0.15	
31		0	0	0	0	0	0.01	0.02	0.06	0.15	0.38	0.38	
32		0	0	0	0	0	0	0	0	0	0	1	
33													

The right-hand side of the matrix equations contains the known (old) temperatures, which are the time-zero solution to the problem. They are entered on the spreadsheet as both the [r] matrix and the solution at time zero:

	A	B	C	D	E	F	G	H	I	J	K	L	M
33													
34	time:	0											
35													
36	[r]	100.0											
37		35.0											
38		35.0											
39		35.0											
40		35.0											
41		35.0											
42		35.0											
43		35.0											
44		35.0											
45		35.0											
46		100.0											
47													

As the integration proceeds, the computed temperature values will be displayed in columns C through L. To keep track of the times corresponding to the integration steps, the time is computed in row 34:

$$\texttt{C34: =B34+\$C\$4} \qquad \text{adds } \Delta t \text{ to the previous time}$$

C34		f_x =B34+$C$4										
	A	B	C	D	E	F	G	H	I	J	K	L	M
33													
34	time:	0	1										
35													
36	[r]	100.0											
37		35.0											
38		35.0											
39		35.0											
40		35.0											
41		35.0											
42		35.0											
43		35.0											
44		35.0											
45		35.0											
46		100.0											
47													

To solve for the temperatures at each grid point at the end of the first integration step, simply multiply the inverted coefficient matrix and the initial right-hand-side vector:

$$[C]^{-1}[r] \qquad\qquad (15.31)$$

The multiplication is handled by Excel's MMULT( ) array function. Because the new matrix result is temperatures at each grid point, the size of the result matrix will be the same as the size of [r]:

$$\{C36:C46\} \ =MMULT(\$B\$22:\$L\$32, B36:B46)$$

Note: Be sure to use an absolute reference (dollar signs) on the [C inv] matrix; it becomes important when the result is copied to integrate over additional time steps.

	C36		▼		f_x	{=MMULT($B$22:$L$32,B36:B46)}						
	A	B	C	D	E	F	G	H	I	J	K	L	M
33													
34	time:	0	1										
35													
36	[r]	100.0	100.0										
37		35.0	59.8										
38		35.0	44.5										
39		35.0	38.7										
40		35.0	36.6										
41		35.0	36.1										
42		35.0	36.6										
43		35.0	38.7										
44		35.0	44.5										
45		35.0	59.8										
46		100.0	100.0										
47													

Completing the Solution This is the step in which the power of Excel's array functions really becomes apparent. To calculate the results at additional times, simply copy cells C34:C46 to the right. The more copies you create, the more time steps you are integrating over. The temperatures at each time step are displayed in rows 36 through 46, and the elapsed time is displayed above each temperature result. Temperature profiles for the first 10 minutes are shown here:

	A	B	C	D	E	F	G	H	I	J	K	L	M
33													
34	time:	0	1	2	3	4	5	6	7	8	9	10	
35													
36	[r]	100.0	100.0	100.0	100.0	100.0	100.0	100.0	100.0	100.0	100.0	100.0	
37		35.0	59.8	71.0	76.9	80.5	83.0	84.9	86.5	87.8	89.0	90.0	
38		35.0	44.5	53.1	59.6	64.6	68.6	71.8	74.5	76.9	79.1	81.0	
39		35.0	38.7	43.8	49.0	53.7	58.0	61.9	65.4	68.5	71.3	73.9	
40		35.0	36.6	39.6	43.4	47.6	51.8	55.8	59.6	63.1	66.4	69.4	
41		35.0	36.1	38.4	41.8	45.7	49.8	53.8	57.7	61.3	64.7	67.8	
42		35.0	36.6	39.6	43.4	47.6	51.8	55.8	59.6	63.1	66.4	69.4	
43		35.0	38.7	43.8	49.0	53.7	58.0	61.9	65.4	68.5	71.3	73.9	
44		35.0	44.5	53.1	59.6	64.6	68.6	71.8	74.5	76.9	79.1	81.0	
45		35.0	59.8	71.0	76.9	80.5	83.0	84.9	86.5	87.8	89.0	90.0	
46		100.0	100.0	100.0	100.0	100.0	100.0	100.0	100.0	100.0	100.0	100.0	
47													

KEY TERMS

Boundary condition
Dependent variable
Euler's technique
Explicit solution
Fictitious point
Grid point
Implicit solution
Independent variable

Initial condition
Integration step size
Method of lines
Ordinary differential equation (ODE)
Parameter
Partial differential equation (PDE)
Residence time
Runge–Kutta methods

Shooting method
Simultaneous ODEs
Sparse matrix
Time scale
Time step
Tridiagonal matrix

SUMMARY

The chief virtue of Euler's method is its simplicity; unfortunately, it can be fairly inaccurate unless you use a very small integration step size. The fourth-order Runge–Kutta method requires that you write a function, but the function is pretty simple, and the Runge–Kutta method yields much better results with larger integration steps. Partial differential equations can be integrated by using either explicit or implicit techniques. Implicit techniques, such as are illustrated in this chapter, tend to be more accurate for a given integration step size.

Note: The fourth-order Runge–Kutta method does work for PDEs, but it is a pretty significant extrapolation from `rk4two()`*, the simultaneous ODE function used in this chapter, to a version for N simultaneous ODEs, where N is the number of grid points used when converting the PDE to a series of ODEs. Math packages such as* MATLAB *and* MathCad *provide fourth-order Runge–Kutta functions for N simultaneous equations.*

Problems

Concentration During a Wash-in

1. Example 15.1 in this chapter was a "washout" problem, in which a tank contained a specified concentration of some chemical species (called component A) in the example and no A in the influent flow. The concentration of A decreases with time as it is washed out of the tank.

    Rework Example 15.1 as a "wash-in" problem. Assume that the initial concentration of A in the tank is zero, but that the influent contains A at a level of 100 mg/ml.

    The analytical result still applies:

    $$C_A = C_{A_{\text{inf}}} + (C_{A_{\text{inf}}} - C_{A_{\text{inf}}})e^{-t/\tau}. \tag{15.32}$$

    Plot the concentration of A calculated by each method (Euler's method and analytical result) as a function of time.

Tank Temperature During a Wash-in

2. One evening, a few friends come over for a soak, but the hot tub has been turned off for days, so it has to be warmed up. As your friends turn on the hot water to warm up the tub, you want to know how long this is going to take and write out a quick energy balance on a well-mixed tank. You end up with a differential equation relating the temperature in the tank T to the temperature of the hot water entering the tank T_{in}, the volume of the tank V, and the volumetric flow rate of hot water $\dot{V}$:

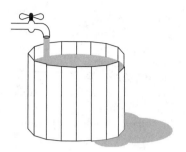

$$\frac{dT}{dt} = \frac{\dot{V}}{V}(T_{\text{in}} - T). \tag{15.33}$$

If the initial temperature is 45°F (7.2°C) and you want to heat it to 100°F (37.8°C), you estimate that you will have plenty of time to solve this equation in Excel.

Suppose the hot water enters the tank at 120°F (54.4°C) at a rate of 30 liters per minute, and the tank contains 3000 liters of water.

a.   Integrate the differential equation, using either Euler's method or the fourth-order Runge–Kutta technique, to obtain tank-temperature values as a function of time.

b.   Graph the temperature and time values.

c.   Approximate how long it will take to heat up the tub to 100°F.

d.   Compute how long it would take to heat the tub to 100°F if it was first drained and then refilled with hot water.

Steady-State Conduction

3.   Example 15.2 used a heat-transfer coefficient of $h = 50$ W/m² K, a value that might represent heat loss to a fast-moving air stream. Rework Example 15.2, but use the following heat-transfer coefficients (the h value is included in the user-written dD2dI() function, which will have to be modified for this problem):

a. $h = 5$ W/m² K (typical of energy transfer to slow-moving air)

b. $h = 300$ W/m² K (typical of energy transfer to a moving liquid)

Unsteady Diffusion

4.   A heat-transfer example was used in Example 15.3. The mass-transfer analog is a mass-diffusion problem, namely,

$$\frac{\partial C_A}{\partial t} = D_{AB}\frac{\partial^2 C_A}{\partial x^2}, \tag{15.34}$$

where

C_A is the concentration of A,

D_{AB} is the diffusivity of A diffusing through B—say, 0.00006 cm²/sec,

t is time, and

x is position.

The units on C_A in this equation are not critical, because they are the same on both sides of the equation. However, D_{AB} is a constant only if the concentration of A is low.

One initial condition and two boundary conditions are needed to solve this problem:

$C_A = 0$ throughout the diffusion region ($0 \leq x \leq 10$ cm) at time zero;

$C_A = 100$ mg A/mL at x = 0 (all times);

$C_A = 0$ mg/mL at x = 10 cm (all times).

a. Use the method of lines to transform the PDE into a series of ODEs (one per grid point).

b. Use the boundary conditions to eliminate the fictitious points in the boundary ODEs.

c. Write out the coefficient matrix and right-hand-side vector required for the implicit method of solution.

d. Solve the systems of ODEs to find out how long it will take for the concentration at $x = 5$ cm to reach 20 mg/mL.

Index